Selected Titles in This Series

200 **V. V. Prasolov and V. M. Tikhomirov,** Geometry, 2001
199 **Shigeyuki Morita,** Geometry of characteristic classes, 2001
198 **V. A. Smirnov,** Simplicial and operad methods in algebraic topology, 2001
197 **Kenji Ueno,** Algebraic geometry 2: Sheaves and cohomology, 2001
196 **Yu. N. Lin′kov,** Asymptotic statistical methods for stochastic processes, 2001
195 **Minoru Wakimoto,** Infinite-dimensional Lie algebras, 2001
194 **Valery B. Nevzorov,** Records: Mathematical theory, 2001
193 **Toshio Nishino,** Function theory in several complex variables, 2001
192 **Yu. P. Solovyov and E. V. Troitsky,** C^*-algebras and elliptic operators in differential topology, 2001
191 **Shun-ichi Amari and Hiroshi Nagaoka,** Methods of information geometry, 2000
190 **Alexander N. Starkov,** Dynamical systems on homogeneous spaces, 2000
189 **Mitsuru Ikawa,** Hyperbolic partial differential equations and wave phenomena, 2000
188 **V. V. Buldygin and Yu. V. Kozachenko,** Metric characterization of random variables and random processes, 2000
187 **A. V. Fursikov,** Optimal control of distributed systems. Theory and applications, 2000
186 **Kazuya Kato, Nobushige Kurokawa, and Takeshi Saito,** Number theory 1: Fermat's dream, 2000
185 **Kenji Ueno,** Algebraic Geometry 1: From algebraic varieties to schemes, 1999
184 **A. V. Mel′nikov,** Financial markets, 1999
183 **Hajime Sato,** Algebraic topology: an intuitive approach, 1999
182 **I. S. Krasil′shchik and A. M. Vinogradov, Editors,** Symmetries and conservation laws for differential equations of mathematical physics, 1999
181 **Ya. G. Berkovich and E. M. Zhmud′,** Characters of finite groups. Part 2, 1999
180 **A. A. Milyutin and N. P. Osmolovskii,** Calculus of variations and optimal control, 1998
179 **V. E. Voskresenskiĭ,** Algebraic groups and their birational invariants, 1998
178 **Mitsuo Morimoto,** Analytic functionals on the sphere, 1998
177 **Satoru Igari,** Real analysis—with an introduction to wavelet theory, 1998
176 **L. M. Lerman and Ya. L. Umanskiy,** Four-dimensional integrable Hamiltonian systems with simple singular points (topological aspects), 1998
175 **S. K. Godunov,** Modern aspects of linear algebra, 1998
174 **Ya-Zhe Chen and Lan-Cheng Wu,** Second order elliptic equations and elliptic systems, 1998
173 **Yu. A. Davydov, M. A. Lifshits, and N. V. Smorodina,** Local properties of distributions of stochastic functionals, 1998
172 **Ya. G. Berkovich and E. M. Zhmud′,** Characters of finite groups. Part 1, 1998
171 **E. M. Landis,** Second order equations of elliptic and parabolic type, 1998
170 **Viktor Prasolov and Yuri Solovyev,** Elliptic functions and elliptic integrals, 1997
169 **S. K. Godunov,** Ordinary differential equations with constant coefficient, 1997
168 **Junjiro Noguchi,** Introduction to complex analysis, 1998
167 **Masaya Yamaguti, Masayoshi Hata, and Jun Kigami,** Mathematics of fractals, 1997
166 **Kenji Ueno,** An introduction to algebraic geometry, 1997
165 **V. V. Ishkhanov, B. B. Lur′e, and D. K. Faddeev,** The embedding problem in Galois theory, 1997
164 **E. I. Gordon,** Nonstandard methods in commutative harmonic analysis, 1997
163 **A. Ya. Dorogovtsev, D. S. Silvestrov, A. V. Skorokhod, and M. I. Yadrenko,** Probability theory: Collection of problems, 1997

(Continued in the back of this publication)

Translations of
MATHEMATICAL MONOGRAPHS

Volume 200

Geometry

V. V. Prasolov
V. M. Tikhomirov

AMERICAN MATHEMATICAL SOCIETY
Providence, Rhode Island

В. В. Прасолов, В. М. Тихомиров

ГЕОМЕТРИЯ

МЦНМО, Москва, 1997

Translated from the Russian by O. V. Sipacheva
Translation editied by A. B. Sossinski

2010 *Mathematics Subject Classification.* Primary 51–01, 51M04, 51M09.

Abstract. This book provides a systematic introduction to various geometries, including Euclidean, affine, projective, elliptic, hyperbolic, and infinite-dimensional geometries. A uniform approach to different geometries is suggested, and the unified origins of different phenomena are traced. All geometric objects are studied from the point of view of duality theory. The theory of conics and quadrics, including the theory of conics for non-Euclidean geometries, is thoroughly developed. The book contains many striking geometric facts and solutions to plenty of beautiful geometric problems.

Numerous pictures help gain a clearer understanding of the presented geometric theorems. Almost all chapters include problem sections, which allows the book to be used as a textbook. The majority of problems are supplied with answers and hints, some with complete solutions.

The purpose of the book is to contribute to the development of research in geometry and improvement of mathematical education. The book is intended for college undergraduate and graduate students, high school mathematics teachers, and researchers in mathematics and physics.

Library of Congress Cataloging-in-Publication Data

Prasolov, V. V. (Viktor Vasil′evich)
[Geometriia. English]
Geometry / V. V. Prasolov, V. M. Tikhomirov ; [translated from the Russian by O. V. Sipacheva].
p. cm. — (Translations of mathematical monographs, ISSN 0065-9282 ; v. 200)
Includes bibliographical references and index.
ISBN 0-8218-2038-9 (acid-free paper)
1. Geometry. I. Tikhomirov, V. M. (Vladimir Mikhailovich), 1934– II. Title. III. Series.
QA445.P7313 2001
516—dc21 2001022063

AMS softcover ISBN: 978-1-4704-2543-2

14 13 12 11 10 9 8 7 6 5 29 28 27 26 25 24

Contents

Preface

Once, Israil' Gelfand said that mathematics has three parts: analysis, geometry, and combinatorics. "What is combinatorics?"—the listeners asked. The answer was: "This is a science not yet created ... "

It is natural to ask: What is geometry?

... In one of his interviews (published in the popular science journal *Kvant*), I. M. Gelfand recollects his past. He spent his childhood in a tiny town not far from Odessa. Very early, he began to think about mathematical problems—alone, because he had nobody to ask: there were neither learned people nor books in the town. He had to invent and devise things on his own. He came to the conclusion that there exists two mathematics separated by an abyss: algebra (actually, algebra and analysis; the second word had not been familiar to the boy at that time) and geometry. Polynomials are algebra, while sines are quite different, they have to do with geometry.

Gelfand said in his interview that, many times in his life, he had been very lucky. For instance, he asked his parents to buy him a book on mathematics, but his request was declined because of lack of money. But (oh, luck!) the boy had had an attack of appendicitis, and it was necessary to go to the city for surgery. Then, he said that he would not go anywhere if his parents did not buy the book. And so the book was bought.

It was an ordinary provincial calculus textbook, where, say, functions were classified into continuous, differentiable, and empirical. But, among other stuff, the book contained the Maclaurin formula and the series

$$\sin x = x - x^3/6 + x^5/120 - \cdots .$$

This led to a revolution in the boy's mind: he understood that mathematics is indivisible!

So, what is the place of geometry in this indivisible mathematics?

About two decades ago, an interesting discovery in physiology was made: Professor Roger Sperry of the California Institute of Technology proved, by inspecting patients with cut "callous body" (which joins the two cerebral hemispheres), that the functions of the hemispheres are somewhat asymmetric. (For this discovery, R. Sperry won the 1981 Nobel Prize in biology and medicine.)

At present, it is accepted as proved that the two hemispheres of the human cerebrum have different functions: the left hemisphere (for right-handers) is responsible for logical analysis. It also controls speech, writing, and other "algorithmic" procedures. This is, so to say, the "algebraic-analytical" hemisphere. The right hemisphere is, on the contrary, synthetic, it "governs" everything "imaginational," such as vision, image perception, imagery ... Thus we can say that the world of the right hemisphere is the world of Geometry!

Would it not be justified to regard Geometry as the "artistic mathematics," the part of mathematics in which we can imagine or picture something with our eyes closed—a curving mobile surface, a plane section of a polyhedron, a network of lines, a chain of linked manifolds going away to infinity? ...

Thus the people are divided into two groups according to their inclinations: some of them prefer calculations, and the others, images. This is why geometry is indispensable in teaching, in the system of education. It provides a unique possibility for simultaneously developing logical thinking and teaching intuitive cognition of truth based on perception. In this latter hypostasis, geometry cannot be compared to anything.

This book is written by two mathematicians. One of them, V. V. Prasolov, is largely a geometer, and the other, V. M. Tikhomirov, an analyst. This determines a certain equilibrium between the imaginational and logical orientations of the book. It contains many illustrations, figures, constructions, cross-sections, and other visual material, but much attention is given also to the logical structure of geometry, models, and algebraic-analytical approaches.

The book is intended for high school students, teachers, college students majoring in mathematics, and for anyone interested in geometry, no matter what their education and occupation. The book is structured accordingly. It is divided into chapters and has an Addendum. The chapters deal with various geometries: Euclidean, affine, projective, non-Euclidean, and with infinite-dimensional generalizations of affine and Euclidean geometries. The latter have to do with linear and convex objects. A separate chapter deals with second-order curves and surfaces.

The contents of each chapter, except for the last one, can be divided into two parts. First, the simplest (one-dimensional and plane) geometries are considered. These fragments are meant for the widest class of readers, including high school students.

The remaining sections of the first five chapters are concerned with finite-dimensional geometries. These sections constitute a basic course in geometry for universities and teacher's colleges.[1] The sixth chapter reveals the geometrical roots of infinite-dimensional analysis.

The general orientation of this course follows the approach to mathematical education in which courses are organized according to the *stage-by-stage* principle: like in climbing a mountain, the same scenery is viewed from different positions and altitudes. At the beginning, the student must be given a chance to acquire basic mathematical knowledge, gain understanding of the general structure of mathematics as a whole, and master the fundamental notions, ideas, and methods. At the first stage, the course must impart the most important skills in the basic fields of mathematics—analysis, geometry, and combinatorics in the broad sense, i.e., as the science of finitization and practical implementation of mathematical concepts that involve the notion of infinity (if Gelfand's words cited at the beginning of this Preface are taken as a first approximation to the truth). Then, the second parts of the main chapters may be regarded as an attempt to constitute something like a university course *Geometry I*. (Our version of the course implements the "continuity" of education; also we try to explain the place of high school geometry in the history of our science.)

---

[1]They correspond to the (somewhat extended) contents of the geometry course taught by V. M. Tikhomirov and V. V. Prasolov at the Independent Moscow University in 1992–96.

The other fragments of the book, collected in the Addendum, are addressed to the experts and true lovers of geometry. These are selected topics and beautiful facts of geometry that appeared in the work of great geometers and still preserve their beauty and freshness today. These sections also describe some applications of geometry to the natural sciences.

A substantial part of the book is devoted to problems. The problems are mostly given at the end of each chapter, but sometimes also in the main text. We tried to select nontrivial and expressive problems that may be useful in the thorough study of the lecture course and as material for the exercise classes.

The book covers the subject matter from the earliest steps due to Thales, Pythagoras, Euclid, Archimedes, and Apollonius to the middle of the twentieth century.

There was one additional motivation for writing this book, which is a tribute to the memory of Andrei Nikolaevich Kolmogorov, to whom the authors feel deeply indebted.

One of Kolmogorov's favorite ideas in the late years of his life was that of creating a school course in geometry which would "gradually prepare grounds for understanding the possibility of different 'geometries' (like Lobachevsky geometry)." Kolmogorov's concept of geometrical education was that it should "include Euclidean geometry as a particular case (of the concept of metric space)." Similar ideas are contained in this book.

We taught the geometry course at Independent Moscow University together with V. O. Bugaenko, M. N. Vyalyi, I. V. Gribkov, and R. A. Sarkisyan. We are grateful to them for detailed discussions of separate parts of this book.

The authors acknowledge the financial support of the Russian Foundation for Basic Research under grants 98-00-555 (V. V. Prasolov) and 96-01-00325 (V. M. Tikhomirov).

Introduction

> In the nineteenth century, geometry developed in many directions, at first glance, bearing no relation to each other. By means of the notion of transformation group, the Erlangen Program knotted together all the varieties of geometry and simultaneously determined the distinguishing features of each; this program states and gives a fundamental answer to the question "What is geometry?"
>
> Hermann Weyl

In the famous paper *L'Architecture des mathématiques*, Nicolas Bourbaki puts the question: "La Mathématique ou les Mathématiques?" In English, this would be "Is there only one mathematics, or are there many?" Exactly the same question can be asked about our field of mathematics—geometry: is there only one geometry, or are there many? Indeed, there are projective, affine, higher, elementary, spherical, hyperbolic, convex, Euclidean, differential, integral, conformal, symplectic, analytic, algebraic, Banach, computer, and a diversity of other geometries, but do they have anything in common?

In the preface, we named one—extrinsic—property that unites them: the geometric objects are imaginable; that is, they can be visualized.

But is there something else that various geometries, such as Euclidean, affine, projective, Riemann, Lobachevsky, have in common?

It turns out that these geometries are tied together by bonds uncovered in the Erlangen Program; these are the "knots" mentioned by H. Weyl (see the epigraph). But what is the Erlangen Program?

The *Erlangen Program* is the traditional Professor's inaugural speech, given by the outstanding German mathematician *Felix Klein* (1849–1924); the speech was then published in the form of a booklet entitled *Vergleichende Betrachtungen über neuere geometrische Forschungen*, which means "Comparative remarks about new geometric investigations."

A prominent contemporary mathematician F. Hirzebruch, Professor at the University of Bonn, writes: "The professors' inauguration ceremony is an old tradition of German universities. After receiving the doctoral degree, a future professor continues his scientific work and, several years later, presents a more extensive text—a *Venia Legendi* thesis—to the faculty for gaining the right to teach students and to lecture. The thesis is reported and discussed at a meeting of the faculty council. In a few weeks, the ceremony concludes by its culmination, an inaugural public lecture. [...]

These lectures are supposed to develop a serious and important topic so that not only mathematicians but also representatives of the other natural sciences and students be able to follow the reasoning. The lecture must shed light on the history of the question and the relationship of various fields of mathematics to each other and to different sciences. In addition, the audience should feel the charm of mathematics, at least a little, and better understand its role in the cultural life of society." (An excellent guideline for authors of mathematical books, too.)

Klein made his speech in 1872, at an age of 23. In the lecture, Klein tried to answer the same sacramental question: What is geometry?

According to Klein's concept, geometry studies *invariants of transformation groups of homogeneous spaces*. A homogeneous space is a set of points on which some transformation group acts transitively (i.e., so that any point can be transformed into any other).

The Euclidean plane can be thought of as the infinitely extended plane of a school blackboard. This plane is a two-dimensional variety of points endowed with a distance. On the variety, the group of isometries (distance-preserving one-to-one maps of the plane onto itself) is introduced. It consists of translations, rotations about points, and reflections about straight lines.

Euclidean geometry studies figures "up to isometry." They include, for example, the fairly rich family of triangles (depending, as we all know, on three parameters), the two-parameter family of ellipses, etc.

The Euclidean plane (precisely the plane, since there is no generalization to spaces of higher dimensions) admits yet another remarkable description: it can be represented as the *complex* plane, as the set of complex numbers. The complex interpretation of the Euclidean plane opens up new possibilities in elementary geometry.

The affine plane consists of the same points as the Euclidean plane, but it admits a wider transformation group, the group of *affine transformations*. In addition to isometries, it contains all dilations in arbitrary directions. The affine transformations can also be defined as one-to-one transformations of the plane that map straight lines into straight lines. Affinely, all triangles are identical; the "affine geometry of the triangle" is much poorer than Euclidean, but it has its own remarkable theorems (such as the Ceva and Menelaus theorems). All ellipses on the affine plane are also equivalent. However, "convex" geometry lives on the affine plane, and it is a discipline exceedingly rich in content.

The projective plane contains not only the points of the affine plane. The affine plane is completed by the "straight line at infinity." The projective transformations are compositions of projections of the plane from some point onto another plane. Topologically, the projective line is a circle, and the projective plane is a "nonorientable" two-dimensional surface obtained from a disk by attaching the "Möbius band" to its boundary.

Affinely, an ellipse is not equivalent to a parabola or hyperbola, while projectively, they are identical. It turns out that in a projective space, a calculus of distances can be introduced, which has led to many remarkable discoveries. On the one hand, this has provided Cayley and Klein with a means for interpreting non-Euclidean geometry (we shall explain this a little later), and on the other, this has made it possible to describe projective geometry in purely algebraic terms. It was found that any division ring (i.e., an algebraic structure with addition and multiplication, where multiplication is not necessarily commutative) generates a unique projective space corresponding to this division ring, and there are only three locally compact division rings, namely, the real numbers, the complex numbers, and the quaternions (this result is due to Kolmogorov and Pontryagin).

Now, we turn to non-Euclidean geometries. The past century was marked by great achievements in non-Euclidean geometries. The dramatic history of the discovery and proof of the consistency of Lobachevsky geometry (which involved great mathematicians such as Gauss, Lobachevsky, Bolyai, Riemann, Beltrami, Cayley, Klein, Poincaré, Hilbert, and others) resulted in a revolution in the understanding of mathematics as a whole.

It remains to mention several points concerning the Riemann and Lobachevsky planes. The Riemann plane can be represented as the ordinary sphere in three-dimensional Euclidean space, with all antipodal points identified pairwise. The distance on this plane is the length of the shorter arc of the great circle joining two points on the sphere. The motions are rotations about the center of the sphere, reflections in the planes passing through its center, and their compositions. The role of straight lines on this plane is played by the great circles.

Through any two points in the Riemann plane, one and only one straight line passes, but this plane contains no parallel lines—all straight lines intersect.

The Lobachevsky plane can be imagined as the upper sheet of the hyperboloid in three-dimensional space specified by the system $x_3^2 - x_1^2 - x_2^2 = 1$, $x_3 > 0$.

The motions are the linear transformations of $\mathbb{R}^3$ that map the cone $x_3 \geqslant \sqrt{x_1^2 + x_2^2}$ onto itself. These transformations have played an important rôle in the history of science because they proved to be related to the Lorentz transformations, which are the basic transformations of relativity theory. A distance can be defined on the hyperboloid, and the Lobachevsky plane is therefore a metric space. The motions of the Lobachevsky plane are the isometries (for this distance).

The geometries of the Euclidean plane, Riemann plane, and Lobachevsky plane have much in common. For instance, each isoscelcs triangle has equal base angles. But there are also many differences between them; say, neither Riemann nor Lobachevsky geometry has any similarity transformation (in both, triangles are determined by their angles), there is no parallelism in Riemann geometry (all lines intersect), while in Lobachevsky geometry, through a point outside some straight line, many parallel lines can be drawn. Nevertheless, all these geometries can be treated in a unified way.

When Klein realized this, he set forth his Erlangen Program. In organizing this book, we largely followed this Program.

CHAPTER 1

The Euclidean World

The Euclidean world is familiar to all of us. The image of a school blackboard infinitely extended on all sides, where we can mark points by thinnest pins and draw straight lines through them by using sharp-pointed pencils and straightedges, and where we can plot circles of any radius with a compass,—this image appears to each of us almost simultaneously with consciousness itself and remains in our imagination forever. This is why we start with Euclidean geometry.

1.1. The Euclidean line and plane

> Although the ideas on which the method of analytic geometry is based on are childishly simple, this method is so powerful that applying it, ordinary seventeen-year-old young people can solve problems that would have stumped the greatest geometers of the past.
>
> E. T. Bell

High school students study Euclidean geometry in the plane. This study is based on the *deduction method* (sometimes carefully hidden). The method describes rather than defines the basic notions. The fundamental properties of the basic elements of the theory are stated in the form of axioms, which are accepted without proof. Other assertions of the theory are logically derived from the axioms.

The first effort to deductively construct geometry that we are aware of was made by Euclid in the third century B.C. For this reason, it is natural to start our acquaintance with great geometers from Euclid.

Euclid (c. 356–c. 300 B.C.), ancient Greek mathematician. He lived in Alexandria in the times of Ptolemy I (one of Alexander the Great's generals, who became the king of Egypt after Alexander died). According to a legend, Ptolemy got interested in geometry, and Euclid began to explain its principles to him. Soon, the king became impatient: "Could you put this somehow simpler?" Euclid proudly answered that *there is no royal road to geometry.*

The main work of Euclid, *Elements*, is one of the greatest achievements of scientific thought; it has served as the foundation for school education in geometry during the entire subsequent history of civilized humanity. No other scientific book enjoyed the same success as the *Elements*: since the first printed edition in 1482, the *Elements* was translated into the majority of languages and has had a run of over 500 editions.

Euclid's *Elements* opens with a description of the basic notions: "*a point is what has no parts,*" [...] "*a straight line is a line equiarranged with respect to its points,*" "*a plane surface is a surface equiarranged with respect to all straight lines on it,*" etc.). Next, there follow axioms (Euclid calls them *postulates*, though). By way of example, we cite three of them:

A.1. "*It is possible to draw a straight line from any point to any point.*" The modern version of this axiom is: *Through any two distinct points, there passes one and only one straight line.*

A.2. "*It is possible to describe a circle from any center with an arbitrary radius.*"

A.3. Euclid's fifth axiom (postulate) is given below in its modern statement: "*Through a point outside a straight line, there passes one and only one line parallel to the given one.*"

After stating the geometric axioms and several general assertions concerning the properties of quantities (such as: "*equals to the same thing are equal to each other*"—it is these assertions that are called axioms in the *Elements*), Euclid proceeds to proofs. His first theorem is stated as follows: "*There exists an equilateral triangle*" (Euclid's statement: "*On a given bounded straight line,*[1] *an equilateral triangle can be constructed*"). But a careful analysis of Euclid's arguments (even the first basic ones) reveals gaps: Euclid's axiomatics does not enjoy the *completeness* necessary to axiomatically construct a theory in the modern sense of the word. For example, to draw an equilateral triangle ABC, Euclid uses the existence of a common point of two circles of radius $|AB|$ centered at A and B, although this does not follow from his axioms.

A complete axiomatics of the Euclidean plane was described by Hilbert in 1899 in his famous treatise *Foundations of Geometry.*

David Hilbert (1862–1943), one of the greatest mathematicians of the twentieth century. The scope of his works included virtually all the mathematics of his time. He made a fundamental contribution to algebra, geometry, number theory, classical and infinite-dimensional analysis, mathematical physics, logic, and the foundations of mathematics.

In constructing the axiomatics, Hilbert followed Euclid's path. His axiomatics includes five groups of axioms; it is based on the notions of point, line, and plane; on the properties of a point to lie on a line and to lie between two other points; and on the properties of figures to be congruent, i.e., to coincide under superposition. These notions and properties are not defined, but they are related to each other by axioms.

After Hilbert, numerous new axiomatics of the Euclidean plane were proposed. Only ten years passed since Hilbert's work had appeared when F. Schur suggested a system of geometry axioms based on the notion of motion (Schur's basic notions were points, straight lines, plane, and motion). He followed the ideas of Klein. One more decade later, H. Weyl created a vector axiomatics of the Euclidean plane and Euclidean space; we describe it in the next section. We shall also tell about one more axiomatics due to A. N. Kolmogorov and based on four notions, those of point, line, plane, and distance.

However, axiomatic geometry is our secondary concern; in this book, we present geometries in the language of geometric models. We shall start with a description of the model of the Euclidean plane that dates back to Descartes and Fermat.

---

[1]This is the name given to line segments by Euclid.

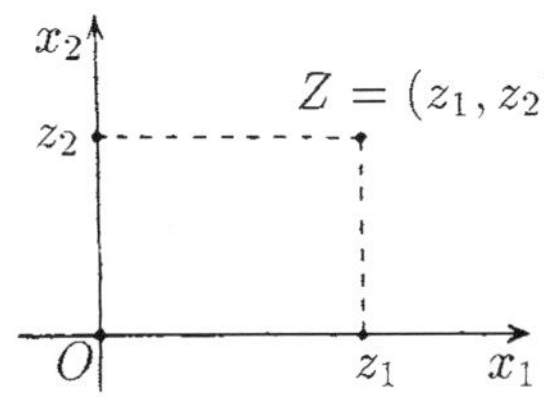

FIGURE 1.1

Cartesian model of the Euclidean straight line and plane. The model of the Euclidean straight line is the set $\mathbb{R}$ of real numbers. Geometrically, the line can be thought of as a rigid infinitely extended rod "without thickness." It cannot be bent, stretched, or compressed. It can only be translated along itself as a solid body and reflected ("rotated") about any point. Having chosen an initial point O, a positive direction, and a scale on such a rod, we can represent our geometric image by its arithmetical model: each point X on the rod is assigned a number x (the coordinate of this point) that equals its distance from the origin with the $+$ sign if the point lies on the positive ray and with the $-$ sign if it lies on the negative ray. After that, we can abandon the geometric image and only deal with numbers. On the real number line, we introduce the distance $d(x, y) = |x - y|$ and the group of motions, i.e., the set of distance-preserving one-to-one transformations of the line. The motions include translations $x \mapsto x + a$, reflections $x \mapsto -x$, and their compositions. The reader can easily prove that all motions can be described in this way.

The Cartesian (the Latin version of Descartes' name is Renatus Cartesius) model of the Euclidean plane is the set of all pairs of real numbers. Geometrically, it can be imagined as the infinitely extended plane of a school blackboard (also "without thickness"). It is assumed that we possess an unbounded straightedge and a compass, which enable us to draw straight lines and arbitrarily centered circles of any radii.

The plane cannot be bent or stretched. It can only be moved as a solid body (parallel to itself or with rotation about some point) or reflected about a straight line; compositions of these transformations are also allowed.

Let us draw two perpendicular straight lines on the plane and denote their intersection point by O (to be definite, we assume that one of these lines, the Ox_1 axis, is horizontal and the other, the Ox_2 axis, is vertical). We call the intersection point O the *origin.* Let us choose a certain scale. If Z is an arbitrary point in the plane, we can project it on the Ox_1 axis by drawing a straight line parallel to the Ox_2 axis until it intersects Ox_1. We denote the distance from this intersection point to the origin by z_1. In a similar manner, we obtain a number z_2. The numbers z_1 and z_2 are called the *coordinates* of the point Z (see Figure 1.1).

Thus in the model described, a *point* is an ordered pair of real numbers and the *plane* is the set of all ordered pairs $X = (x_1, x_2)$, where x_i are real numbers. We shall denote points by uppercase letters, $X, Y, Z, \dots$, and their coordinates by the respective lowercase letters: $X = (x_1, x_2)$, $Y = (y_1, y_2)$, $Z = (z_1, z_2)$, etc. The set of all real numbers is denoted by $\mathbb{R}$ and the set of all pairs of real numbers, by $\mathbb{R}^2$. We have defined points and the plane. It remains to define two more basic concepts, those of a line and distance. A *straight line* is the set of points satisfying a linear equation $a_1x_1 + a_2x_2 = b$, where the numbers a_1 and a_2 are not both zero.

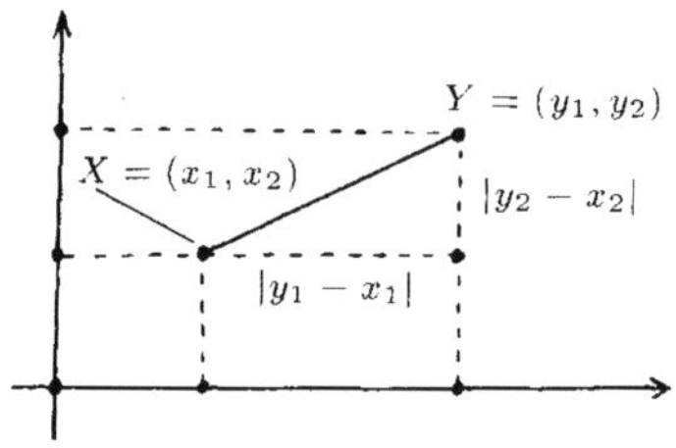

FIGURE 1.2

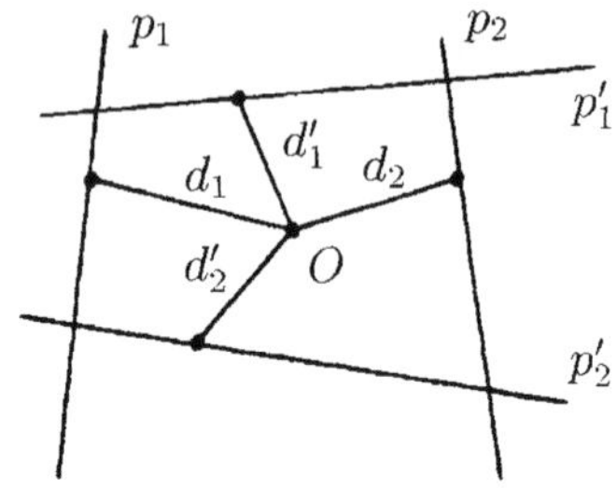

FIGURE 1.3

The triples (a_1, a_2, b) and $(\lambda a_1, \lambda a_2, \lambda b)$ with $\lambda \neq 0$ determine the same line. We shall denote lines by lowercase letters from the middle of the alphabet $(l, m, n, \dots)$.

In this book, we shall encounter various geometries such as affine, projective, Riemann, and Lobachevsky geometries. The Euclidean plane (unlike the affine plane and like the Riemann and Lobachevsky planes) is a *metric space*: its most important feature is that for two arbitrary points, a *distance* between them is defined. In our arithmetical model, the distance $d(X, Y)$ between the points $X = (x_1, x_2)$ and $Y = (y_1, y_2)$ is, by the Pythagorean theorem, $\sqrt{(x_1 - y_1)^2 + (x_2 - y_2)^2}$ (Figure 1.2). A *motion* of the Euclidean plane is a one-to-one map of the plane onto itself that preserves the distances between pairs of points. The motions include parallel translations ($x_1' = x_1 + a_1$, $x_2' = x_2 + a_2$), rotations about the origin through an angle φ ($x_1' = x_1 \cos\varphi + x_2 \sin\varphi$, $x_2' = -x_1 \sin\varphi + x_2 \cos\varphi$), reflections about the Ox_1 axis ($x_1' = x_1$, $x_2' = -x_2$), and all combinations of these transformations. (There are no other motions; we shall prove this later on.)

We obtained the object mentioned in Klein's Erlangen Program, namely, a set of points $X = (x_1, x_2)$ on which the specified group of motions acts transitively.

We have described the arithmetical (or *Cartesian*) model of the Euclidean plane; the Euclidean plane represented in this way is denoted by $\mathbb{E}^2$.

Historical comments. Rectilinear coordinates and the derivation of the equations of straight lines and second-order curves appeared in Pierre de Fermat's treatise *Ad locus planos et solidos isagoge* (Introduction to the theory of plane and spatial figures), which was written in 1636 but was not published until 1679, after Fermat had died. In 1637, René Descartes' famous *Geometry* appeared; in this book, too, the model of the Euclidean plane described above was virtually constructed. At present, this model is called Cartesian.

Interestingly, Fermat and Descartes arrived at the idea of analytic geometry independently, but they were solving the same problem from *Mathematical Collection* of Pappus of Alexandria (third century A.D.). Here is the statement of the problem. Suppose given n straight lines (in Figure 1.3, $n = 4$) divided into two sets: if $n = 2k$, the sets are formed by the lines $p_1, \dots, p_k$ and $p_1', \dots, p_k'$, and if $n = 2k + 1$, by $p_1, \dots, p_k$ and $p_1', \dots, p_{k+1}'$. From an arbitrary point O, we can draw line segments to the given lines at given angles. Let d_i and d_i' be the lengths of the segments drawn to the lines p_i and p_i'. It is required to describe the set of points O for which the ratio of the product of the lengths d_i to the product of the lengths d_i' is constant. Pappus communicates that Apollonius has solved the problem for $n = 3$ and 4. In both cases, the answers are conic sections.

Descartes (1596–1650), philosopher, mathematician, and physicist, the greatest scientist of France. Fermat's rival in many questions of mathematics and natural philosophy. His treatise *Geometry* is a masterpiece of scientific literature. In *Geometry*, Descartes introduced the notion of variable and gave an algebraic description of geometric objects.

In addition to the coordinate system, a folium, an oval and other geometric figures were also named after Descartes.

The name of *Fermat* (1601–1665) is certainly known to everybody thanks to Fermat's Last (or Lost) Theorem. Over centuries, Fermat's Last Theorem caused incredible commotion. Quite recently, it has been proved by A. Wiles.

However, the name of Fermat says a lot to anyone interested in mathematics even irrespective of his great Conjecture. He undoubtedly was one of the most sagacious men of his time, the time of Giants. He was destined to become the creator of number theory and unraveled many of its mysteries. Simultaneously with Descartes, he laid the foundation of analytic geometry; he made fundamental contributions to mathematical analysis and combinatorics.

Shortly before his death Fermat had written: "Perhaps, posterity will be grateful to me for having shown that the Ancients did not know everything." Galileo, Kepler, Descartes, and Fermat laid the foundations for the Science of the New Times.

Now we can, as we did for the line, only deal with the model of the plane rather than its geometric image (but we shall refer to both). Importantly, *all the assertions of Euclidean geometry, in particular, all its axioms, become theorems of algebra and arithmetic in the Cartesian model.*

Our first theorem is a set of assertions about the simplest properties of lines and distances that were cited above as Euclidean postulates.

The statement of the theorem involves two new notions. Although they certainly are well known to the reader, we give their formal definitions.

DEFINITION. A *circle* is the set of points equidistant from some point. A *straight line parallel to a given straight line* is either this line itself or a straight line having no common points with the given one.

THEOREM 1 (properties of lines and circles). *In the Cartesian model of the Euclidean plane,*

(i) *through any two different points, there passes one and only one straight line;*
(ii) *it is possible to draw a circle from any center with an arbitrary radius;*
(iii) *through a point outside a straight line, there passes one and only one line parallel to the given one.*

Proof. (i) Let $Y = (y_1, y_2)$ and $Z = (z_1, z_2)$ be two different points. If $y_1 = z_1$, then the required straight line is the line specified by the equation $x_1 = y_1$. If the first coordinates of the points do not coincide, then the required line is the set of points (x_1, x_2) such that

$$x_2 - \frac{z_2 - y_2}{z_1 - y_1}(x_1 - y_1) = y_2.$$

(ii) The circle of radius $r > 0$ centered at $Z = (z_1, z_2)$ is the set of points (x_1, x_2) such that

$$(x_1 - z_1)^2 + (x_2 - z_2)^2 = r^2.$$

(iii) The straight line passing through the point $Z = (z_1, z_2)$ parallel to the line l specified by the equation $a_1x_1 + a_2x_2 = b$ is described by the equation

$$a_1x_1 + a_2x_2 = a_1z_1 + a_2z_2.$$

We have proved the *existence* of the objects required. The proof of their uniqueness is left to the reader. □

Every notion of Euclidean geometry can be given an algebraic definition in a similar way. The *incidence* relation between a point and a line (in Hilbert's axiomatics) means precisely that the coordinates of the point satisfy the equation of the line; a point $X = (x_1, x_2)$ lies between $A = (a_1, a_2)$ and $B = (b_1, b_2)$ if there exists a number θ, $0 \leq \theta \leq 1$, such that

$$\theta(a_1, a_2) + (1 - \theta)(b_1, b_2) = (x_1, x_2);$$

two figures are called *congruent* if one of them is transformed into the other by a motion.

Let us discuss the proofs of several assertions among the first propositions of Euclid's *Elements* (we give their modern formulations).

PROPOSITION 1. *There exists an equilateral triangle of an arbitrary side length.*

To prove this, it suffices to produce three points at a distance a from each other, say, $(0,0)$, $(0,a)$, and $(a/2, a\sqrt{3}/2)$. (We mentioned that the proof given by Euclid was incomplete.)

PROPOSITION 5. *The base angles of an isosceles triangle are equal.*

Euclid does not give a definition, anywhere near precise, of an angle between the sides of a triangle, and he uses several unmotivated assertions; but we can define the angle between sides BA and CA as

$$\arccos \frac{(b_1 - a_1)(c_1 - a_1) + (b_2 - a_2)(c_2 - a_2)}{d(A,B) \cdot d(A,C)},$$

which obviously implies Proposition 5.

We shall see that Proposition 5 is also valid in Riemann and Lobachevsky geometries. It is believed to be due to Thales.

Thales of Miletus (c. 625–547 B.C.), ancient Greek mathematician. In the history of science he is considered to be the first mathematician. He is given credit for the first proofs in geometry, such as the equality of vertical angles, the equality of base angles in an isosceles triangle, the equality of an angle at a point on a circle subtended by the diameter of the circle to the right angle, etc. His proofs employed motions and the related compositions. Thales became the first "applied mathematician," for he calculated the altitude of a pyramid and the distance to a ship in the sea. He predicted the solar eclipse of May 28, 585 B.C.

Let us also mention a simple formula that describes the situation where three points $X = (x_1, x_2)$, $Y = (y_1, y_2)$, and $Z = (z_1, z_2)$ are collinear, i.e., lie on one straight line. By $\det\{A, B\}$ for $A = (a_1, a_2)$ and $B = (b_1, b_2)$, we denote the expression $a_1 b_2 - a_2 b_1$ (it is called the *determinant* of the matrix $\begin{pmatrix} a_1 & a_2 \\ b_1 & b_2 \end{pmatrix}$). Points X, Y, and Z are collinear if and only if

$$\det\{X, Y\} + \det\{Y, Z\} + \det\{Z, X\} = 0. \tag{1}$$

Indeed, $\det\{A, B\}$ is nothing but the (oriented) *area of the parallelogram spanned by the vectors A and B*. (The reader can either prove this or read the proof on p. 24.) Thus formula (1) has a natural geometric meaning, namely (see Figure 1.4),

area of triangle OXY + area of triangle OYZ + area of triangle OZX

equals zero.

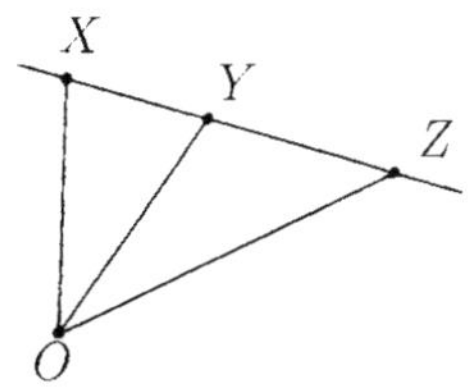

FIGURE 1.4

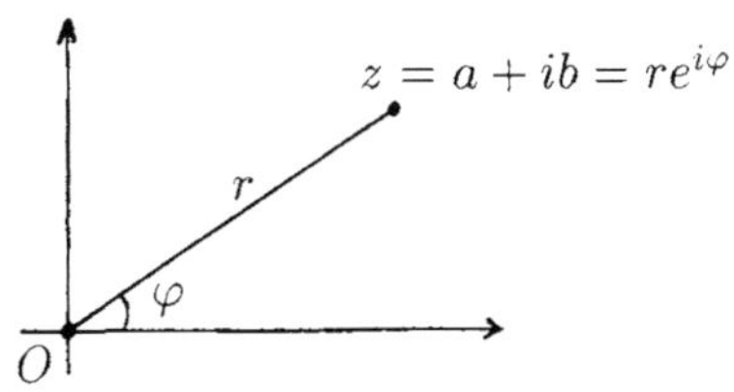

FIGURE 1.5

In Euclidean geometry, any assertion, any problem, and any formula can be proved, solved, or derived algebraically, just as the theorem on properties of lines and circles was proved. As to problems, we shall solve many interesting problems of plane geometry even in this section; to gain more success in solving them, we need to make one more brief excursion into algebra and describe another ("complex") model of the Euclidean plane.

The Euclidean plane and complex numbers. A *complex number* is an expression of the form $a + bi$, where a and b are real numbers and i is a symbol satisfying the relation $i^2 = -1$. The numbers a and b are called, respectively, the *real* and *imaginary* parts of the complex number $z = a + bi$ (the notation is $a = \operatorname{Re} z$ and $b = \operatorname{Im} z$).

Complex numbers are multiplied according to the usual rules for opening parentheses and collecting similar terms; i^2 is always replaced by -1, i.e.,

$$(a + bi)(c + di) = (ac - bd) + (ad + bc)i.$$

Complex numbers can be divided by one another (except by zero, of course):

$$(a + bi) : (c + di) = \frac{(ac + bd) + (bc - ad)i}{c^2 + d^2}.$$

The number $\bar{z} = a - bi$ is called *complex conjugate to* $z = a + bi$.

In the Cartesian coordinate system, there is a one-to-one correspondence $(a, b) \longleftrightarrow (a + bi)$ between the complex numbers and the points in the plane. The multiplication by a complex number z can be given the following geometric interpretation. Let r be the distance from z to zero, and let φ be the angle through which we must rotate the ray containing the positive real half-axis $\{ z \mid z = a + 0i,\ a \geqslant 0 \}$ to obtain a ray passing through z (see Figure 1.5). The numbers r and φ are called the *absolute value* and the *argument* of the number z, respectively (the notation is $r = |z|$ and $\varphi = \arg z$). Multiplication of complex numbers has the following geometric interpretation: *when complex numbers are multiplied, their absolute values are multiplied, and their arguments are added up.* The number z can be written in the form $z = re^{i\varphi}$.

We have obtained one more model of the Euclidean plane, where the points are complex numbers z, the plane itself is the set of all complex numbers (denoted by $\mathbb{C}$), the lines are the sets of points specified by the equations

$$\bar{a}z + a\bar{z} = \beta, \quad \text{where } \beta = \bar{\beta} \text{ and } a \neq 0 \tag{2}$$

(thus each straight line is determined by a pair (a, β), where a is a complex number and β is a real number; for any nonzero real number λ, the pairs (a, β) and $(\lambda a, \lambda\beta)$ determine the same line), and finally, the distance between points z and ζ is $d(z, \zeta) = |z - \zeta|$.

The motions of the Euclidean plane in the complex model are described as follows. A translation is specified by $z' = z + a$, a rotation about the origin by $z' = bz$ with $|b| = 1$, and the reflection about the real axis by $z' = \bar{z}$.

In this complex model, Theorem 1 has a complete analog. The equation of the line through points ζ and ζ' (we denote this line by $\zeta\zeta'$) can be written as $\zeta(\bar{\zeta}' - \bar{z}) - \zeta'(\bar{\zeta} - \bar{z}) + z(\bar{\zeta} - \bar{\zeta}') = 0$ (it easily reduces to the form (2) if multiplied by i); the equation of the circle of radius r centered at ζ has the form $(z - \zeta)(\bar{z} - \bar{\zeta}) = r^2$; and the equation of the line parallel to the line $\bar{a}z + a\bar{z} = \beta$ and passing through a point ζ is $\bar{a}z + a\bar{z} = \bar{a}\zeta + a\bar{\zeta}$.

When you forget a theorem or formula of elementary geometry, there is no need to refer to a textbook: most likely, you can easily derive it by using complex numbers (you only have to apply basic trigonometric formulas and the Euler formula $e^{i\varphi} = \cos\varphi + i\sin\varphi$). By way of example, let us derive two important formulas of Euclidean geometry.

We start with the *law of cosines*, which expresses the length of a side of a triangle in terms of the lengths of the two other sides and the angle between them.

To derive the required formula, we position a triangle ABC whose sides have lengths a and b and the angle between them is φ in such a way that $C = 0$, $A = be^{i\varphi}$, and $B = a$. Let $|AB|$ be the length of the side AB. Then

$$\begin{aligned} |AB|^2 = c^2 = |a - be^{i\varphi}|^2 &= (a - be^{i\varphi})(a - be^{-i\varphi}) \\ &= a^2 + b^2 - 2ab\cos\varphi = |AC|^2 + |BC|^2 - 2|AC||BC|\cos\varphi. \end{aligned}$$

This is the law of cosines. In particular, if $\varphi = \pi/2$, we obtain the *Pythagorean theorem* $c^2 = a^2 + b^2$.

Next, let us prove the *law of sines*, which allows us to evaluate the lengths of two sides of a triangle knowing the length of the third side and the two angles at this side. To derive the formula, we circumscribe a circle about the triangle ABC under consideration (let its radius be R) and place the origin at the center of the circle. Then $A = Re^{i\varphi_1}$, $B = Re^{i\varphi_2}$, $C = Re^{i\varphi_3}$, and therefore,

$$\begin{aligned} |BC|^2 = R^2|e^{i\varphi_2} - e^{i\varphi_3}|^2 &= R^2(e^{i\varphi_2} - e^{i\varphi_3})(e^{-i\varphi_2} - e^{-i\varphi_3}) \\ &= 2R^2(1 - \cos(\varphi_2 - \varphi_3)) = 4R^2\sin^2\frac{\varphi_2 - \varphi_3}{2} = 4R^2\sin^2 A \end{aligned}$$

(the inscribed angle equals half the corresponding central one). Thus $|BC| = 2R\sin A$; similarly, $|AC| = 2R\sin B$ and $|AB| = 2R\sin C$; this implies the law of sines

$$\frac{\sin A}{a} = \frac{\sin B}{b} = \frac{\sin C}{c}.$$

The statements of the law of cosines and the law of sines in spherical, Riemann, and Lobachevsky geometries will be given in the fifth chapter, devoted to these remarkable geometries.

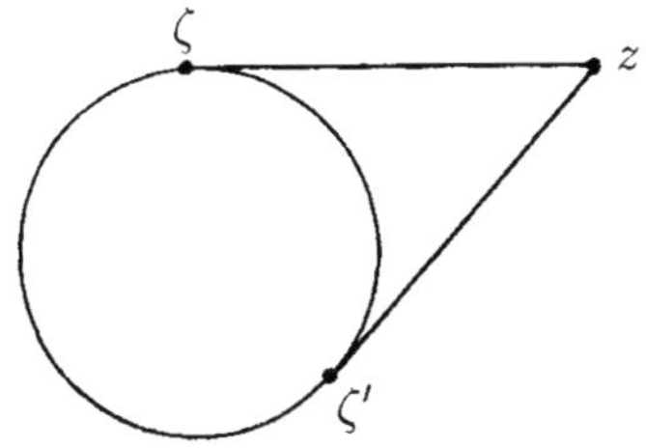

FIGURE 1.6

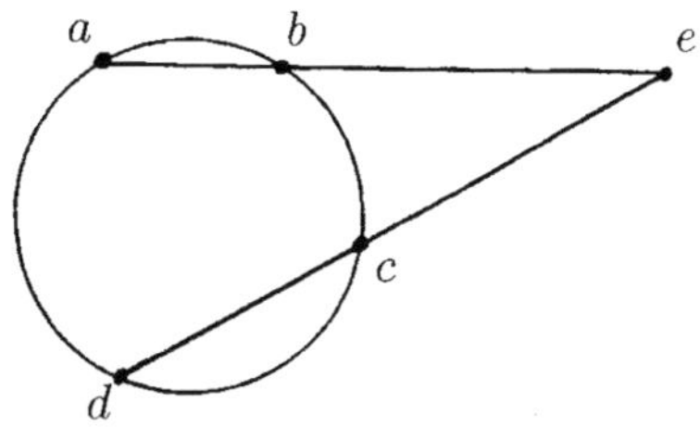

FIGURE 1.7

For $a = b$, the law of sines yields *Thales' theorem*: *the base angles in an isosceles triangle are equal.*

We have gained one more, "complex," key to solving any geometric problems. But before demonstrating how to use our two keys, Cartesian and complex, we shall write out several relations. Their proofs are quite elementary, and we leave them to the reader. First, we introduce two notations. Let z_1, z_2, and z_3 be three complex numbers. The expression

$$(z_1, z_2, z_3) := \frac{z_1 - z_3}{z_2 - z_3}$$

is called the *affine ratio* of these numbers. If z_1, z_2, z_3, and z_4 are four different complex numbers, then the expression

$$(z_1, z_2, z_3, z_4) := \frac{z_1 - z_3}{z_2 - z_3} : \frac{z_1 - z_4}{z_2 - z_4}$$

(i.e., the ratio of the affine ratios (z_1, z_2, z_3) and (z_1, z_2, z_4)) is called the *cross ratio* of these numbers.

The following relations and formulas are valid.

1. *Points z_1, z_2, and z_3 lie on one line if and only if the affine ratio (z_1, z_2, z_3) is real.*

2. *Four points $\{z_1, z_2, z_3, z_4\}$ lie on one line or on one circle if and only if the cross ratio (z_1, z_2, z_3, z_4) is real.*

3. *If z is the intersection point of tangents ζ and ζ' to the unit circle, then*

$$z = \frac{2}{1/\zeta + 1/\zeta'} = \frac{2}{\bar{\zeta} + \bar{\zeta}'}$$

(*in other words, z is the harmonic mean of ζ and ζ'; we denote it by $\gamma(\zeta, \zeta')$*) (Figure 1.6).

4. *If a, b, c, and d are four points on the unit circle $|z| = 1$, then the intersection point e of the lines ab and cd can be found by the formula*

$$e = \frac{(\bar{a} + \bar{b}) - (\bar{c} + \bar{d})}{\bar{a}\bar{b} - \bar{c}\bar{d}}$$

(Figure 1.7).

Some problems. In this section, we solve several problems and, in passing, get acquainted with some of the greatest geometers of all times.

NEWTON'S PROBLEM. *The midpoints of the diagonals in a quadrilateral circumscribed about a circle are collinear with the center of the circle* (Figure 1.8).

Sir Isaac Newton (1643–1727), the greatest genius in the history of mankind. In his 1687 treatise *Philosophiae Naturalis Principia Mathematica* (Mathematical Principles of

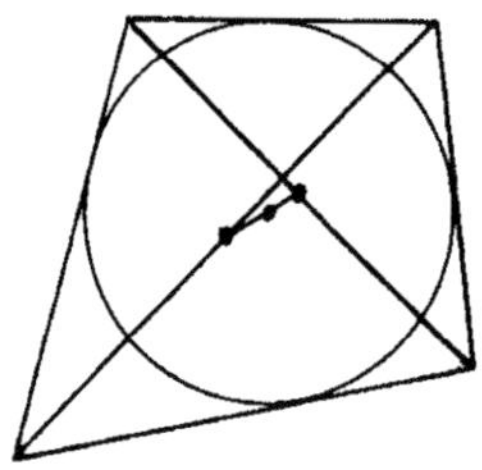

FIGURE 1.8

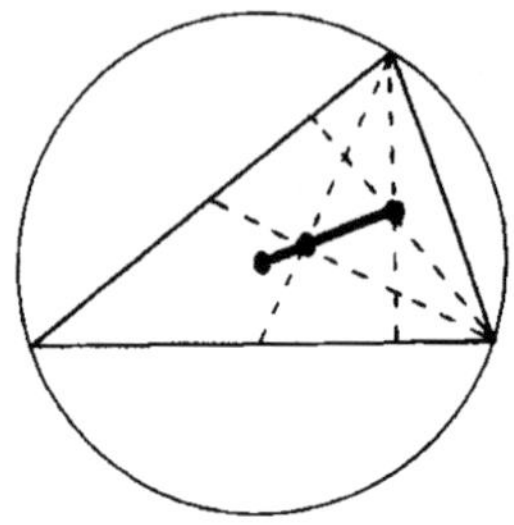

FIGURE 1.9

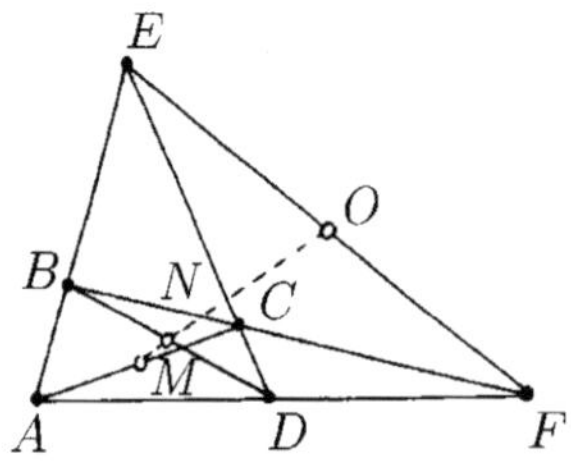

FIGURE 1.10

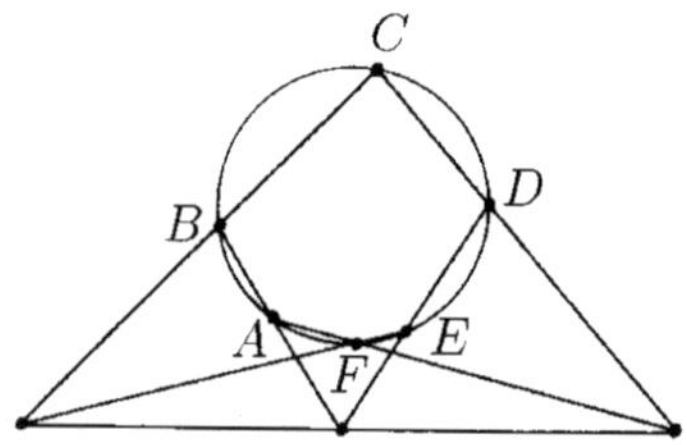

FIGURE 1.11

Natural Philosophy), Newton laid the foundations for the modern natural science and created the System of the Universe. Lagrange called this work "the greatest creation of the human mind." Together with *Leibniz*, Newton is the procreator of the calculus. He made a significant contribution to geometry, too; in particular, he classified third-order curves.

EULER'S LINE. *The center of mass, the orthocenter, and the center of the circumscribed circle of an arbitrary triangle are collinear* (Figure 1.9).

Leonhard Euler (1707–1783), mathematician, physicist, expert in mechanics, and astronomer; the greatest and noblest worker in the history of science. The author of more than 800 treatises and scientific papers that fill up over 90(!) volumes. His correspondence contains about 4000 letters to hundreds of addressees; only a person of outstanding personal qualities could be capable of this. He made great contributions to all fields of mathematics, including geometry. There is abundant evidence that his contemporaries and successors admired him; here is one of them: "Read Euler, he is the teacher of us all ... " (Laplace).

GAUSS' LINE. *Consider an arbitrary quadrilateral $ABCD$. Let us denote the intersection point of the lines AB and CD by E, the intersection point of the lines BC and AD by F, the midpoint of $[A,C]$ by M, the midpoint of $[B,D]$ by N, and the midpoint of $[E,F]$ by O. Then the points M, N, and O are collinear* (Figure 1.10).

Carl Friedrich Gauss (1777–1855) made great contributions to algebra, analysis, and geometry, as well as to physics, astronomy, and geodesy. People used to call him Princeps Mathematicorum (the King of Mathematicians). He discovered the principles of non-Euclidean geometry in parallel with Bolyai and Lobachevsky (Gauss did it earlier, although he never published his studies). Gauss was among the first to suggest a geometric interpretation of complex numbers; he laid the foundations of differential geometry.

PASCAL'S THEOREM. *The intersection points of the lines containing the opposite sides of an inscribed hexagon are collinear* (Figure 1.11).

Blaise Pascal (1623–1662), mathematician, physicist, and philosopher. Pascal (in parallel with Fermat) laid the foundations of combinatorics (recall "the Pascal triangle") and probability theory. He was among those who had developed the principles of the calculus. The famous theorem of Pascal (stated above) is the corner-stone of the building of projective geometry. Pascal proved it at the age of 16. Pascal is the author of the aphorism (we are free to disagree with it, though): "It is not the vastness of the World that arouses admiration, but the human being who measured it."

What a brilliant constellation of names; what a splendid collection of geometric masterpieces! Try to solve these problems geometrically, without resorting to algebra. If you succeed, there will be no end to your delight.

But the knowledge that we have acquired in such a short time and formulas that fit in one page are enough to figure out at once how to solve all these problems, and bringing the job to an end is merely a matter of technique.

You don't believe us? Look!

To solve Newton's problem, it is natural to apply the relations

$$\zeta_1 = \gamma(z_1, z_4), \quad \zeta_2 = \gamma(z_1, z_2), \quad \zeta_3 = \gamma(z_2, z_3), \quad \zeta_4 = \gamma(z_3, z_4)$$

(see assertion 3 on p. 13) and assertion 1 on p. 13 to the points $(\zeta_1 + \zeta_3)/2$, 0, and $(\zeta_2 + \zeta_4)/2$.

To solve Euler's problem, we position the triangle ABC so that its vertices lie at $C = (0,0)$, $B = (a, 0)$, and $A = (b\cos\varphi, b\sin\varphi)$. You can find the coordinates of the center of mass, the center of circumscribed circle, and the orthocenter without much effort:

$$\left(\frac{b\cos\varphi + a}{3}, \frac{b\sin\varphi}{3}\right), \quad \left(\frac{a}{2}, \frac{b - a\cos\varphi}{2\sin\varphi}\right), \quad (b\cos\varphi, (a - b\cos\varphi)\cot\varphi),$$

respectively. After that, it only remains to verify that all the centers are collinear (this can be done by using the theorem about the properties of lines and circles or using formula (1) on p. 10).

Gauss' problem is solved by applying formula (1) on p. 10 three times (to the points M, N, and O); then, the same formula should be applied to the four triples of collinear points.

Let us explain this in more detail. Writing expression (1) for the points M, N, and O, we obtain

$$\begin{aligned} \det\left\{\frac{A+C}{2}, \frac{B+D}{2}\right\} &+ \det\left\{\frac{B+D}{2}, \frac{E+F}{2}\right\} + \det\left\{\frac{E+F}{2}, \frac{A+C}{2}\right\} \\ = \frac{1}{4}\big(&\det\{A,B\} + \det\{C,D\} + \det\{A,D\} + \det\{C,B\} \\ &+ \det\{B,E\} + \det\{D,E\} + \det\{B,F\} + \det\{D,F\} \\ &+ \det\{E,A\} + \det\{F,A\} + \det\{E,C\} + \det\{F,C\}\big); \end{aligned}$$

next, we must apply formula (1) to the triples

$$\{A,B,E\}, \ \{C,B,F\}, \ \{A,D,F\}, \ \{C,D,E\},$$

each belonging to one straight line.

Finally, to prove Pascal's theorem, we use assertion 4 on p. 13. According to this assertion,

$$\bar{h} = \frac{a+b-(d+e)}{ab-de}, \quad \bar{k} = \frac{b+c-(e+f)}{bc-ef}, \quad \bar{g} = \frac{c+d-(f+a)}{cd-fa}.$$

Therefore,

$$\frac{\bar{h}-\bar{k}}{\bar{k}-\bar{g}} \in \mathbb{R},$$

because $\bar{a} = 1/a$, etc. To complete the proof of Pascal's theorem, it remains to apply assertion 1 on p. 13.

All our moves suggest themselves, don't they? Here is the confirmation of the thought expressed in the epigraph to this section.

To conclude, we give solutions of two more problems.

First, we shall describe one "historical" problem.

All the proofs described above, which use the coordinate method, compete with remarkable geometric proofs that have preceded the analytical solutions. The problem suggested below was given an analytic solution at once. The theorem belongs to Princess *Elizabeth* (1618–1680), a student of Descartes.

THE PRINCESS ELIZABETH THEOREM. *All points such that the segments of the tangent lines drawn from these points to two given nonconcentric circles have equal lengths lie on one straight line.* (This line is called the *radical axis* of the two circles.)

Proof. To prove the theorem, it suffices to note that the squared length of the segment of the tangent line drawn from a point (x, y) to the circle of radius R centered at (ξ, η) is

$$(x-\xi)^2 + (y-\eta)^2 - R^2,$$

and therefore the coordinates of all the points under consideration satisfy the equation

$$(x-x_1)^2 + (y-y_1)^2 - R_1^2 = (x-x_2)^2 + (y-y_2)^2 - R_2^2,$$

i.e.,

$$2(x_2-x_1)x + 2(y_2-y_1)y = R_1^2 - R_2^2 + x_2^2 - x_1^2 + y_2^2 - y_1^2. \quad \square$$

Descartes was fascinated by this result. He claimed that he would not undertake the job of proving this theorem even in a month! Of course, this might have been a recherché instance of French courtliness.

Finally, a little story about a butterfly.

Consider a circle, its chord AC with midpoint B, and two of its points P and Q on one side of AC. Take the intersection points S and R of the circle with the lines PB and QB, respectively, draw the line segments $[Q, S]$ and $[P, R]$, and denote their intersection points with AC by N and M. Prove that $|MB| = |NB|$ (Figure 1.12).

The polygon $RPBSQ$ reminds us of a butterfly, doesn't it? For this reason, this problem is known as the *butterfly problem*.

The Butterfly is a lady of a fairly respectable age. Her exact date and place of birth are unknown (she keeps her age secret, as a lady should), but anyway, her first "official birth certificate" appeared in 1815 in the English journal *Gentlemen's Diary* (No. 1029 on pp. 39–40). During all the succeeding years, Butterfly has been giving

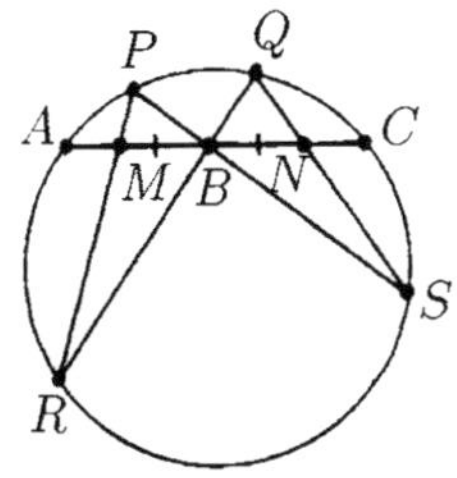

FIGURE 1.12

real pleasure to almost everybody who has happened to make her acquaintance. Butterfly's lovers have devoted many essays to her and written her diverse messages; their stream never runs dry. As time passes, whole surveys appear. The reader may verify this by looking through the paper [**Ba**] of L. Bankoff, for example.

There exist a countless number of solutions to this remarkable problem. One of the "simplest" geometric solutions is given below.

Let us draw a chord PT parallel to AC. Then $|BP| = |BT|$ (i), and the angles PBA, TBC, BPT, and PTB are equal (ii). The angles TPS and TQS are equal or complementary to each other; therefore, we can circumscribe a circle about the quadrilateral $BTQN$, and hence the angles BTN, BQN, and MPB are equal (iii).

It follows from (i)–(iii) that the triangles MPB and BTN are equal, and therefore, $|MB| = |BN|$, as required!

Now try to solve this problem analytically! We shall return to it again (see p. 70).

1.2. n-dimensional Euclidean space

The vector space $\mathbb{R}^n$. Let us generalize the arithmetical model of the Euclidean plane described in the preceding section to the n-dimensional case.

DEFINITION 1. The *vector space* $\mathbb{R}^n$ is the set of all vectors $x = (x_1, \dots, x_n)$, where x_i are real numbers ($x_i \in \mathbb{R}$).

The numbers x_i are called the *coordinates* of the vector x. (We denote vectors by lowercase letters to distinguish them from points.) Two vectors $x = (x_1, \dots, x_n)$ and $y = (y_1, \dots, y_n)$ are regarded as equal if and only if $x_i = y_i$ for $1 \leqslant i \leqslant n$.

There are two natural operations in the space $\mathbb{R}^n$, the addition of vectors and the multiplication of a vector by a number. The *sum of vectors* $x = (x_1, \dots, x_n)$ and $y = (y_1, \dots, y_n)$ is the vector $x + y := (x_1 + y_1, \dots, x_n + y_n)$. The *product of a number* $\lambda \in \mathbb{R}$ *by a vector* $x = (x_1, \dots, x_n)$ is the vector $\lambda x := (\lambda x_1, \dots, \lambda x_n)$.

It is easy to verify that the following relations hold:

(i) $x + y = y + x$;
(ii) $(x + y) + z = x + (y + z)$;
(iii) there exists a zero vector 0 such that $x + 0 = x$ for any vector x;
(iv) any vector x has an opposite vector y such that $x + y = 0$;
(v) $1x = x$;
(vi) $\alpha(\beta x) = (\alpha\beta)x$;
(vii) $\alpha(x + y) = \alpha x + \alpha y$;
(viii) $(\alpha + \beta)x = \alpha x + \beta x$.

A set X with operations of addition of elements and of multiplication of elements by real numbers satisfying conditions (i)–(viii) is called a *vector space*, or *linear space*.

The vector $0 = (0, \dots, 0)$ plays the role of zero vector in $\mathbb{R}^n$.

First examples of vector spaces are the real line $\mathbb{R}$ itself and $\mathbb{R}^n$.

Other examples are the space $\mathcal{P}_{n-1}$ of polynomials of degree at most $n-1$ (we shall explain why $n-1$ rather than n a little later) and the space $C([a,b])$ of continuous functions on the interval $[a,b]$.

Let X and Y be vector spaces. A map $A\colon X \to Y$ from X to Y is said to be *linear* if $A(\lambda_1 x_1 + \lambda_2 x_2) = \lambda_1 A(x_1) + \lambda_2 A(x_2)$ for any $\lambda_1, \lambda_2 \in \mathbb{R}$ and $x_1, x_2 \in X$.

A linear function on $\mathbb{R}^n$ (i.e., a linear map from $\mathbb{R}^n$ to $\mathbb{R}$) is specified by the equality

$$f(x) = a_1x_1 + \cdots + a_nx_n.$$

DEFINITION 2. Vector spaces X and Y are called *isomorphic* if there exists a bijective linear map $\Lambda : X \to Y$. The map Λ is then called an *isomorphism* of X onto Y.

DEFINITION 3. A subset L of $\mathbb{R}^n$ is called a *linear subspace* if, for any vectors $x, y \in L$ and any real numbers $\lambda, \mu \in \mathbb{R}$, the vector $\lambda x + \mu y$ lies in L.

To put it differently, a subset $L \subset X$ of vectors which is a linear space with respect to the actions defined on X is called a linear subspace. The most important examples of linear subspaces in $\mathbb{R}^n$ are the straight lines and the hyperplanes (passing through the origin). A *straight line* is determined by a vector $\bar{x} \neq 0$ and consists of all vectors $\lambda\bar{x}$, where $\lambda \in \mathbb{R}$. A *hyperplane* consists of the vectors whose coordinates satisfy one linear equation $a_1x_1 + \cdots + a_nx_n = 0$, where a_i are not all zero.

Let $a^1, \dots, a^n$ be vectors in $\mathbb{R}^n$. The vector $\lambda_1 a^1 + \cdots + \lambda_n a^n$, where $\lambda_i \in \mathbb{R}$, is called a *linear combination* of the vectors $a^1, \dots, a^n$.

The vectors $a^1, \dots, a^n$ are called *linearly independent* if $\lambda_1 a^1 + \cdots + \lambda_n a^n = 0$ implies that $\lambda^1 = \cdots = \lambda^n = 0$. A family of linearly independent vectors in a subspace $L \subseteq \mathbb{R}^n$ is called a *basis* of L if every vector of the subspace is a linear combination of vectors of this family. All bases consist of the same number of vectors; this number is called the *dimension* of the subspace L. An example of a basis in $\mathbb{R}^n$ is the *canonical basis*, which consists of the n vectors

$$e_1 = (1, 0, \dots, 0), \dots, e_n = (0, \dots, 0, 1).$$

The dimension of $\mathcal{P}_{n-1}$ equals n (this is why we took the degrees of the polynomials to be no greater than $n-1$); the monomials $\{t^k\}_{k=0}^{n-1}$ can be taken as a basis in this space. The space $C([a,b])$ is infinite-dimensional; it contains infinitely many linearly independent elements, for example, the functions $x_n = t^n$, where $n = 0, 1, \dots$.

DEFINITION 4. The *scalar product of vectors* $a = (a_1, \dots, a_n)$ and $b = (b_1, \dots, b_n)$ is the number

$$(a, b) = a_1b_1 + \cdots + a_nb_n.$$

The vector space $\mathbb{R}^n$ with scalar product is known as the *Euclidean* (vector) space.

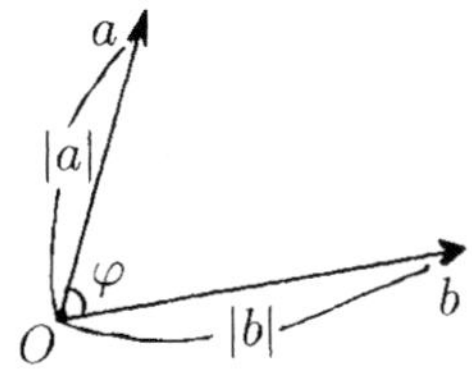

FIGURE 1.13

The scalar product generates the *distance*

$$d(x,y) = \sqrt{(x-y,x-y)} = \big((x_1-y_1)^2+\dots+(x_n-y_n)^2\big)^{1/2}.$$

By definition, $d(x,y) = d(x-y,0)$.

The distance introduced has the following properties:

(i) $d(x,y) \geqslant 0$ for any $x,y \in \mathbb{R}^n$, and $d(x,y) = 0$ if and only if $x = y$;

(ii) $d(x,y) = d(y,x)$;

(iii) $d(x,z) \leqslant d(x,y) + d(y,z)$ (the *triangle inequality*).

The first two properties are obvious. We shall start with the proof of the triangle inequality for $y = 0$. It is based on the well-known Cauchy–Bunyakovskii–Schwarz inequality

$$\Big|\sum_{i=1}^n x_i z_i\Big| \leqslant \Big(\sum_{i=1}^n x_i^2\Big)^{1/2}\Big(\sum_{i=1}^n z_i^2\Big)^{1/2}. \tag{1}$$

This inequality gives

$$\begin{aligned} d^2(x,z) &= \sum_{i=1}^n (x_i - z_i)^2 = \sum_{i=1}^n (x_i^2 + z_i^2) - 2\sum_{i=1}^n x_i z_i \\ &\overset{(1)}{\leqslant} \sum_{i=1}^n x_i^2 + 2\Big(\sum_{i=1}^n x_i^2\Big)^{1/2}\Big(\sum_{i=1}^n z_i^2\Big)^{1/2} + \sum_{i=1}^n z_i^2 \\ &= (d(x,0) + d(z,0))^2. \end{aligned}$$

Now we obtain

$$d(x,z) = d(x-y,z-y) \leqslant d(x-y,0) + d(z-y,0) = d(x,y) + d(z,y),$$

as required.

Thus we have made $\mathbb{R}^n$ a metric space. The number

$$|x| := d(x,0) = \Big(\sum_{i=1}^n x_i^2\Big)^{1/2}$$

is called the *length* of the vector x.

Vectors a and b are called *orthogonal* if $(a,b) = 0$. The *angle* between nonzero vectors a and b is the number $\arccos\big((a,b)/|a||b|\big)$.

Thus the scalar product of two vectors a and b equals the product of the lengths of these vectors by the cosine of the angle between them (see Figure 1.13).

Vectors $f_1,\dots,f_k$ in $\mathbb{R}^n$ are called *orthonormal* if

$$(f_i,f_i) = 1, \quad (f_i,f_j) = 0 \quad \text{for} \quad i \neq j.$$

It is easy to show that n orthonormal vectors form a basis in $\mathbb{R}^n$. Indeed, if $\lambda_1 f_1 + \dots + \lambda_n f_n = 0$, then $\lambda_1(f_1,f_i) + \dots + \lambda_n(f_n,f_i) = 0$, i.e., $\lambda_i = 0$.

An example of an orthonormal basis is the canonical basis introduced above. At this point, we shall digress a bit.

The affine space $\mathbb{A}^n$ and the Euclidean space $\mathbb{E}^n$.

The Euclidean (vector) space $\mathbb{R}^n$ with $n = 2$ is not a generalization of the school plane geometry as yet. The true generalization is formed in two stages. At the first stage, *points* and *vectors* arise. The generalization of such a structure is the notion of *affine space*. Affine spaces are obtained from vector spaces by ignoring the origin and even coordinates in general. Next, this structure is endowed with a metric, and we obtain the n-dimensional Euclidean space generalizing the Euclidean plane studied at school.

DEFINITION 5. The *affine space over* $\mathbb{R}^n$ consists of the points $A = (a_1, \dots, a_n)$, where $a_i \in \mathbb{R}$. Each ordered pair of points $A = (a_1, \dots, a_n)$ and $B = (b_1, \dots, b_n)$ is assigned a vector $\overrightarrow{AB} = (b_1 - a_1, \dots, b_n - a_n)$ with initial point A and endpoint B. The point $O = (0, \dots, 0)$ is the *origin of coordinates* in this space.

Thus the affine space over $\mathbb{R}^n$ (we denote it by $\mathbb{A}^n$) is a set of points; each point $A \in \mathbb{A}^n$ is assigned a copy of the linear space $\mathbb{R}^n$, namely, the space of vectors $\overrightarrow{AB}$, where A and B are points of $\mathbb{A}^n$.

Axiomatically, an affine space $\mathbb{A}^n$ is a triple $(\mathbb{A}^n, X, +)$, where $\mathbb{A}^n$ is a set of points, X is a linear space, and "+" is the operation of applying a vector to a point; these objects must satisfy the following three axioms:

(i) for each point $A \in \mathbb{A}^n$ and any $x, x' \in X$, we have $(A+x)+x' = A+(x+x')$;

(ii) $A + 0 = A$ for all $A \in \mathbb{A}^n$;

(iii) for any $A, B \in \mathbb{A}^n$, there exists a unique vector $x \in X$ such that $A + x = B$.

The "standard" n-dimensional affine space (or merely the affine n-space) $\mathbb{A}^n$ over $\mathbb{R}^n$ is the triple $(\mathbb{R}^n, \mathbb{R}^n, +)$ formed by the points $A = (a_1, \dots, a_n)$, the vectors $x = (x_1, \dots, x_n) \in \mathbb{R}^n$, and the operation $+$ of applying a vector to a point. Setting the origin at the point $O = (0, \dots, 0)$, we can identify the points of this affine space with the vectors of $\mathbb{R}^n$; namely, a point $A = (a_1, \dots, a_n)$ is identified with the vector $a = (a_1, \dots, a_n) = \overrightarrow{OA}$. For this reason, we sometimes denote points and the corresponding vectors by the same letters (although, the "correct" notation assumes the use of capitals for points and lowercase letters for vectors).

It is pertinent here to explain how a line passing through two points and a line segment joining two points are described in the affine space.

The line passing through two points A and B with coordinates $a = (a_1, \dots, a_n)$ and $b = (b_1, \dots, b_n)$ is the set of points $\lambda_1 a + \lambda_2 b$, where $\lambda_1 + \lambda_2 = 1$; the line segment $[A, B]$ is the set of points $\lambda_1 a + \lambda_2 b$, where $\lambda_i \geqslant 0$ and $\lambda_1 + \lambda_2 = 1$.

In more detail, affine spaces are considered in Chapter 2. Here, we shall pass to the n-dimensional generalization of the Euclidean plane.

DEFINITION 6. The *n-dimensional Euclidean space (or merely the Euclidean n-space)* $\mathbb{E}^n$ *over* $\mathbb{R}^n$ with scalar product $(x, x') = \sum_{i=1}^n x_i x_i'$ is the point set $\mathbb{A}^n$ specified in the definition of affine space endowed with the distance between points defined as $d(A, A') = \left(\sum_{i=1}^n (a_i' - a_i)^2\right)^{1/2}$ (here $A = (a_1, \dots, a_n)$ and $A' = (a_1', \dots, a_n')$).

We denote this space by $\mathbb{E}^n$.

Thus the Euclidean space $\mathbb{E}^n$ is the triple $(\mathbb{A}^n, (\mathbb{R}^n, (\cdot,\cdot)), +)$ consisting of the points of the affine space $\mathbb{A}^n$, the vector n-space, and the operation of applying a vector to a point. The space $\mathbb{E}^n$ is a metric space.

We have described points and distances between points. Now is the right time to say something about motions in Euclidean spaces. As in the case of the plane, motions are distance-preserving one-to-one maps.

Let $X \mapsto f(X)$ be a motion of the Euclidean space $\mathbb{E}^n$ with origin O, and let $A = f(O)$. Then $f(X) = A + g(\overrightarrow{OX})$, where g is a transformation of the vector space such that $g(0) = 0$. We have

$$|g(\overrightarrow{OX}) - g(\overrightarrow{OY})| = d(f(X), f(Y)) = d(X, Y) = |\overrightarrow{XY}|.$$

Thus $|g(x) - g(y)|^2 = |x - y|^2$, and therefore $(g(x), g(y)) = (x, y)$.

Put $\varepsilon_i = g^{-1}(e_i) = a_{i1}e_1 + \cdots + a_{in}e_n$, where $e_1, \ldots, e_n$ is the canonical basis. Then $g(\varepsilon_i) = e_i$; therefore,

$$(g(x), e_i) = (x, \varepsilon_i) = x_1a_{i1} + \cdots + x_na_{in},$$

i.e., $g(x) = x_1'e_1 + \cdots + x_n'e_n$, where $x_i' = x_1a_{i1} + \cdots + x_na_{in}$. In particular, g is a linear transformation. The vectors $\varepsilon_1, \ldots, \varepsilon_n$ satisfy the relation $(\varepsilon_i, \varepsilon_j) = (e_i, e_j)$, i.e., $\sum_{k=1}^n a_{ik}^2 = 1$ and $\sum_{k=1}^n a_{ik}a_{jk} = 0$ for $i \neq j$. The matrices (a_{ij}) whose elements satisfy these relations are called *orthogonal*, and the linear transformations g with orthogonal matrices (a_{ij}) are also called *orthogonal*.

Thus every motion of the Euclidean space is the composition of a translation $x \mapsto a+x$ and an orthogonal transformation. It is easy to verify that the converse is also true, i.e., both the translations and the orthogonal transformations are motions.

Historical comments. Fermat and Descartes, the creators of analytic geometry on the plane, introduced the correspondence between points and pairs of numbers in a somewhat obscure form; otherwise it would not have taken a century to pass from the plane to three-dimensional space, while the principles of three-dimensional analytic geometry were laid down by Clairaut only in 1731.

Alexis-Claude Clairaut (1713–1765), French mathematician and astronomer. Began his scientific activity at the age of 12. At 18, published a work on the analytic and differential geometry of three-dimensional space.

The arithmetical model of n-dimensional space appeared only a century later. The main honor of its creation is owed to *W. R. Hamilton* (1805–1865).

Hamilton made important contributions to geometry. He was the first to introduce the notion of "vector" and to define vector spaces. He also discovered a new numerical system, quaternions.

1.3. Introduction to the multidimensional world of Euclidean geometry

At school, we first study Euclidean plane geometry and then, space geometry.

The basic object of plane geometry is a line, and that of space geometry, a plane. After these notions are introduced, the world of geometry is filled with triangles, circles, squares, balls, pyramids, and other figures. First, we shall describe the generalization of lines and planes to Euclidean space. A more detailed description of these subsets of $\mathbb{A}^n$ can be found in linear algebra courses.

Affine varieties. Euclidean and affine n-spaces are constructed similarly to the two-dimensional ones. But to feel at home in the multidimensional world, we must acquire some knowledge of linear algebra. Below, relevant information and (when possible) its geometric interpretation are presented.

We start with the generalization of lines and planes. A straight line in the arithmetical model can be defined in two ways. First, it is the set of points $X = (x_1, x_2)$ that satisfy an equation of the form $a_1x_1 + a_2x_2 = b$; secondly, this is the set of points $X + \lambda e$, where e is a fixed vector in $\mathbb{R}^2$ and X is a fixed point in the plane.

The second definition also works in the multidimensional case: the set of points $X + \lambda e$, where X is a point in the Euclidean n-space and e is a vector in $\mathbb{R}^n$, is a straight line in $\mathbb{E}^n$. The first definition gives another important notion, namely, the notion of hyperplane. A *hyperplane* in the Cartesian model of the Euclidean n-space is given by the equation

$$a_1x_1 + a_2x_2 + \cdots + a_nx_n = b, \tag{1}$$

i.e., it is the level set of a linear function. In our "ordinary" 3-space, the sets so specified are precisely the ordinary planes.

A straight line is a one-dimensional object and a hyperplane, an $(n-1)$-dimensional one. What does this mean? and what lies between 1 and $n-1$? Our immediate purpose is to define the k-dimensional analogs of straight lines and planes in n-space; these analogs are called k-dimensional affine varieties. The straight lines and planes have the following property: together with any two points, they contain the entire line passing through these points. Sets with this property in n-space are called *affine varieties*.

One-dimensional affine varieties (straight lines) are sets of points $X+\lambda e$, where e is a fixed n-dimensional vector and $\lambda \in \mathbb{R}$. It is natural to define the two-dimensional affine varieties—two-dimensional affine planes—by analogy, as the sets of points $X+\lambda_1e_1+\lambda_2e_2$, where e_1 and e_2 are two fixed linearly independent vectors of $\mathbb{R}^n$ and λ_1 and λ_2 are arbitrary real numbers.

The k-dimensional affine variety is the set of points $X + \sum_{i=1}^k \lambda_i e_i$, where $\{e_i\}_{i=1}^k$ are fixed linearly independent vectors of $\mathbb{R}^n$ and $\{\lambda_i\}_{i=1}^k$ are arbitrary real numbers. Clearly, the k-dimensional affine varieties coincide with the translated k-dimensional linear subspaces. It is proved in linear algebra that the k-dimensional affine varieties can be described in a different ("dual") way as the intersections of $n-k$ hyperplanes. A hyperplane has dimension $n-1$. For example, the hyperplane specified by the equation $x_n = 0$ is the set of points

$$O + \sum_{i=1}^{n-1} \lambda_i e_i,$$

where

$$e_1 = (1, 0, \ldots, 0), e_2 = (0, 1, \ldots, 0), \ldots, e_{n-1} = (0, \ldots, 1, 0).$$

Linear algebra studies the theory of linear (and then, quadratic) functions and linear equations in the space $\mathbb{R}^n$.

Linear functions are functions of the form

$$f(x) = a_1x_1 + \cdots + a_nx_n.$$

A system of m equations in n unknowns is given by m linear equations

$$f_1(x) = b_1, \dots, f_m(x) = b_m$$

or, in expanded form, by

$$\begin{gathered} a_{11}x_1 + \cdots + a_{1n}x_n = b_1, \\ \dots\dots\dots\dots\dots\dots \\ a_{m1}x_1 + \cdots + a_{mn}x_n = b_m. \end{gathered}$$

Thus, geometrically, the theory of linear equations is the description of intersections of families of hyperplanes in $\mathbb{R}^n$.

In comparatively recent times it has become clear that an important role in practice is played by problems that involve *linear inequalities* and have the following form: find the minimum of the linear function

$$f_0(x) = c_1x_1 + \cdots + c_nx_n$$

subject to the constraints

$$f_i(x) = a_{i1}x_1 + \cdots + a_{in}x_n \leqslant b_i, \quad 1 \leqslant i \leqslant m.$$

Such problems are called *linear programming* % problems.

A table of the form

$$A = \begin{pmatrix} a_{11} & \dots & a_{1n} \\ \dots & \dots & \dots \\ a_{m1} & \dots & a_{mn} \end{pmatrix} = (a_{ij})_{\substack{1\leqslant i\leqslant m \\ 1\leqslant j\leqslant n}}$$

is said to be a *matrix* of size m-by-n (or an $m \times n$ matrix).

Matrices are multiplied by the rule:

$$A = (a_{ij})_{\substack{1\leqslant i\leqslant m, \\ 1\leqslant j\leqslant n}} \quad B = (b_{jk})_{\substack{1\leqslant j\leqslant n \\ 1\leqslant k\leqslant s}} \implies AB = (c_{ik})_{\substack{1\leqslant i\leqslant m, \\ 1\leqslant k\leqslant s}}$$

where

$$c_{ik} = \sum_{j=1}^{n} a_{ij}b_{jk}.$$

A matrix has m rows, each of which can be interpreted as an n-dimensional row vector, and n columns, each of which is interpreted as an m-dimensional column vector.

Among matrices, the square matrices, i.e., matrices of size $n \times n$, are of special importance.

The system of m equations in n unknowns considered above can be written in the abbreviated form $Ax = b$, where the vectors x and b are represented as column vectors, or $m \times 1$ matrices.

Two very important notions have to do with matrices—the notions of *determinant* and *rank.* Both of them have geometric interpretations. We start with determinants.

Determinants and volumes. Let us represent a parallelogram $A_0A_1A_2A_3$ in the plane in vector form. We take the point A_0 as the origin and denote $a^1 = \overrightarrow{A_0A_1}$ and $a^2 = \overrightarrow{A_0A_2}$. The vertices of the parallelogram are then the origin and the three vectors a^1, a^2, and $a^1 + a^2$, and the parallelogram itself consists of the points $x = \lambda_1a^1 + \lambda_2a^2$, where $0 \leqslant \lambda_i \leqslant 1$. Similarly, a parallelepiped in 3-space (with a vertex at the origin) consists of the points $\sum_{i=1}^3 \lambda_ia^i$, where $0 \leqslant \lambda_i \leqslant 1$.

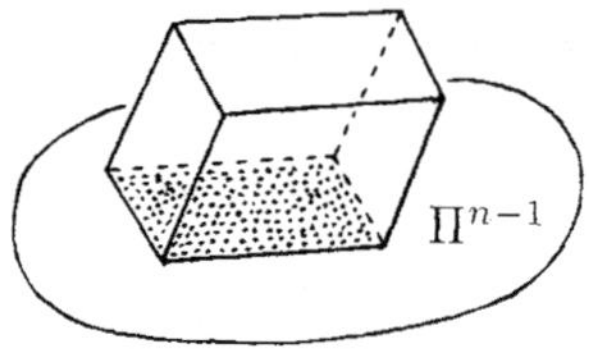

FIGURE 1.14

This is very easily extended to the k-dimensional case. Again, we take A_0 as the origin and consider the vectors $a^i = \overrightarrow{A_0 A_i}$. The set of "endpoints" of the vectors $\sum_{i=1}^{k} t_i a^i$, where $0 \leqslant t_i \leqslant 1$, is called the *n-dimensional parallelepiped*, or simply n-parallelepiped, in $\mathbb{R}^n$.

If the vectors $\{a^i\}_{i=1}^n$ are pairwise orthogonal, the parallelepiped is called *rectangular*. The lengths of the vectors $a^1, \ldots, a^n$ are the edge lengths of the parallelepiped. A rectangular parallelepiped with equal edges is called a *cube*. The volume of a unit n-cube is usually taken to be the unit volume; on this basis, measure theory in the space $\mathbb{R}^n$ can be constructed.

Let us define a k-face of an n-parallelepiped in $\mathbb{R}^n$. We start with an $(n-1)$-face. A hyperplane Π^{n-1} is said to be a face hyperplane of an n-parallelepiped $M^n \subset \mathbb{R}^n$ if M^n lies "on one side" of Π^{n-1} and the intersection of M^n with Π^{n-1} is an $(n-1)$-parallelepiped (see Figure 1.14). The intersection of a face hyperplane with an n-parallelepiped M^n is called an *$(n-1)$-dimensional face*, or an *$(n-1)$-face*, of the parallelepiped M^n. For the n-parallelepiped $X + \sum_{i=1}^{n} t_i a^i$, each face of dimension $n-1$ consists of the vectors $X + \varepsilon a^k + \sum_{i \neq k} t_i a^i$, where $\varepsilon = 0$ or 1 and $0 \leqslant t_i \leqslant 1$.

Thus an $(n-1)$-face of an n-parallelepiped is an $(n-1)$-parallelepiped. This allows us to define an *$(n-2)$-face* as an $(n-2)$-face of an $(n-1)$-face, etc. The 0-faces are called *vertices* and the 1-faces, *edges* of the parallelepiped.

Each $(n-k)$-face of an n-parallelepiped consists of the endpoints of the vectors

$$X = \sum_{j=1}^{k} \varepsilon_j a^{m_j} + \sum_{i \neq m_1, \ldots, m_k} t_i a^i;$$

therefore, the number of $(n-k)$-faces in an n-parallelepiped is $2^k \binom{n}{n-k}$.

Next, let us define the *volume of a parallelepiped*.

Let us return for a short while to our familiar Euclidean plane. Choose an orthogonal Cartesian coordinate system and consider the parallelogram $ABCO$ (see Figure 1.15) generated by the vectors $a_1 = (a_{11}, a_{21})$ and $a_2 = (a_{12}, a_{22})$. Evaluating the area of the triangle OAC and doubling the result, we obtain the area of the parallelogram, which is

$$a_{12}a_{22} + (a_{11} - a_{12})(a_{21} + a_{22}) - a_{11}a_{21} = a_{11}a_{22} - a_{12}a_{21}.$$

More precisely, the area of the parallelogram equals this number if the vectors a_1 and a_2 are arranged as shown in Figure 1.15. In the general case, the area is $|a_{11}a_{22} - a_{21}a_{12}|$.

The columns of the coordinates of the vectors a_1 and a_2 form the matrix

$$A = \begin{pmatrix} a_{11} & a_{12} \\ a_{21} & a_{22} \end{pmatrix}.$$

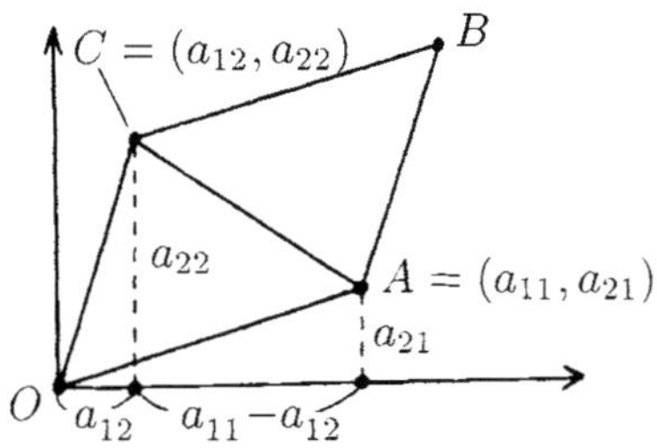

FIGURE 1.15

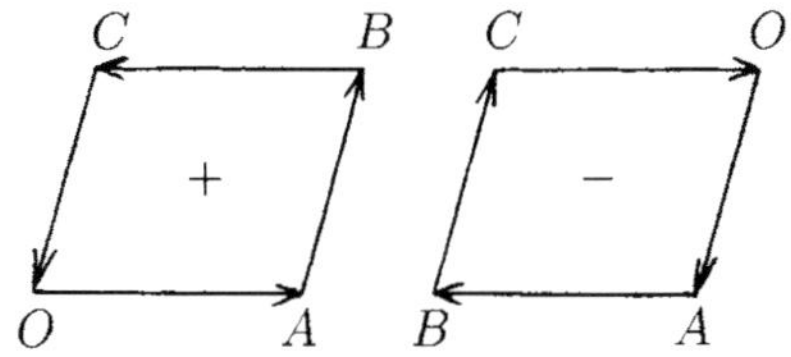

FIGURE 1.16

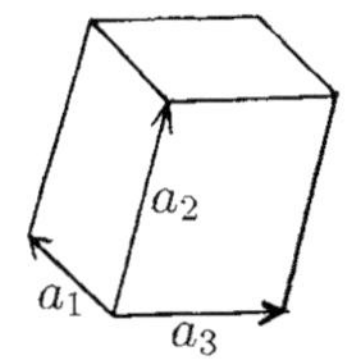

FIGURE 1.17

The *determinant* of the matrix A is the number $a_{11}a_{22} - a_{12}a_{21}$. This number can be called the *oriented area of the parallelogram* $ABCO$. The orientation of the parallelogram is defined by specifying the direction of traversing its vertices. A change of the parallelogram orientation results in a change of the sign of the oriented area (see Figure 1.16). The determinant of the matrix A is denoted by $\det A$ or by $\begin{vmatrix} a_{11} & a_{12} \\ a_{21} & a_{22} \end{vmatrix}$; A is said to be *nonsingular* if $\det A \neq 0$. The nonsingular matrices form a group denoted by $GL(2, \mathbb{R})$.

Next, let us define the volume of the three-parallelepiped generated by vectors $\{a_1, a_2, a_3\}$ (see Figure 1.17). We start with the unoriented volume. It can be defined as $V = hS$, where S is the area of the face $\{a_1, a_2\}$ and h is the length of the projection of a_3 on the normal to the plane of this face. Now, let us pass to the definition of the oriented volume $V(a_1, a_2, a_3)$ of the parallelepiped generated by vectors a_1, a_2, a_3. The definition must be consistent with the relation $|V(a_1, a_2, a_3)| = h|S(a_1, a_2)|$. We define the number $V(a_1, a_2, a_3)$ by the following four conditions:

(i) $V(\dots a_i, a_{i+1} \dots) = -V(\dots a_{i+1}, a_i \dots)$;
(ii) $V(\dots \lambda a_i \dots) = \lambda V(\dots a_i \dots)$;
(iii) $V(\dots a_i + b_i \dots) = V(\dots a_i \dots) + V(\dots b_i \dots)$;
(iv) $V(e_1, e_2, e_3) = 1$, where $e_1 = (1, 0, 0)$, $e_2 = (0, 1, 0)$, and $e_3 = (0, 0, 1)$.

Condition (i) is consistent with the definition of unoriented volume, because

$$h_1 S_{23} = h_2 S_{13} = h_3 S_{12}.$$

Conditions (ii) and (iv) do not require special comments, and condition (iii) comes from the observation that the projection of the sum of two vectors on some unit vector equals the sum of their projections (the projections are taken with signs).

First, we verify that (i)–(iv) determine the function $V(a_1, a_2, a_3)$ uniquely. Conditions (ii) and (iii) show that

$$V\Big(\sum_i a_{i1}e_i, \sum_j a_{j2}e_j, \sum_k a_{k3}e_k\Big) = \sum_{i,j,k} a_{i1}a_{j2}a_{k3}V(e_i, e_j, e_k).$$

Condition (i) implies that $V(e_i, e_j, e_k)$ equals $\pm V(e_1, e_2, e_3)$ if the numbers i, j, and k are distinct, and vanishes otherwise. The sign is plus if transforming the triple (i, j, k) into the triple $(1, 2, 3)$ requires an even number of transpositions (permutations of two neighboring elements), and the sign is minus if the number of transpositions is odd.

The application of (iv) completes the evaluation of $V(a_1, a_2, a_3)$. As a result, we obtain

$$V(a_1, a_2, a_3) = \sum (-1)^\sigma a_{i1} a_{j2} a_{k3},$$

where $(-1)^\sigma$ is the sign of the permutation $\sigma = \left(\begin{smallmatrix} 1 & 2 & 3 \\ i & j & k \end{smallmatrix}\right)$. It is easy to verify that the function $V(a_1, a_2, a_3)$ specified by this formula does satisfy conditions (i)–(iv), i.e., these conditions are consistent and uniquely determine the function $V(a_1, a_2, a_3)$. The consistency of conditions (i)–(iv) with the relation $V = hS$ implies that $|V(a_1, a_2, a_3)|$ coincides with the unoriented volume of the parallelepiped generated by the vectors a_1, a_2, and a_3.

DEFINITION. The *oriented volume* $V(a_1, \dots, a_n)$ of the parallelepiped generated by vectors $a_1, \dots, a_n$ in the space $\mathbb{R}^n$ is the number

$$V(a_1, \dots, a_n) = \sum (-1)^\sigma a_{i_1 1} \cdots a_{i_n n},$$

where $(-1)^\sigma$ is the sign of the permutation $\sigma = \left(\begin{smallmatrix} 1 & \dots & n \\ i_1 & \dots & i_n \end{smallmatrix}\right)$ and a_{ij} are the coordinates of the vector a_j.

The number $V(a_1, \dots, a_n)$ specified by this formula is called the *determinant of the matrix* $A = (a_{ij})_{1 \leqslant i,j \leqslant n}$ and denoted by $\det A$. The nonsingular $n \times n$ matrices form the group $GL(n, \mathbb{R})$.

For $b_{ij} = a_{ji}$, $A^T = (b_{ij})_{1 \leqslant i,j \leqslant n}$ is the transposed matrix. The elements of the matrix $A^T A$ are the scalar products of the vectors

$$a_i = (a_{1i}, \dots, a_{ni}), \qquad a_j = (a_{1j}, \dots, a_{nj}).$$

The determinant of this matrix is called the *Gramian*. It equals the squared determinant of the matrix A, i.e., the squared volume of the parallelepiped spanned by the vectors $a_1, \dots, a_n$.

We have considered determinants. Now, let us go to the rank of a matrix. We begin with an algebraic definition.

Let $A = (a_{ij})_{\substack{1 \leqslant i \leqslant m \\ 1 \leqslant j \leqslant n}}$ be an $m \times n$ matrix. The determinant of the matrix formed by the elements in the intersections of p rows and p columns of A is called a pth-order *minor* of the matrix A. The maximum order of a nonzero minor is called the *rank* of the matrix. It is proved in linear algebra that the rank of a matrix coincides with the maximum number of linearly independent columns and the maximum number of linearly independent rows of the matrix.

This definition can be given the following geometric interpretation. The correspondence $x \mapsto Ax$, where x is an n-dimensional column vector and A is an $m \times n$ matrix, is interpreted as a linear map from $\mathbb{R}^n$ to $\mathbb{R}^m$. The rank of the matrix is the dimension of the image of $\mathbb{R}^n$ under this map.

One more additional remark. We have been talking about the real Euclidean world. But we might similarly construct the complex world based on the complex vector space $\mathbb{C}^n$ with the scalar product $(z, \xi) = z_1 \bar{\xi}_1 + \dots + z_n \bar{\xi}_n$. (Moreover, the theory can be constructed over an arbitrary field, but it is difficult to give a geometric interpretation in the general case.)

Simplices and balls. A set C in an affine space is said to be *convex* if, together with any two points $X, Y \in C$, it contains the entire line segment $[X, Y]$.

The *convex hull* of points $A_1, \dots, A_N$ in an affine space is the smallest convex set containing these points (i.e., the intersection of all convex sets that contain $A_1, \dots, A_N$).

Alternatively, the convex hull can be defined as follows. Choose an arbitrary point O and consider the set C consisting of the endpoints of the vectors $\lambda_1\overrightarrow{OA_1} + \dots + \lambda_N\overrightarrow{OA_N}$, where $\lambda_i \geqslant 0$ and $\sum \lambda_i = 1$. It is easy to verify that the set C does not depend on the choice of O. For two points A_1 and A_2, this is the segment $[A_1, A_2]$. After that, it is easy to show that in the general case, C is the convex hull of the points $A_1, \dots, A_N$.

A set of points X_i with $0 \leqslant i \leqslant k$ in the affine n-space is called *affinely independent* if the vectors $\overrightarrow{X_0X_i}$ with $1 \leqslant i \leqslant k$ are linearly independent. The convex hull of $n+1$ affinely independent points in the affine n-space is called an *n-dimensional simplex*, or an *n-simplex*. A one-dimensional simplex is a segment, a two-dimensional simplex is a triangle, and a three-dimensional simplex is a tetrahedron. Among all simplices in Euclidean n-space, especially remarkable ones are the *regular simplices*, namely those whose edges X_iX_j are all of equal length.

The *ball* of radius r centered at a point O in a Euclidean space is the set of points whose distance from O does not exceed r; the set of points whose distance from O is exactly r is called the *sphere* of radius r centered at O.

The multidimensional world of Euclidean geometry is discussed in detail in the Addendum.

Problems

Multidimensional spaces.

1.1. Prove that if the vertices of a triangle have rational coordinates, then the center of the circumscribed circle has rational coordinates.

1.2. Find the angle between the planes

$$a_1x_1 + a_2x_2 + a_3x_3 + a_4 = 0 \quad \text{and} \quad b_1x_1 + b_2x_2 + b_3x_3 + b_4 = 0$$

in the space $\mathbb{R}^3$.

1.3. Find the coordinates of the vector parallel to the intersection line of the planes

$$a_1x_1 + a_2x_2 + a_3x_3 + a_4 = 0 \quad \text{and} \quad b_1x_1 + b_2x_2 + b_3x_3 = 0$$

in $\mathbb{R}^3$.

1.4. Specify two 2-dimensional planes in $\mathbb{R}^4$ that intersect at exactly one point.

1.5. Specify two skew (i.e., nonparallel and disjoint) 2-dimensional planes in $\mathbb{R}^4$.

1.6. A point A lies inside a sphere in the space $\mathbb{R}^3$; the latter is embedded in $\mathbb{R}^4$. Is it possible to move A in $\mathbb{R}^4$ so that A never meets the sphere and at the end A is again in $\mathbb{R}^3$ but outside the sphere under consideration?

1.7. Two circles are linked in $\mathbb{R}^3$ (Figure 1.18 (a)), and $\mathbb{R}^3$ is embedded in $\mathbb{R}^4$. Is it possible to move one of the circles in $\mathbb{R}^4$ so that it never meets the second circle and at the end both circles are again in $\mathbb{R}^3$, but they are no longer linked (Figure 1.18 (b))?

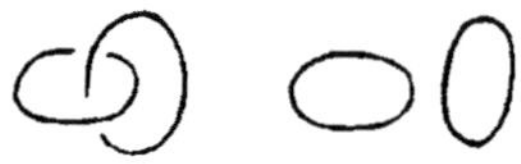

(a) (b)

FIGURE 1.18

1.8. Given that the sum of the squared lengths of the projections of vectors $a_1, \dots, a_n \in \mathbb{R}^n$ on any straight line is constant, prove that for some $\lambda \in \mathbb{R}$ the vectors $\lambda a_1, \dots, \lambda a_n$ form an orthonormal basis.

1.9. (a) Is it possible to arrange nine points in a cube with edge d so that the pairwise distances between them will be no shorter than d?

(b) Is it possible to arrange 17 points in a four-dimensional cube with edge d so that the pairwise distances between them will be no shorter than d?

1.10. Given an orthonormal basis $e_1, \dots, e_n$ and a set of vectors $a_1, \dots, a_n$ such that the angle between the vectors e_i and a_i equals α_i for each i, prove that if

$$\cos\alpha_1 + \dots + \cos\alpha_n > \sqrt{n(n-1)},$$

then the vectors $a_1, \dots, a_n$ are linearly independent.

1.11. In $\mathbb{R}^n$, balls of radii $R_1, \dots, R_m$ centered at $A_1, \dots, A_m$ have a common point. Prove that if $B_1, \dots, B_m$ are points such that $|B_iB_j| \leqslant |A_iA_j|$, then the balls of radii $R_1, \dots, R_m$ centered at $B_1, \dots, B_m$ also have a common point.

1.12. (a) Given vectors a_1, a_2, a_3 in $\mathbb{R}^3$ such that the angle between any two of them is not obtuse, prove that there exists a Cartesian coordinate system such that the coordinates of the given vectors are nonnegative.

(b) Specify vectors $a_1, \dots, a_5$ in the space $\mathbb{R}^n$ with $n \geqslant 5$ such that the angle between any two of them is not obtuse, but there exists no Cartesian coordinate system such that the coordinates of the given vectors are nonnegative.

Determinants.

1.13. Prove that the points (x_1, y_1), (x_2, y_2), (x_3, y_3) are collinear if and only if

$$\begin{vmatrix} x_1 & y_1 & 1 \\ x_2 & y_2 & 1 \\ x_3 & y_3 & 1 \end{vmatrix} = 0.$$

1.14. Prove that

$$\begin{vmatrix} x_1 & y_1 & z_1 \\ x_2 & y_2 & z_2 \\ x_3 & y_3 & z_3 \end{vmatrix}^2 \leqslant (x_1^2 + x_2^2 + x_3^2)(y_1^2 + y_2^2 + y_3^2)(z_1^2 + z_2^2 + z_3^2).$$

1.15. Vectors a_1, a_2, a_3 are of lengths a, b, c, respectively; α, β, γ are the angles between a_2 and a_3, a_1 and a_3, a_1 and a_2; V is the volume of the parallelepiped spanned by the vectors a_1, a_2, a_3. Prove that

$$V^2 = a^2b^2c^2(1 + 2\cos\alpha\cos\beta\cos\gamma - \cos^2\alpha - \cos^2\beta - \cos^2\gamma).$$

1.16. Given a tetrahedron whose ith and jth faces form a dihedral angle φ_{ij}, prove that

$$\begin{vmatrix} -1 & \cos\varphi_{12} & \cos\varphi_{13} & \cos\varphi_{14} \\ \cos\varphi_{21} & -1 & \cos\varphi_{23} & \cos\varphi_{24} \\ \cos\varphi_{31} & \cos\varphi_{32} & -1 & \cos\varphi_{34} \\ \cos\varphi_{41} & \cos\varphi_{42} & \cos\varphi_{43} & -1 \end{vmatrix} = 0.$$

1.17. (a) Prove that the volume V_n of a simplex $A_1 \dots A_{n+1}$ in Euclidean space $\mathbb{E}^n$ equals hV_{n-1}/n, where V_{n-1} is the volume of the simplex $A_1 \dots A_n$ and h is the distance from the vertex A_{n+1} to the face $A_1 \dots A_n$.

(b) Prove that the volume V_n of a simplex $A_1 \dots A_{n+1}$ equals

$$\frac{1}{n!}|V(a_1, \dots, a_n)|,$$

where $a_i = \overrightarrow{A_{n+1}A_i}$.

1.18. Given a simplex of volume V_n with vertices $A_1, \dots, A_{n+1}$ such that $d_{ij} = |A_iA_j|$, prove that

$$V_n^2 = \frac{(-1)^{n+1}}{(n!)^2 2^n} \begin{vmatrix} 0 & d_{12}^2 & \dots & d_{1,n+1}^2 & 1 \\ d_{21}^2 & 0 & \dots & d_{2,n+1}^2 & 1 \\ \dots & \dots & \dots & \dots & \dots \\ d_{n+1,1}^2 & d_{n+1,2}^2 & \dots & 0 & 1 \\ 1 & 1 & \dots & 1 & 0 \end{vmatrix}.$$

1.19. Given a simplex with vertices $A_1, \dots, A_{n+1}$ such that $d_{ij} = |A_iA_j|$ and the circumscribed sphere has radius R, prove that

$$R^2 = -\frac{\Delta}{2D}, \quad \text{where} \quad \Delta = \begin{vmatrix} 0 & d_{12}^2 & \dots & d_{1,n+1}^2 \\ d_{21}^2 & 0 & \dots & d_{2,n+1}^2 \\ \dots & \dots & \dots & \dots \\ d_{n+1,1}^2 & d_{n+1,2}^2 & \dots & 0 \end{vmatrix}$$

and D is the determinant defined in the preceding problem.

1.20. Prove that points $A_1, \dots, A_{n+2}$ of the space $\mathbb{R}^n$ lie on one $(n-1)$-sphere or in one hyperplane if and only if

$$\begin{vmatrix} 0 & d_{12}^2 & \dots & d_{1,n+2}^2 \\ d_{21}^2 & 0 & \dots & d_{2,n+2}^2 \\ \dots & \dots & \dots & \dots \\ d_{n+2,1}^2 & d_{n+2,2}^2 & \dots & 0 \end{vmatrix} = 0,$$

where $d_{ij} = |A_iA_j|$.

1.21. (a) Given spheres of radii $r_1, \dots, r_{n+1}$ in $\mathbb{R}^{n-1}$ pairwise externally tangent to each other, prove that

$$\left(\frac{1}{r_1} + \dots + \frac{1}{r_{n+1}}\right)^2 = (n-1)\left(\frac{1}{r_1^2} + \dots + \frac{1}{r_{n+1}^2}\right).$$

(b) Given spheres of radii $r_1, \dots, r_{n+1}$ in $\mathbb{R}^{n-1}$ making the same angle φ with each other, prove that

$$\left(\frac{1}{r_1} + \dots + \frac{1}{r_{n+1}}\right)^2 = \left(n + \frac{1}{\cos\varphi}\right)\left(\frac{1}{r_1^2} + \dots + \frac{1}{r_{n+1}^2}\right).$$

(The angle between spheres centered at O_1 and O_2 and meeting at A is, by definition, the angle between the vectors $\overrightarrow{AO_1}$ and $\overrightarrow{AO_2}$.)

1.22. Given numbers $d_{ij} > 0$, where $1 \leqslant i < j \leqslant n+1$, prove that there exists a simplex $A_1 \dots A_{n+1}$ with edges d_{ij} in $\mathbb{R}^n$ if and only if

$$(-1)^k \begin{vmatrix} 0 & d_{i_1 i_2}^2 & \dots & d_{i_1 i_k}^2 & 1 \\ d_{i_2 i_1}^2 & 0 & \dots & d_{i_2 i_k}^2 & 1 \\ \dots & \dots & \dots & \dots & \dots \\ d_{i_k i_1}^2 & d_{i_k i_2}^2 & \dots & 0 & 1 \\ 1 & 1 & \dots & 1 & 0 \end{vmatrix} > 0$$

for all $k = 2, \dots, n+1$ and any set of different points $A_{i_1}, \dots, A_{i_k}$ (in other words, if and only if the volumes of the simplices $A_{i_1} \dots A_{i_k}$ evaluated by the formula given in Problem 1.18 are positive).

The geometry of a simplex and a cube.

1.23. Find the angle between two $(n-1)$-faces of a regular n-simplex.

1.24. Given a $(2n+1)$-dimensional cube in $\mathbb{R}^{2n+1}$ with integer coordinates of all vertices, prove that the length of its edge is an integer.

1.25. Prove that there exists a two-dimensional plane that intersects a four-dimensional simplex but is disjoint with all its edges.

1.26. Prove that there exists a section of an n-dimensional cube by a two-dimensional plane that is a regular $2n$-gon.

1.27. (a) Prove that through any interior point of a three-dimensional cube, there passes a two-dimensional plane that intersects all edges of the cube at interior points (i.e., not at vertices).

(b) Prove that through any interior point of an n-dimensional cube, there passes a three-dimensional plane that intersects all $(n-1)$-dimensional faces of the cube at interior points.

1.28. A simplex with vertices $A_1, \dots, A_{n+1}$ is called *orthocentric* if its altitudes meet at one point. Prove that a simplex is orthocentric if and only if one of the following conditions holds.

(a) The value $(\overrightarrow{A_1A_i}, \overrightarrow{A_1A_j})$ does not depend on i and j whenever $i \neq j$ and $i, j \neq 1$.

(b) The edges A_iA_j and A_kA_l are orthogonal if i, j, k, and l are pairwise distinct.

(c) One of the $(n-1)$-faces is an orthocentric simplex, and the altitude to this face passes through the orthocenter.

(d) If i, j, k, and l are pairwise distinct, then

$$|A_iA_j|^2 + |A_kA_l|^2 = |A_iA_k|^2 + |A_jA_l|^2 = |A_iA_l|^2 + |A_jA_k|^2. \qquad (2)$$

(e) There exist numbers $a_1, \dots, a_{n+1} \in \mathbb{R}$ such that $|A_iA_j|^2 = a_i + a_j$.

(f) There exists a point A_0 (orthocenter) such that (2) holds for any pairwise distinct i, j, k, l.

Show that then there exist numbers $a_0, \dots, a_{n+1} \in \mathbb{R}$ such that $|A_iA_j|^2 = a_i + a_j$, and these numbers are related by

$$a_0^{-1} + \dots + a_{n+1}^{-1} = 0.$$

1.29. Given an orthocentric simplex $A_1 \dots A_{n+1}$ with orthocenter H and center of circumscribed sphere O, prove that

$$\overrightarrow{OA}_1 + \dots + \overrightarrow{OA}_{n+1} = (n-1)\overrightarrow{OH}.$$

CHAPTER 2

The Affine World

The affine space consists of the same points as Euclidean space, but it has a larger transformation group; the geometry of the affine world is therefore poorer than the geometry of the Euclidean world. This is the world of linear equations, linear inequalities, and convexity.

We begin this chapter, as usual, with the simplest, one- and two-dimensional, affine spaces: the affine line and the affine plane.

2.1. The affine line and the affine plane

Arithmetical model of the affine line. The model of the affine line is, as in the Euclidean case, the straight line $\mathbb{R}$. Geometrically, the affine line can be thought of as an unbending rod that, however, can be stretched.

On this rod, we choose a point (denoted by O), a positive direction, and a scale; this allows us to assign a coordinate x to each point X on the line. So far, everything is similar to the Euclidean case. But in the affine case, the scale is not fixed, and we do not introduce distances; all possible homotheties (arbitrary dilations or contractions) centered at any point are admitted. This means that the transformations of the affine line have the form $x' = \alpha x + \beta$, where $\alpha \neq 0$ and $\alpha, \beta \in \mathbb{R}$. We have described the model of the affine line. As we said, on the affine line, the notion of distance is not defined, but that of *passage to the limit* is: $X_n \underset{n\to\infty}{\longrightarrow} X$ if $x_n \to x$, where x_n and x are the coordinates of the points X_n and X in some coordinate system; the result does not depend on the choice of the system. Thus the affine line does not have a metric structure, but it does have a topological structure.

Arithmetical model of the affine plane. The model of the affine plane is, as in the Euclidean case, the space $\mathbb{R}^2$ of ordered pairs (x_1, x_2) of real numbers. These are points. The set of points is the affine plane proper. In addition to points, the affine plane contains very important objects of another kind: straight lines. As before, these are the level lines of linear functions, or, equivalently, the sets of points satisfying the equations $a_1x_1 + a_2x_2 = b$, where $a_1^2 + a_2^2 \neq 0$. The motions of the affine plane are one-to-one maps of $\mathbb{R}^2$ onto itself of the form

$$f(x_1, x_2) = (a_{11}x_1 + a_{12}x_2 + a_1, a_{21}x_1 + a_{22}x_2 + a_2).$$

Those properties of geometric figures that are preserved by affine transformations are called the *affine properties*. They include parallelism between straight lines, convexity of figures, the ratio in which a point divides a line segment, the property to be a median of a triangle, the ratio of lengths of parallel line segments, etc.

An arbitrary triangle is the image of an equilateral triangle under an affine transformation. For this reason, to prove that all triangles have a certain affine property, it suffices to prove that some particular triangle has this property. For instance, the medians of an equilateral triangle obviously intersect at one point and are divided by this point in the ratio 1 : 2. Therefore, the medians of any triangle intersect at one point and are divided by this point in the ratio 1 : 2.

The abstract definition of affine n-space was given in Chapter 1. Taking the particular case of this definition in dimension two, we obtain the triple consisting of the points and vectors of $\mathbb{R}^2$ and the operation of applying a vector to a point.

We have described the arithmetical model of affine geometry. As to the notion of affine geometry as such, its appearance followed that of Klein's Erlangen Program.

The key role in affine geometry is played by the *affine varieties.* On the plane, these are the points, the straight lines, and the entire plane. In a coordinate system, a point is a pair (x_1, x_2), and a straight line is specified by an equation of the form $a_1x_1 + a_2x_2 = b$, where $a_1^2 + a_2^2 \neq 0$; the line specified by this equation coincides with the line given by $\lambda a_1x_1 + \lambda a_2x_2 = \lambda b_2$ with $\lambda \neq 0$. Note that any straight line divides the affine plane $\mathbb{A}^2$ into three parts—two open half-planes and the line itself; in coordinates, these are $a_1x_1 + a_2x_2 > b$, $a_1x_1 + a_2x_2 < b$, and $a_1x_1 + a_2x_2 = b$. Each point belongs to one of these three sets, and if X_1 belongs to the first half-plane and X_2 to the second, then the line segment $[X_1, X_2]$ necessarily intersects the line separating the two half-planes.

Now, we pass to two basic theories of affine geometry, those of linear equations and convex sets.

Linear equations on the plane. Our objective is to describe the set of solutions to the system of equations

$$a_{i1}x_1 + a_{i2}x_2 = b_i, \quad 1 \leqslant i \leqslant N.$$

One such equation determines a straight line. The next and most important stage is to examine two equations in two unknowns:

$$\begin{aligned} a_{11}x_1 + a_{12}x_2 &= b_1, \\ a_{21}x_1 + a_{22}x_2 &= b_2. \end{aligned} \tag{1}$$

We disregard the case in which all a_{ij}, b_i are zero (any pair (x_1, x_2) is then a solution). Each of the two equations under consideration determines a straight line. There are three possibilities: either these lines coincide, or they are disjoint (have no common points), or they intersect at one point.

THEOREM 1 (on the solution of a system of two equations in two unknowns). *System* (1) *has a solution if and only if either the triples* $(a_{11}, a_{12}, -b_1)$ *and* $(a_{21}, a_{22}, -b_2)$ *are proportional or the determinant* $\det A = \det(a_{ij})_{1\leqslant i,j\leqslant 2}$ *of the system is nonzero. In the latter case,* (1) *has a unique solution, which is*

$$x_1 = \frac{\det A_1}{\det A}, \quad x_2 = \frac{\det A_2}{\det A}, \quad \textit{where } A_1 = \begin{pmatrix} b_1 & a_{12} \\ b_2 & a_{22} \end{pmatrix} \textit{ and } A_2 = \begin{pmatrix} a_{11} & b_1 \\ a_{21} & b_2 \end{pmatrix}$$

(Cramer's formulas).

Theorem 1 can be reformulated as follows: System (1) is solvable if and only if the rank of the matrix

$$A = \begin{pmatrix} a_{11} & a_{12} \\ a_{21} & a_{22} \end{pmatrix}$$

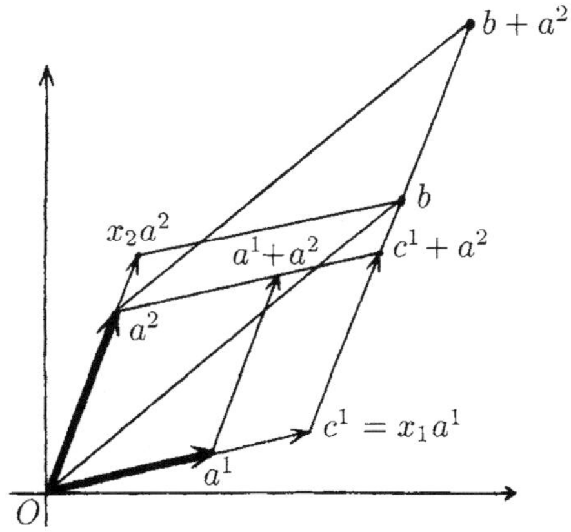

FIGURE 2.1

coincides with the rank of the augmented matrix

$$A' = \begin{pmatrix} a_{11} & a_{12} & -b_1 \\ a_{21} & a_{22} & -b_2 \end{pmatrix}.$$

If the rank of A is two, then there is only one solution, and it is specified by Cramer's formulas.

Proof. We give two proofs, algebraic and geometric.

Algebraic proof. Let us multiply the first equation in (1) by a_{22} and the second by a_{12}. Subtracting the second resulting equation from the first, we obtain

$$(a_{11}a_{22} - a_{12}a_{21})x_1 = b_1a_{22} - b_1a_{12}.$$

Similarly,

$$(a_{11}a_{22} - a_{12}a_{21})x_2 = b_2a_{11} - b_2a_{21}.$$

Therefore, if $\det A = a_{11}a_{22} - a_{12}a_{21} \neq 0$, then the solution is given by Cramer's formulas. If $\det A = 0$, then $a_{11}/a_{21} = a_{12}/a_{22}$, and a solution exists if and only if b_1/b_2 has the same value, what was to be proved.

Geometric proof. Many beautiful and important geometric facts lie on the surface, so to say; all that is needed to find them is to draw a picture and say the magic word "look!" This is the case here. Remember: the determinant of a matrix is the area of the parallelogram constructed on the vectors specified by the columns of the matrix. Now, Cramer's formulas follow directly from the picture presented in Figure 2.1; all that we need is to carefully look at it and study the formulas below. So—look!

Let us write our system of equations in the form $x_1a^1 + x_2a^2 = b$. We have

$$x_1 = \frac{S_{Oc^1(c^1+a^2)a^2}}{S_{Oa^1(a^1+a^2)a^2}},$$

$$S_{Oc^1(c^1+a^2)a^2} = S_{Ob(b+a^2)a^2} = \det \begin{pmatrix} b_1 & a_{12} \\ b_2 & a_{22} \end{pmatrix} = \det A_1,$$

$$S_{Oa^1(a^1+a^2)a^2} = \det A \implies x_1 = \frac{\det A_1}{\det A}. \quad \square$$

COROLLARY 1. *Two distinct intersecting straight lines have exactly one intersection point.*

COROLLARY 2. *Through any point, there passes exactly one straight line parallel to a given one.*

Corollary 1 follows directly from Theorem 1. Let us prove Corollary 2. Let $a_1x_1 + a_2x_2 = b$ be the given line. Consider an arbitrary point (ξ_1, ξ_2). The line $a_1x_1 + a_2x_2 = a_1\xi_1 + a_2\xi_2$ is parallel to the given one, and if $a_1'x_1 + a_2'x_2 = b'$ is another line parallel to the given one, then, by Theorem 1, $a_1' = \lambda a_1$ and $a_2' = \lambda a_2$, i.e., $b' = \lambda a_1\xi_1 + \lambda a_2\xi_2$; therefore, these two lines coincide.

To complete the theory of linear equations, it remains to consider the case of a family $\{a_{i1}x_1 + a_{i2}x_2 = b_i,\ i \in J\}$ of more than two equations.

It may happen that all lines specified by equations of this family coincide. This occurs if and only if $(a_{i1}, a_{i2}, -b_i) = \lambda_i(a_{11}, a_{12}, -b_1)$. It may also happen that all the lines pass through one point. This occurs if and only if the values

$$\frac{b_i a_{j2} - b_j a_{i2}}{a_{i1}a_{j2} - a_{j1}a_{i2}} \quad \text{and} \quad \frac{b_j a_{i1} - b_i a_{j1}}{a_{i1}a_{j2} - a_{j1}a_{i2}}$$

do not depend on i and j. In all other cases, the "solution" is the empty set.

We have proved the following assertion: *a solution of a system of linear equalities is an affine set* (*i.e., the empty set, a point, a straight line, or the entire plane*).

Convex geometry on the plane and the theory of linear inequalities.

The notion of convexity dates back to ancient times: for instance, Archimedes used the notion of convex surface. But the theory of convex figures appeared only in the nineteenth century. Cauchy and later Blaschke, Weyl, and others developed the theory of convex surfaces. The principles of the theory of convex figures were developed in the late nineteenth–early twentieth century by Minkowski. At approximately the same time, the relationship between the geometry of convex figures and the theory of linear spaces was discovered. But the rigorous mathematical theory of such inequalities was only constructed in the late 1930s–early 1940s.

Recall that a set C in an affine space is said to be *convex* if together with any points X and Y, it contains the entire line segment $[X, Y]$.

On the affine plane, the notion of passing to the limit is defined: $X_n \to X$ if $x_{n1} \to x_1$ and $x_{n2} \to x_2$, where x_{n1}, x_{n2} are the coordinates of X_n and x_1, x_2 are the coordinates of X in some coordinate system. Clearly, the limit does not depend on the coordinate system. With this definition, the affine plane becomes a topological space, and hence the notions of continuity and open and closed sets are defined.

A set C is said to be *closed* if it contains all its limit points. In other words, $X \in C$ whenever there is a sequence $X_i \in C$ converging to X.

The sets with closed complements are called *open*. Another definition: A set C is said to be open if for any point $A \in C$ with coordinates (a_1, a_2) in some coordinate system, we can find an $r > 0$ such that C contains the entire open disk $\{ X \mid (x_1 - a_1)^2 + (x_2 - a_2)^2 < r^2 \}$ of radius r centered at A.

The *convex closure* of a set C is the intersection of all closed convex sets containing C.

Linear inequalities in two variables are inequalities of the form $a_1x_1 + a_2x_2 \leqslant b$ (or $\geqslant b$, or $< b$, or $> b$). We can write all inequalities with $\leqslant$ or $<$ by changing sign where necessary. In this section, we largely consider nonstrict inequalities (with $\leqslant$).

One of our objectives is to describe the set of solutions to the system of linear inequalities $a_{1\alpha}x_1 + a_{2\alpha}x_2 \leqslant b_\alpha$ with $\alpha \in \mathcal{A}$, where $\mathcal{A}$ is some index set, not necessarily finite or countable. The set $a_1x_1+a_2x_2 \leqslant b$ is a closed half-plane. Thus we want to describe intersections of closed half-planes.

The theory of convex sets and the theory of linear inequalities are based on the following two separation theorems.

THEOREM 2 (First separation theorem on the plane). *A nonempty convex (not necessarily closed) set C in the affine plane can be separated from any point A that does not belong to this set (i.e., we can draw a straight line so that C lies entirely in one of the two closed half-planes determined by this line and A lies either in the other open half-plane or on the line itself).*

THEOREM 3 (Second separation theorem on the plane). *A closed convex (nonempty) set C in the affine plane can be strictly separated from any point A that does not belong to this set (i.e., we can draw a straight line so that C lies in one of the two closed half-planes determined by this line and A lies in the other open half-plane).*

We give an analytical proof of Theorem 3 and a sketch of a geometric proof of Theorem 2.

The proof of Theorem 3 relies on the well-known Weierstrass theorem: *A continuous function on $\mathbb{R}^2$ attains its greatest lower bound on a bounded closed subset of $\mathbb{R}^2$.* Our continuous function is the distance to a fixed point.

Proof. By assumption, the set C is nonempty; choose a point Y in C. Let C_1 be the intersection of C with the closed ball B of radius $|AY|$ centered at A. The set C_1 is closed and bounded; therefore, by the Weierstrass theorem, the distance from A to the points of C_1 attains its minimum at some point $O \in C_1$, and the minimum distance from A to the points of C equals AO. Indeed, if $Z \in C$ and $Z \notin C_1$, then Z lies outside the ball B, and $|AZ| > |AY| \geqslant |AO|$.

Choose a Cartesian coordinate system with origin at O. Denote the coordinates of the point A by (a_1, a_2). Let us show that the line $x_1a_1 + x_2a_2 = 0$ strictly separates A from C, i.e., $z_1a_1 + z_2a_2 \leqslant 0$ for any point $Z = (z_1, z_2) \in C$. Suppose that $Z \in C$ and $z_1a_1 + z_2a_2 > 0$. The set C is convex and contains the points $Z = (z_1, z_2)$ and $O = (0, 0)$; therefore, it contains all points $Z_t = (tz_1, tz_2)$ with $0 \leqslant t \leqslant 1$. In addition,

$$\begin{aligned}|AZ_t|^2 &= (a_1 - tz_1)^2 + (a_2 - tz_2)^2 = a_1^2 + a_2^2 - 2t(a_1z_1 + a_2z_2) + t^2(z_1^2 + z_2^2)\\ &< a_1^2 + a_2^2 \qquad \text{if } 0 < t < 2\frac{a_1z_1 + a_2z_2}{z_1^2 + z_2^2}.\end{aligned}$$

Thus for small $t > 0$, the points $Z_t \in C$ are closer to A than O is. This contradiction completes the proof of Theorem 3. □

Sketch of the proof of Theorem 2. If the set C lies on a straight line, then everything is simple: this line is the required one. If C does not lie on a straight line, then it has "interior" points (a point $X \in C$ is called *interior* if there exists an open set that contains this point and is contained in C). This is the first item to be proved.

Next, we take an interior point X in C, draw a straight line through A and X, and "rotate" this line about A until it meets interior points of C "for the first time."

(We must define what "rotate" and "for the first time" mean.) The straight line thus obtained separates C from A (this claim should also be substantiated). □

The proof can be formalized, but this would involve coordinates and calculus theorems. Yet, the proof as presented is fairly evident and convincing, isn't it?

Next, we shall describe solutions of systems of linear inequalities.

THEOREM 4 (Description of solutions to systems of linear inequalities on the plane). *The solution of any system of nonstrict linear inequalities is a closed convex set, and any closed convex set is the solution of such a system.*

Proof. As explained above, it suffices to consider only nonstrict inequalities $\leqslant$. Consider the system of nonstrict inequalities

$$a_{1\alpha}x_1 + a_{2\alpha}x_2 \leqslant b_\alpha, \quad \alpha \in \mathcal{A}. \tag{2}$$

Suppose that some points (x_1, x_2) and (x'_1, x'_2) satisfy all the inequalities from this system. Then the point $(tx_1+(1-t)x'_1,\ tx_2+(1-t)x'_2)$ also satisfies all inequalities in (2). Each inequality specifies a closed half-plane, and any intersection of closed sets is closed; therefore, (2) specifies some closed convex (possibly empty) set.

Now, let C be a closed convex set. If C coincides with the entire plane, then C is the solution of the inequality $0x_1 + 0x_2 \leqslant 0$. Suppose that C does not coincide with the plane. Then we can find a point $\hat{X} \notin C$. According to the second separation theorem, there exists a closed half-plane containing C.

Consider the intersection of *all* closed half-planes containing C. This is a closed convex set C' containing C. The assumption about the existence of $Z \in C' \setminus C$ (this means that $Z \in C'$ and $Z \notin C$) leads to a contradiction (it suffices to apply the second separation theorem once more and construct a half-plane that contains C and does *not* contain Z). Thus C is the intersection of closed half-planes, and hence (in some coordinate system) a solution of a system of nonstrict inequalities. □

Theorem 4 gives a dual description of convexity: a convex set is the union of its points and the intersection of the half-planes containing it.

Historical comments. The theorem that each convex set is the intersection of half-planes and its generalization to the n-dimensional case are due to Hermann Minkowski.

H. Minkowski (1864–1909), a remarkable German mathematician and physicist, a close friend of David Hilbert. Constructed the basics of the theory of convex sets (in *Geometry of Numbers*, which appeared in 1886). Implemented geometrization ideas in physics, namely, gave a geometric interpretation of special relativity by introducing the four-dimensional space with hyperbolic metric (it will be discussed later on). Taught at the universities of Königsberg, Berlin, Zurich, and Göttingen. One of his Zurich students was Einstein. Einstein did not make a strong impression on Minkowski, who found him too slow-witted. (It is helpful to remember that not all people predestined to contribute to science are quick at seminars and science olympiads.)

The foundations of the theory of linear inequalities were also laid in the nineteenth century (by Fourier); then, much work was done in the early twentieth century (by Vallée-Poussin, Voronoi, Weyl, and others). But the most extensive development of the theory dates to the 1940s; it was due to the discovery of the fact that the theory is applicable to economics problems, made by Kantorovich in the USSR (1939) and Koopmans in the USA (the early 1940s). They were both awarded the 1975 Nobel Prize in economics.

We would like to mention another important theorem of convex geometry, the Minkowski theorem on extreme points.

A point X in a set C is said to be *extreme* if it is not an interior point of a line segment with endpoints in C.

THEOREM 5 (Minkowski). *A bounded closed convex plane set is the convex closure of its extreme points.*

Recall that the convex closure of a set is the intersection of all the closed convex sets that contain this set.

Proof. First, we give a geometric proof of this theorem and then formalize it. For sets lying on a straight line, the assertion is obvious: a bounded closed convex set on a line is an interval, its extreme points are its endpoints, and the interval itself is their convex hull (which coincides with their convex closure).

Suppose that the set is planar. We take an interior point of this set, draw a line through it, and shift this line parallel to itself in some direction until the line "leaves" the set. At this last instant, the line still intersects C (in an interval), and C lies on one side of the line. It is easy to see that the endpoints of the interval are extreme for C, i.e., the set of extreme points of C is nonempty. Let C' be the convex hull of the set of extreme points. Clearly, $C' \subset C$. If C' does not coincide with C, we choose a point $X \in C \setminus C'$, separate it from C' by a straight line, and move this line "in the direction of X" again until the "last" moment, at which one more extreme point is found. This point belongs to $C \setminus C'$. On the other hand, all extreme points are contained in C' by construction. This contradiction completes the proof of the theorem. □

Let us formalize the above arguments. We take $(\lambda_1, \lambda_2) \neq (0, 0)$ and choose a point $A = (a_1, a_2) \in C$ for which the value $f(X) = \lambda_1 x_1 + \lambda_2 x_2$ is minimal (such a point A exists by the Weierstrass theorem). The intersection of the line $\lambda_1 x_1 + \lambda_2 x_2 = f(A)$ with C is convex, closed, and nonempty (it contains the point A). Therefore, this intersection is an interval $[S, T]$ (or one point; S and T then coincide with A). Let us prove that T is an extreme point for C. Suppose that T is the midpoint of an interval $[Y, Z]$, where $Y, Z \in C$. Then $2f(T) = f(Y) + f(Z)$. Since $f(Y), f(Z) \geqslant f(A) = f(T)$, we have $f(Y) = f(Z) = f(A)$, i.e., $Y, Z \in [S, T]$. The contradiction obtained shows that the set of extreme points of C is nonempty.

Let C' be the convex closure of the set of extreme points of C. Clearly, $C' \subset C$. Suppose that there exists $B = (b_1, b_2) \in C \setminus C'$. According to the second separation theorem, we can find a line $\mu_1 x_1 + \mu_2 x_2 = b$ such that

$$\mu_1 b_1 + \mu_2 b_2 > b \quad \text{and} \quad \mu_1 x_1 + \mu_2 x_2 \leqslant b \quad \text{for any} \quad (x_1, x_2) \in C'.$$

Let $A \in C$ be a point for which the value $g(X) = \mu_1 x_1 + \mu_2 x_2$ is maximal. On the one hand, $g(A) \geqslant g(B) > b$, and hence the line $\mu_1 x_1 + \mu_2 x_2 = g(A)$ contains no points of C'. On the other hand, this line contains extreme points of C. We have arrived at a contradiction. □

The fundamental theorem of affine geometry. The following assertion is often called the *fundamental theorem of affine geometry* because of its importance (and nontriviality).

LEMMA. *Let $f\colon \mathbb{R}^2 \to \mathbb{R}^2$ be a one-to-one map such that the images of any three collinear points are again collinear. Then f is an affine transformation.*

Proof. First, note that the map f is a one-to-one transformation of each straight line into a straight line. Indeed, let A_1 and B_1 be the images of two different points

A and B. Then the image of the line AB lies on the line A_1B_1. It remains to show that, if C_1 is a point on A_1B_1, then its preimage C belongs to AB. Suppose that C does not lie on AB. Then the lines AC and BC are distinct, and their images lie on the line A_1B_1. Through an arbitrary point X in the plane, we can draw a line intersecting AC and BC at different points A' and B'. The images of A' and B' lie on the line A_1B_1; hence the image of X also lies on A_1B_1. This contradicts the assumption that the image of f is the entire plane.

So, suppose that $f\colon \mathbb{R}^2 \to \mathbb{R}^2$ is a one-to-one map taking each straight line to a straight line. We shall successively formulate various properties of this map and prove them; each time we shall use the results obtained at the preceding steps.

Step 1. The map f takes parallel lines to parallel lines.

Indeed, the images of two disjoint lines cannot intersect because f is one-to-one.

Step 2. The action of f on the vectors is well defined, i.e., if $\overrightarrow{AB} = \overrightarrow{CD}$, then $\overrightarrow{A_1B_1} = \overrightarrow{C_1D_1}$, where A_1, B_1, C_1, and D_1 are the images of the points A, B, C, and D, respectively.

Suppose that the point C does not lie on the line AB. Consider the parallelogram (more precisely, two pairs of parallel lines) $ABDC$. The first property (Step 1) implies that these two pairs are mapped to two pairs of parallel lines. If the point C lies on AB, then we choose two auxiliary points X and Y such that $\overrightarrow{XY} = \overrightarrow{AB}$ and X does not lie on AB.

In what follows, we treat f as a map of vectors. It is required to prove that f is a linear transformation.

Step 3. $f(0) = 0$, where 0 is the zero vector.

This is so because f is one-to-one.

Step 4. $f(a+b) = f(a) + f(b)$.

We can assume that $a = \overrightarrow{AB}$ and $b = \overrightarrow{BC}$; then

$$f(a+b) = f(\overrightarrow{AC}) = \overrightarrow{A_1C_1} = \overrightarrow{A_1B_1} + \overrightarrow{B_1C_1} = f(a) + f(b).$$

Step 5. $f(ka) = kf(a)$ for $k \in \mathbb{Q}$.

For a positive integer n, the fourth property (Step 4) implies that

$$f(na) = f(a + \cdots + a) = f(a) + \cdots + f(a) = nf(a).$$

In addition, $f(na) + f(-na) = f(na - na) = f(0) = 0$. For $k = m/n$, where m and n are integers, we obtain

$$nf\left(\frac{m}{n}a\right) = f(ma) = mf(a), \qquad \text{i.e.,} \quad f(ka) = \frac{m}{n}f(a) = kf(a).$$

If we would have assumed f to be continuous, then the proof of the theorem would have been completed because any number $k \in \mathbb{R}$ can be approximated by rational numbers. But we made no continuity assumption, and the most difficult part of the proof only begins.

Let $a = \overrightarrow{OA}$ and $b = \overrightarrow{OB}$ be basis vectors. The map f takes them to vectors $a_1 = \overrightarrow{O_1A_1}$ and $b_1 = \overrightarrow{O_1B_1}$. Choose points X and Y on the lines OA and OB, respectively. They are mapped to points X_1 and Y_1 lying on O_1A_1 and O_1B_1. This means that $f(xa) = \varphi(x)a_1$ and $f(yb) = \psi(y)b_1$, where φ and ψ are some transformations of $\mathbb{R}$.

Step 6. $\varphi(t) = \psi(t)$.

Indeed, if $\overrightarrow{OX} = t\,\overrightarrow{OA}$ and $\overrightarrow{OY} = t\,\overrightarrow{OB}$, then the lines XY and AB are parallel, and hence X_1Y_1 and A_1B_1 are parallel also, i.e., $\varphi(t) = \psi(t)$.

We have proved that $f(xa + yb) = \varphi(x)a_1 + \varphi(y)b_1$. It remains to prove that $\varphi(x) = x$ for all $x \in \mathbb{R}$. Recall that $\varphi(x) = x$ for $x \in \mathbb{Q}$ (Step 5). Therefore, it is sufficient to show that $\varphi(x) < \varphi(y)$ whenever $x < y$.

Step 7. $\varphi(xy) = \varphi(x)\varphi(y)$ for all $x, y \in \mathbb{R}$.

Consider the proportional vectors $xa + yb$ and $\frac{x}{y}a + b$. Their images $\varphi(x)a_1 + \varphi(y)b_1$ and $\varphi\left(\frac{x}{y}\right)a_1 + b_1$ are also proportional; hence $\varphi\left(\frac{x}{y}\right) = \frac{\varphi(x)}{\varphi(y)}$. In particular,

$$\varphi\left(\frac{1}{y}\right) = \frac{\varphi(1)}{\varphi(y)} = \frac{1}{\varphi(y)} \quad \text{and} \quad \varphi\left(\frac{x}{1/y}\right) = \frac{\varphi(x)}{\varphi(1/y)} = \varphi(x)\varphi(y).$$

Step 8. If $x < y$, then $\varphi(x) < \varphi(y)$.

The fourth property implies $\varphi(y) = \varphi(y - x + x) = \varphi(y - x) + \varphi(x)$. Therefore, it suffices to verify that $\varphi(t) > 0$ if $t = y - x > 0$. Any positive number t can be represented as $t = s^2$, where $s \in \mathbb{R}$; hence $\varphi(t) = \varphi(s)^2 > 0$.

This concludes the proof of the theorem. □

COROLLARY. *If a one-to-one map* $f\colon \mathbb{R}^2 \to \mathbb{R}^2$ *takes each circle to a circle, then* f *is an affine transformation.*

Proof. The transformation f^{-1} maps any three collinear points $f(A)$, $f(B)$, and $f(C)$ into three points A, B, and C, also collinear. Indeed, if the points A, B, and C are not collinear, then they are pairwise distinct, and we can draw a circle through them. The points $f(A)$, $f(B)$, and $f(C)$ are then also pairwise distinct and lie on a circle; therefore, these three points do not lie on a straight line. Thus the transformation f^{-1} is affine; hence f is affine as well. □

REMARK. Note that the converse statement is not true because an affine map can take a circle to an ellipse.

The fundamental theorem of affine geometry means that to describe an affine plane, it suffices to specify a set of points, a set of lines, and an intersection point for each pair of intersecting lines. Certainly, such a list is immense and less convenient in many respects than the arithmetical model of the affine plane. But it suggests another—axiomatic—approach to geometry. Namely, we can formulate axioms that uniquely specify sets of points and lines and then deduce theorems from these axioms (which involve only points, lines, and intersection points of lines).

An affine plane can be defined not only over $\mathbb{R}$, but also over an arbitrary (even noncommutative) field; see **[A]** for more details. The proof of the fundamental theorem of affine geometry remains unchanged, except Step 8 (and Step 5 for fields of nonzero characteristic). In particular, for the affine space over the field $\mathbb{C}$ of complex numbers, a repetition of the argument from the proof of the fundamental theorem leads to a transformation $\varphi\colon \mathbb{C} \to \mathbb{C}$ such that $\varphi(xy) = \varphi(x)\varphi(y)$ and $\varphi(x + y) = \varphi(x) + \varphi(y)$. Such transformations are called automorphisms of the field $\mathbb{C}$. We have seen that the field $\mathbb{R}$ has only the identity automorphism. As to $\mathbb{C}$, it has many automorphisms, though only two of them are continuous (namely, the identity automorphism and the conjugation $z \mapsto \bar{z}$). Without the continuity requirement, we can construct an automorphism $\varphi\colon \mathbb{C} \to \mathbb{C}$ that maps an arbitrary transcendental number to an arbitrary other transcendental number; there also exists an automorphism that maps a given root of an irreducible polynomial over $\mathbb{Q}$

to any other root of this polynomial (this is proved in [**Bo1**], Chapter IV, Section 6, Proposition 3).

2.2. Affine space. Theory of linear equations and inequalities

The definition of an affine n-space was given in Chapter 1. Surprisingly, the passage from 2 to n dimensions involves no new ideas; it only requires some comments. Let us formulate the n-dimensional analogs of Theorems 1–5 proved in the preceding section.

THEOREM 1′. *The system of equations in n unknowns*

$$\begin{aligned} a_{11}x_1 + \cdots + a_{1n}x_n &= b_1, \\ \cdots\cdots\cdots&\cdots\cdots \\ a_{n1}x_1 + \cdots + a_{nn}x_n &= b_n \end{aligned} \tag{1}$$

is solvable if and only if the rank of the matrix $A = \begin{pmatrix} a_{11} & \dots & a_{1n} \\ \dots & \dots & \dots \\ a_{n1} & \dots & a_{nn} \end{pmatrix}$ *coincides with the rank of the augmented matrix*

$$A' = \begin{pmatrix} a_{11} & \dots & a_{1n} & b_1 \\ \dots & \dots & \dots & \dots \\ a_{n1} & \dots & a_{nn} & b_n \end{pmatrix}.$$

If the rank of A equals n, then (1) *has a unique solution, which is specified by Cramer's formulas*

$$x_i = \frac{\det A_i}{\det A}, \qquad i = 1, \dots, n, \tag{2}$$

where A_i is obtained from A by replacing the ith column with the column of free terms.

The first part of the theorem follows from the definition of rank and is proved in linear algebra courses. The second part, which is also rigorously proved in linear algebra courses, has essentially the same geometric interpretation as Theorem 1.

In the n-dimensional case, the role of straight lines is played by hyperplanes. In coordinates, they are determined by equations of the form

$$a_1x_1 + \cdots + a_nx_n = b.$$

Each such hyperplane determines two *open half-spaces*

$$a_1x_1 + \cdots + a_nx_n < b \quad \text{and} \quad a_1x_1 + \cdots + a_nx_n > b.$$

(These are open subsets in $\mathbb{A}^n$.) The half-spaces

$$a_1x_1 + \cdots + a_nx_n \leqslant b \qquad \text{and} \qquad a_1x_1 + \cdots + a_nx_n \geqslant b$$

are called closed half-spaces.

THEOREM 2′ (First separation theorem in affine n-space). *A nonempty convex (not necessarily closed) set C in affine n-space can be separated from any point A that does not belong to this set (i.e., there exists a hyperplane such that C lies entirely in one of the two closed half-spaces determined by this hyperplane, and A lies either in the other open half-space or in the hyperplane itself).*

The geometric proof uses induction: in 3-space, we "rotate" a plane passing through A and an interior point of C about the line that separates the section of C by this plane from A, etc.

THEOREM 3′ (Second separation theorem in affine n-space). *A nonempty convex closed set C in the affine n-space can be strictly separated from any point A that does not belong to this set (i.e., there exists a hyperplane such that C lies in one of the two closed half-spaces determined by this hyperplane and A lies in the other open half-space).*

The proof of Theorem 3′ coincides almost literally with that of Theorem 3. We choose an arbitrary point Y in C and consider the closed ball B of radius $|AY|$ centered at A. Next, we take a point O in $C_1 = C \cap B$ at which the function $f(X) = d(X, A)$ attains its minimum on the set C_1. The hyperplane passing through O perpendicularly to the line AO strictly separates A from C.

THEOREM 4′ (Description of solutions to linear equations and inequalities).

(a) *The solution of a system of linear homogeneous equations $\sum_{i=1}^n a_{i\alpha} x_i = 0$ with $\alpha \in \mathcal{A}$ is either a k-dimensional subspace, where $0 \leqslant k \leqslant n-1$, or the entire space $\mathbb{R}^n$.*

(b) *The solution of the system of linear homogeneous inequalities $\sum_{i=1}^n a_{i\alpha} x_i \leqslant 0$ with $\alpha \in \mathcal{A}$ is a closed convex cone in $\mathbb{R}^n$.*

(c) *The solution of the system of linear nonhomogeneous equations $\sum_{i=1}^n a_{i\alpha} x_i = b_\alpha$ with $\alpha \in \mathcal{A}$ is either a k-dimensional affine set k, where $0 \leqslant k \leqslant n-1$, or $\mathbb{R}^n$, or the empty set.*

(d) *The solution of the system of linear nonhomogeneous inequalities is either the empty set, or the entire space $\mathbb{R}^n$, or a closed convex subset of $\mathbb{R}^n$.*

This theorem has only one difference from its plane version, namely, the description of solutions of homogeneous inequalities. Recall that a set in a linear space is a cone if, together with its every point X, it contains the entire ray $\{\, tX \mid t > 0 \,\}$. A closed convex cone is a cone which is simultaneously a closed convex set. On the plane, assertion (b) degenerates: the closed convex cones on the plane are merely angles.

THEOREM 5′ (Minkowski). *A bounded closed convex set in the affine n-space coincides with the convex closure of its extreme points.*

Theorems 4′ and 5′ are proved similarly to Theorems 4 and 5; we shall return to them in Chapter 6.

2.3. Introduction to finite-dimensional convex geometry

The Carathéodory and Radon lemmas. Let $A_1, \dots, A_m$ be points in affine space. Choose an arbitrary point O and consider the vector $\overrightarrow{OA} = \sum \lambda_i \overrightarrow{OA_i}$. The point A does not depend on the choice of O if and only if $\sum \lambda_i = 1$. Thus if $\sum \lambda_i = 1$, then the linear combination $\sum \lambda_i A_i$ of points is well defined. If all numbers λ_i are nonnegative, the point $A = \sum \lambda_i A_i$ is said to be a *convex linear combination* of the points A_i. A point A is a convex linear combination of $A_1, \dots, A_m$ if and only if A belongs to their convex hull.

LEMMA 1 (Carathéodory). *Let $X \subset \mathbb{R}^n$ be a nonempty set. Then any point A in the convex hull of X can be represented as a convex linear combination of no more than $n+1$ points from X.*

Proof. By assumption, $A = \sum_{i=1}^s \lambda_i A_i$, where $\sum \lambda_i = 1$, $\lambda_i \geqslant 0$, and $A_i \in X$. We can assume that all numbers λ_i are positive. If $s \leqslant n+1$, then we have nothing to prove; for this reason, we assume that $s \geqslant n+2$. The vectors $\overrightarrow{A_1A_2}, \dots, \overrightarrow{A_1A_s}$ are then linearly dependent, i.e., $\sum_{i=2}^{s} \mu_i \overrightarrow{A_1A_i} = 0$, where μ_i are not all zero. If O is an arbitrary point, then $\sum_{i=1}^{s} \nu_i \overrightarrow{OA_i} = 0$, where $\nu_1 = -(\mu_2 + \dots + \mu_s)$ and $\nu_i = \mu_i$ for $i \geqslant 2$. Therefore, for any $t \in \mathbb{R}$,

$$A = \sum_{i=1}^{s} (\lambda_i + t\nu_i) A_i \qquad \text{and} \qquad \sum (\lambda_i + t\nu_i) = \sum \lambda_i = 1,$$

because $\sum \nu_i = 0$.

For $t = 0$, all numbers $\lambda_i + t\nu_i$ are positive, and as $t \to +\infty$, there is at least one negative number among them because the ν_i are not all zero. Take a minimal $t > 0$ for which at least one number $\lambda_i + t\nu_i$ is equal to zero. Then all numbers $\lambda_i + t\nu_i$ in the representation $A = \sum_{i=1}^{s} (\lambda_i + t\nu_i) A_i$ are nonnegative, but one of the points A_i can be removed because it has zero coefficient.

Repeating this procedure several times, we obtain a representation of the point A in the required form. $\square$

LEMMA 2 (Radon). *An arbitrary set of points $A_1, \dots, A_s$ in $\mathbb{R}^n$ with $s \geqslant n+2$ can be divided into two disjoint subsets so that the convex hulls of these subsets have nonempty intersection.*

Proof. In the proof of the Carathéodory lemma, it was shown that

$$\sum_{i=1}^{s} \nu_i \overrightarrow{OA_i} = 0, \qquad \text{where} \quad \sum_{i=1}^{s} \nu_i = 0$$

and the numbers ν_i are not all zero. In the first set, we include the points A_i for which $\nu_i > 0$. Without loss of generality, we can assume that the first set contains the points $A_1, \dots, A_t$ and the second, $A_{t+1}, \dots, A_s$. Let $a = (\nu_1 + \dots + \nu_t)^{-1} > 0$. The convex hulls of the sets $\{A_1, \dots, A_t\}$ and $\{A_{t+1}, \dots, A_s\}$ share the point

$$a\nu_1 A_1 + \dots + a\nu_t A_t = (-a\nu_{t+1}) A_{t+1} + \dots + (-a\nu_s) A_s. \quad \square$$

Helly's theorem.

THEOREM 1 (Helly). *If $\mathcal{A}$ is an arbitrary index set, $\{X_\alpha\}_{\alpha \in \mathcal{A}}$ is a family of convex sets in $\mathbb{R}^n$ one of which is compact, and each $(n+1)$-element subfamily has nonempty intersection, then the entire family has nonempty intersection.*

Proof. First, we prove by induction on s that any family of s sets, where $s \geqslant n+1$, has nonempty intersection. Suppose that $s \geqslant n+2$ and each $(s-1)$-element family has nonempty intersection. Consider s sets $X_1, \dots, X_s$. Let us remove X_j. By the induction hypothesis, the remaining sets have nonempty intersection. Let A_j be a point from this intersection. By the Radon lemma, the points $A_1, \dots, A_s$ can be divided into two sets so that the convex hulls of these two sets share a point A.

Suppose that $j_1, \dots, j_k$ are the numbers of points in the first set, and $j_{k+1}, \dots, j_s$ are the numbers of points in the second set. By definition, $A_j \in X_i$ for all $i \neq j$. Therefore, if a point A belongs to the convex hull of the points $A_{j_1}, \dots, A_{j_k}$, then $A \in X_i$ for $i \neq j_1, \dots, j_k$, i.e., $A \in X_{j_{k+1}} \cap \dots \cap X_{j_s}$. Similarly, $A \in X_{j_1} \cap \dots \cap X_{j_k}$, whence $A \in X_1 \cap \dots \cap X_s$.

We have proved Helly's theorem for finite families of convex sets. To prove it for an infinite family, we apply the finite intersection lemma (if each finite subsystem of a system of closed subsets of a compact space has nonempty intersection, then the system itself has nonempty intersection). □

Problems

2.1. Specify the following sets by systems of linear inequalities:

(a) the set $x_2 \geqslant x_1^2$ (the "epigraph" of a parabola);

(b) the equilateral triangle centered at zero with vertex at $(1, 0)$;

(c) the set $\{ (x_1, x_2) \mid x_1 x_2 \geqslant 1,\ x_1 \geqslant 0,\ x_2 \geqslant 0 \}$.

2.2. Apply affine transformations to solve the following problems.

(a) Through each vertex of a triangle, two lines dividing the opposite side into three equal parts are drawn. Prove that the diagonals joining the opposite sides of the hexagon formed by these lines intersect at one point.

(b) Given a parallelogram $ABCD$ with points K, L, and M lying on its sides AB, BC, and CD, respectively, and dividing them in the same ratio, prove that the lines b, c, and d passing through the vertices B, C, and D parallel to KL, KM, and ML, respectively, meet at one point.

(c) Given a triangle ABC with points M, N and P lying on its sides AB, BC, and CA and dividing them in the same ratio, prove that the medians of the triangles ABC and MNP intersect at the same point O.

(d) Given a triangle ABC with points M, N, P lying on its sides AB, BC, CA, respectively, and M_1, N_1, P_1 symmetric to M, N, P with respect to the midpoints of the respective sides, prove that the triangles MNP and $M_1N_1P_1$ have equal areas.

2.3. Prove that an arbitrary convex pentagon $ABCDE$ with sides parallel to its diagonals (i.e., such that $AB \| CE$, $BC \| DA$, etc.) can be affinely transformed into a regular pentagon.

2.4. Prove that an arbitrary convex quadrilateral which is not a trapezoid can be affinely transformed into a quadrilateral with two right opposite angles.

2.5. Prove that an arbitrary convex hexagon $ABCDEF$ in which each side is parallel to the opposite side can be affinely transformed into a hexagon with equal diagonals AD, BE, and CF.

2.6. Given three vectors $\vec{a}$, $\vec{b}$, and $\vec{c}$ in the plane such that $\alpha\vec{a} + \beta\vec{b} + \gamma\vec{c} = \vec{0}$, prove that these vectors can be affinely transformed into vectors of equal lengths if and only if there exists a triangle with sides $|\alpha|$, $|\beta|$, and $|\gamma|$.

2.7. Prove that the affine transformations of the n-space preserve the ratios between volumes.

2.8. Prove that the diagonals of an n-parallelepiped intersect at one point and are bisected by this point.

2.9. The centroid of an n-simplex $A_1 \ldots A_{n+1}$ is a point M for which

$$\overrightarrow{MA}_1 + \cdots + \overrightarrow{MA}_{n+1} = \overrightarrow{0}.$$

(a) Prove that the centroid exists and is unique.

(b) Prove that the affine transformations map centroids to centroids.

2.10. Let $(a_{ij})_1^n$ be a matrix with determinant $\Delta \neq 0$. In n-space, consider the figure defined by the inequalities

$$|a_{i1}x_1 + \cdots + a_{in}x_n| \leqslant b_i,$$

where $b_1, \dots, b_n$ are positive numbers. Prove that this figure has volume $2^n b_1 \cdots b_n / |\Delta|$.

2.11. Given convex polygons M_1 and M_2 and positive numbers λ_1 and λ_2 with sum equal to 1, consider

$$\lambda_1 M_1 + \lambda_2 M_2 = \{\lambda_1 x_1 + \lambda_2 x_2 \mid x_1 \in M_1,\ x_2 \in M_2\}.$$

(a) Prove that $\lambda_1 M_1 + \lambda_2 M_2$ is a convex polygon whose number of sides does not exceed the total number of sides of M_1 and M_2.

(b) Prove that the perimeter of the polygon $\lambda_1 M_1 + \lambda_2 M_2$ equals $\lambda_1 P_1 + \lambda_2 P_2$, where P_1 and P_2 are the perimeters of M_1 and M_2, respectively.

(c) Prove that the area $S(\lambda_1, \lambda_2)$ of the polygon $\lambda_1 M_1 + \lambda_2 M_2$ equals

$$\lambda_1^2 S_1 + 2\lambda_1\lambda_2 S_{12} + \lambda_2^2 S_2,$$

where S_1 and S_2 are the areas of M_1 and M_2, respectively, and S_{12} depends only on M_1 and M_2.

CHAPTER 3

The Projective World

> The Cayley principle: Projective geometry is all geometry.
>
> F. Klein

Let us go a step further. Recall that we started with linear spaces, the main example of which was the space $\mathbb{R}^n$; then, we constructed the affine space over $\mathbb{R}^n$, a space of points without a fixed origin. Now, we shall complete the affine space by points at infinity and thus construct the projective space $\mathbb{R}P^n$.

3.1. The projective line and the projective plane

A model and some facts of the geometry of projective line. The projective line can be thought of as the usual line completed by a point at infinity, and the model of the projective line is the line $\mathbb{R}$ completed by the point x_∞ at infinity. The motions of the completed line are the linear-fractional transformations.

The same model can be obtained otherwise. Consider the plane $\mathbb{R}^2$ with coordinate axes Ox_1 and Ox_2. The line $\mathbb{R}$ can be represented as the set of points $(x, 1)$, i.e., as the Ox_1 axis "raised" by one (see Figure 3.1). To each straight line $l(x)$ through the origin, except the axis Ox_1, corresponds its "trace" $(x, 1)$ on the line representing $\mathbb{R}$. To Ox_1, we assign a "point at infinity" x_∞. As we have already mentioned, the line $\mathbb{R}$ completed by the point at infinity is what is called the projective line $\mathbb{R}P^1$. The motions of the projective line are the transformations induced on the set of lines $l(x)$ by arbitrary linear transformations of $\mathbb{R}^2$. To be more precise, consider an arbitrary nonsingular matrix $\left(\begin{smallmatrix} a & b \\ c & d \end{smallmatrix}\right)$ (such matrices form the group $GL(2, \mathbb{R})$ of linear transformations of the plane $\mathbb{R}^2$). The points of a line $l(x)$, which correspond to the point $(x, 1)$ of the projective line, have the coordinates $(\lambda x, \lambda)$. The linear transformation under consideration maps them to the points

$$\begin{pmatrix} a & b \\ c & d \end{pmatrix} \begin{pmatrix} \lambda x \\ \lambda \end{pmatrix} = \lambda \begin{pmatrix} ax + b \\ cx + d \end{pmatrix};$$

hence, if $cx+d \neq 0$, then the point $(x, 1)$ is mapped to the point $\big((ax+b)/(cx+d), 1\big)$. The point $(-d/c, 1)$ is mapped to the point at infinity. Every line through the

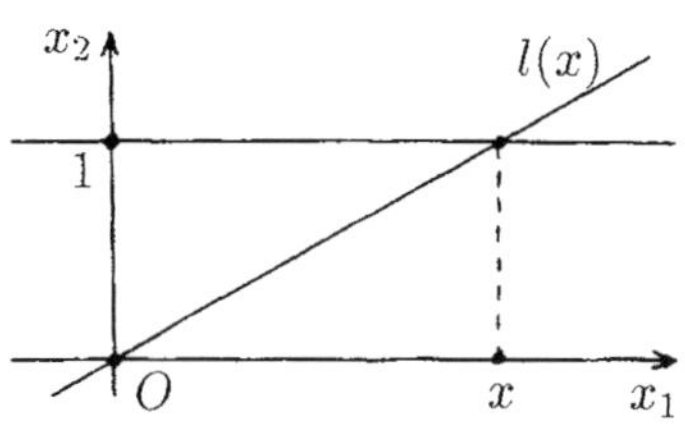

FIGURE 3.1

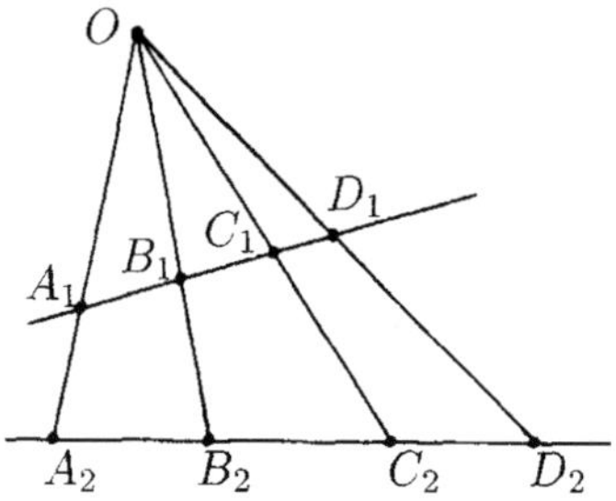

FIGURE 3.2

origin in $\mathbb{R}^2$ is determined by a pair of numbers (ξ_1, ξ_2), where $\xi_i \in \mathbb{R}$ and $i = 1, 2$, which are not both zero; the pairs (ξ_1, ξ_2) and $(\lambda\xi_1, \lambda\xi_2)$, where $\lambda \in \mathbb{R}$ and $\lambda \neq 0$, determine the same line. In other words, points of the projective plane can be specified as ratios $(\xi_1 : \xi_2)$. These coordinates of points of the projective plane are called *homogeneous*. In homogeneous coordinates, the motions are described by

$$(\xi_1 : \xi_2) \mapsto ((a_{11}\xi_1 + a_{12}x_2) : (a_{21}\xi_1 + a_{22}x_2)),$$

where $a_{ij} \in \mathbb{R}$ and $\det(a_{ij}) \neq 0$.

Thus the projective plane belongs to the company of geometries described by the Erlangen Program.

On the projective line, convergence is defined in the natural way, so that the projective line becomes a compact topological space homeomorphic to the circle (this space can even be endowed with a metric; it is defined on p. 88). However, unlike affine geometry, projective geometry has no betweenness relation.

The reader may ask the question: Why is the projective line called projective? The reason for this is that the group of projective transformations of the projective line coincides with the group of compositions of projections of straight lines onto straight lines from points. Let us take a closer look at projections.

Take two straight lines l_1 and l_2 on the (Euclidean) plane and a point O outside the lines. The projection of l_1 onto l_2 from O is shown in Figure 3.2. We can immediately state one important property of projections.

LEMMA 1. *Let points A_2, B_2, C_2, and D_2 of a line l_2 correspond to points A_1, B_1, C_1, and D_1 of a line l_1 under projection from some point O. Then*

$$\frac{|C_1A_1|}{|C_1B_1|} : \frac{|D_1A_1|}{|D_1B_1|} = \frac{|C_2A_2|}{|C_2B_2|} : \frac{|D_2A_2|}{|D_2B_2|}.$$

Proof. Let us denote the area of the triangle ABC by S_{ABC}. We have

$$\frac{|C_1A_1|}{|C_1B_1|} = \frac{S_{OC_1A_1}}{S_{OC_1B_1}} = \frac{|OC_1|\,|OA_1| \sin A_1OC_1}{|OC_1|\,|OB_1| \sin B_1OC_1} = \frac{|OA_1| \sin A_1OC_1}{|OB_1| \sin B_1OC_1}.$$

Therefore,

$$\frac{|C_1A_1|}{|C_1B_1|} : \frac{|D_1A_1|}{|D_1B_1|} = \left(\frac{|OA_1| \sin A_1OC_1}{|OB_1| \sin B_1OC_1}\right) : \left(\frac{|OA_1| \sin A_1OD_1}{|OB_1| \sin B_1OD_1}\right).$$

Reducing the right-hand side, we can express this value in terms of angles only. Since $\sin A_1OC_1 = \sin A_2OC_2$, etc., this implies the required assertion. □

DEFINITION 1. Let A, B, C, D be points on a line l, and let a, b, c, d be their respective coordinates on l. The value

$$[A, B, C, D] := \frac{c-a}{c-b} : \frac{d-a}{d-b}$$

is called the *cross ratio* of these four points.

DEFINITION 2. The *cross ratio* of four lines lying in one plane and passing through one point is the cross ratio of the four points at which these lines intersect an arbitrary line l. Lemma 1 shows that the cross ratio does not depend on the choice of l.

Lemma 1 has three corollaries.

COROLLARY 1. *Any projection of a line onto another one preserves the cross ratios of quadruples of points.*

This follows directly from Lemma 1.

COROLLARY 2. *Any projective transformation of a line is a linear-fractional map.*

Proof. Let a_1, b_1, c_1 be the images of points a, b, c under a projective map, and let x_1 be the image of a point x under this map. By Corollary 1, $[a, b, c, x] = [a_1, b_1, c_1, x_1]$. This equality can be rewritten as

$$\frac{c-a}{c-b} : \frac{x-a}{x-b} = \frac{c_1-a_1}{c_1-b_1} : \frac{x_1-a_1}{x_1-b_1} \Longrightarrow x_1 = \frac{\alpha x+\beta}{\gamma x+\delta}. \quad \square$$

This readily implies the following assertion.

COROLLARY 3. *If a projective transformation of a line has three fixed points, then it is the identity map.*

Projective geometry appeared in the seventeenth century, and its earliest creators were artists—painters and architects who were interested in imaging, perspective, etc. Initially, projective geometry was the geometry "of projections;" this explains its name. The approaches related to linear-fractional transformations appeared in the nineteenth century.

Now, we are ready to prove the main result that combines the old and new approaches.

THEOREM 1. (a) *The following two definitions of the projective transformations of a line are equivalent:*

(i) *the projective transformations are the compositions of projections;*

(ii) *the projective transformations are the linear-fractional transformations.*

(b) *A projective transformation of a line is determined by the images of three points.*

Proof. Corollary 2 of Lemma 1 implies that the compositions of projections are linear-fractional. Next, the linear-fractional maps preserve the cross ratios. Indeed,

$$\text{if} \quad y_i = \frac{\alpha x_i+\beta}{\gamma x_i+\delta}, \quad \text{then} \quad y_i - y_j = \frac{(\alpha\gamma-\beta\delta)(x_i-x_j)}{(\gamma x_i+\delta)(\gamma x_j+\delta)}.$$

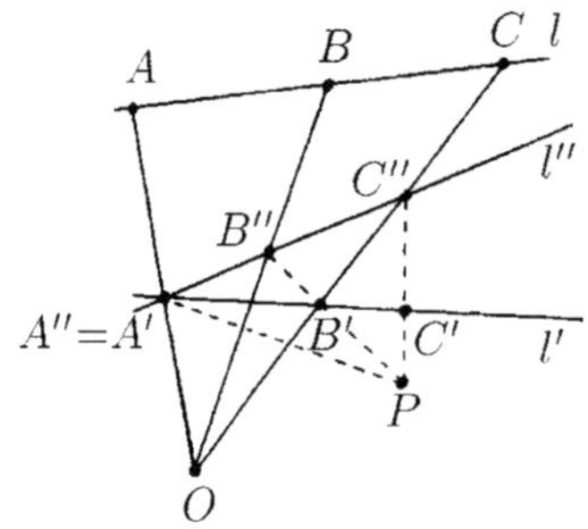

FIGURE 3.3

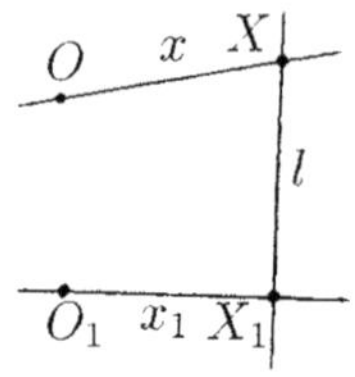

FIGURE 3.4

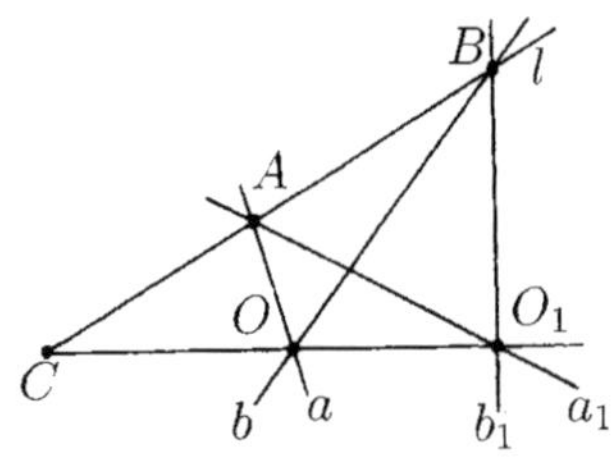

FIGURE 3.5

Therefore, a linear-fractional transformation is determined by the images of three points because the relation $[a, b, c, x] = [a', b', c', x']$ allows us to express x_1 in terms of x. To complete the proof, we must show that there exists a composition of projections that maps three arbitrary points A, B, and C to three given points A', B', and C'. For this purpose, we draw a line l'' different from AA' through the point A', take a point O on the line AA', and project l onto l'' from this point (see Figure 3.3). Let P be the intersection point of the lines $B''B'$ and $C''C'$. Projecting l'' onto l' from P, we obtain the required map. (If $B'B'' \parallel C'C''$, we must consider a parallel projection; if A lies on l' but does not coincide with A', an additional preliminary projection should be performed.) □

Suppose that to each straight line x passing through a point O we assign a line x_1 passing through a point O_1. The correspondence $x \mapsto x_1$ is said to be *projective* if it preserves the cross ratios of quadruples of lines; in other words, the map $X \mapsto X_1$, where $X = x \cap l$ and $X_1 = x_1 \cap l$ (Figure 3.4), must be projective for an arbitrary line l.

THEOREM 2. *Let $x \mapsto x_1$ be a projective correspondence under which a line $x = OO_1$ is assigned a line $x_1 = O_1O$. Then the intersection points of all pairs of lines corresponding to each other lie on one straight line.*

Proof. Let a, a_1 and b, b_1 be two pairs of lines corresponding to each other, and let A and B be their intersection points (Figure 3.5). The line $l = AB$ meets OO_1 at a point C. If OX is assigned O_1X_1, where X and X_1 are points on l, then $[A, B, C, X] = [A, B, C, X_1]$, and hence $X = X_1$. Therefore, all intersection points of pairs of lines corresponding to each other lie on the line l. □

In the statement of Theorem 2, points can be replaced by lines and lines by points. More precisely, the following theorem is valid.

THEOREM 3. *Let a projective map $X \mapsto X_1$ of a line o to a straight line o_1 leave the intersection point of these lines fixed. Then all straight lines passing through pairs of points corresponding to each other under this map meet at one point.*

Proof. Suppose that A, A_1 and B, B_1 are two pairs of lines corresponding to each other, a and b are the lines passing through them, L is the intersection point of a and b, and c is the line passing through L and the intersection point of o and o_1. If a point X of the line o corresponds to the point X_1 of the line o_1, then $[a, b, c, x] = [a_1, b_1, c_1, x_1]$, where $x = XL$ and $x_1 = LX_1$. Therefore, $x = x_1$, i.e., the line XX_1 passes through the point L. □

The cross ratio of four points depends on the order of the points. Under permutations of the points, it changes according to fairly simple rules. Obviously,

$$[A, B, C, D] = [B, A, C, D]^{-1} = [A, B, D, C]^{-1}.$$

It is also easy to see that

$$[A, B, C, D] = 1 - [A, C, B, D];$$

indeed, this is equivalent to the equality

$$(c-a)(d-b) = (c-b)(d-a) + (b-a)(d-c).$$

The three permutations considered above generate all other permutations. Therefore, if the cross ratio of four points is λ, then the cross ratios of the same points taken in a different order can assume the values λ^{-1}, $1-\lambda$, $(1-\lambda)^{-1}$, $1-\lambda^{-1}$, and $(1-\lambda^{-1})^{-1}$. In certain cases, some of these values may coincide. Equating λ to the five other values, we obtain the following degenerate sets of λ values: $\{-1, 2, -1/2\}$, $\{1, 0, \infty\}$, and $\{-\varepsilon, -\varepsilon^2\}$, where $\varepsilon^2 + \varepsilon + 1 = 0$.

For unordered sets of four points, it is more convenient to consider a function $J(\lambda)$ invariant with respect to the replacements of λ by $1-\lambda$ and by λ^{-1} rather than the cross ratios. For such a function, we can use

$$J_1(\lambda) = \lambda^2 + \frac{1}{\lambda^2} + (1-\lambda)^2 + \frac{1}{(1-\lambda)^2} + \left(\frac{\lambda}{1-\lambda}\right)^2 + \left(\frac{\lambda-1}{\lambda}\right)^2$$

or functions obtained from J_1 by linear transformations, say,

$$J_2(\lambda) = \frac{J_1(\lambda)+3}{2} = \frac{(\lambda^2-\lambda+1)^2}{\lambda^2(1-\lambda)^2},$$

$$J_3(\lambda) = J_2(\lambda) - \frac{27}{4} = \left(\frac{(\lambda+1)(\lambda-2)(\lambda-1/2)}{\lambda(1-\lambda)}\right)^2.$$

Each of these functions J has the property that $J(\lambda) = J(\lambda_1)$ if and only if λ_1 is obtained from λ by applying a composition of the transformations $\lambda \mapsto 1-\lambda$ and $\lambda \mapsto \lambda^{-1}$.

The *cross ratio* of four planes sharing a line in $\mathbb{R}^3$ can be defined as the cross ratio of the four lines obtained by intersecting these planes with a plane that does not contain the shared line. The ratio does not depend on the secant plane. Indeed, if two secant planes have a common line l, then the cross ratio determined by either one is equal to the cross ratio of the points at which l intersects the four given planes.

The projective plane. We could have passed from $\mathbb{R}P^1$ directly to the definition of the n-dimensional projective space $\mathbb{R}P^n$. Instead, we devote a separate section to the two-dimensional case—first, following our general outline, and secondly, bearing in mind the historical role played by the projective plane.

We construct the projective plane $\mathbb{R}P^2$ similarly to the projective line $\mathbb{R}P^1$ constructed in the preceding section.

As the model of the projective plane, we can take the set of all lines through the origin in $\mathbb{R}^3$, the motions being all nonsingular linear transformations of $\mathbb{R}^3$.

To each straight line passing through the origin (except the lines in the plane Ox_1x_2), we assign its trace on the plane $x_3 = 1$. To the lines lying in the plane Ox_1x_2, we assign various points at infinity, and call the set of points at infinity the *line at infinity*.

Every line through the origin in $\mathbb{R}^3$ is determined by a triple of numbers (ξ_1, ξ_2, ξ_3), where $\xi_i \in \mathbb{R}$ for $i = 1, 2, 3$ and not all ξ_i are zero. The triples (ξ_1, ξ_2, ξ_3) and $(\lambda\xi_1, \lambda\xi_2, \lambda\xi_3)$, where $\lambda \in \mathbb{R}$ and $\lambda \neq 0$, determine the same line. In other words, every line through the origin is uniquely assigned a ratio $(\xi_1 : \xi_2 : \xi_3)$. These coordinates of points in the projective plane are called *homogeneous*. In homogeneous coordinates, the motions are described as

$$\begin{aligned}&(\xi_1 : \xi_2 : \xi_3)\\ &\quad\mapsto \big((a_{11}\xi_1 + a_{12}x_2 + a_{13}x_3) : (a_{21}\xi_1 + a_{22}x_2 + a_{23}x_3) : (a_{31}\xi_1 + a_{32}x_2 + a_{33}x_3)\big),\end{aligned}$$

where $a_{ij} \in \mathbb{R}$ and $\det(a_{ij}) \neq 0$.

We have made the projective plane consistent with the Erlangen ideology.

Topologically, $\mathbb{R}P^2$ is the set of all straight lines in $\mathbb{R}^3$ that pass through the point O. This is a nonorientable manifold. Other equivalent topological descriptions of $\mathbb{R}P^2$ are as follows:

- the sphere $\mathbb{S}^2$ in which all antipodal points are pairwise identified;
- the disk in which all antipodal boundary points are pairwise identified;
- the Möbius band with a disk attached to its boundary.

So, what objects does projective geometry study? In addition to *points* (which correspond to straight lines through the origin), there are *lines*, which correspond to the planes containing the origin. They are specified by equations of the form

$$a_1x_1 + a_2x_2 + a_3x_3 = 0, \tag{1}$$

where at least one of the numbers a_1, a_2, and a_3 is nonzero. To the plane $x_3 = 0$ corresponds the straight line "at infinity." If $a_3 \neq 0$, then the line (1) can be "seen": this is the straight line in the plane $x_3 = 1$ determined by the equation

$$\frac{a_1}{a_3}x_1 + \frac{a_2}{a_3}x_2 = -1.$$

In the projective plane, through any two points there passes exactly one line, and any two lines intersect (there is no parallelism in the projective world).

Relation (1) establishes a duality between lines and points in the projective plane: the dual to the line (1) is the point corresponding to the straight line with direction vector (a_1, a_2, a_3). (This and other dualities are given much attention later.)

In addition to points and lines, projective geometry studies *conics*, i.e., curves determined by equations of the form

$$\sum_{i,j=1}^{3} a_{ij}x_ix_j = 0. \tag{2}$$

For instance, the right circular cone $x_3^2 = x_1^2 + x_2^2$ "cuts out" the circle $x_1^2 + x_2^2 = 1$ on the plane $x_3 = 1$, the cone $x_3^2 + x_2^2 = x_1^2$ cuts out the hyperbola $x_1^2 - x_2^2 = 1$, and the cone $x_1^2 - x_2x_3 = 0$, the parabola $x_2 = x_1^2$.

It remains to mention several features of the transformation group of the projective plane. Similarly to the projective line, we consider the transformations of $\mathbb{R}P^2$ induced by the transformations of the general linear group $GL(3,\mathbb{R})$ and declare that two figures on the projective plane are equivalent if one of them can be mapped onto the other by a transformation in $GL(3,\mathbb{R})$. In particular, the circle, the hyperbola, and the parabola (see the preceding paragraph) are equivalent as projective objects.

Pappus' and Desargues' theorems. We have already mentioned the remarkable phenomenon of *duality between lines and points* on the projective plane. In all the geometries that have or will be considered in this book, duality is one of the most important phenomena; historically, the duality in projective geometry was discovered first. In this section, we shall speak of the two most renowned theorems of projective geometry and discuss their dual nature.

The duality principle was first stated by Poncelet.

General *Jean-Victor Poncelet* (1788–1867), French engineer and mathematician. Laid the foundations of projective geometry in his *Traité des propriétés projectives des figures* (Treatise on the Projective Properties of Figures, 1822), most of which was written when he was a prisoner of war in Russia after Napoleon's disastrous campaign of 1812.

Let us look again at relation (1) from the preceding section. It implies that the straight line generated by the vector $a = (a_1, a_2, a_3)$, i.e., the set of points in $\mathbb{R}^3$ that can be represented as $\lambda(a_1, a_2, a_3)$ with $\lambda \in \mathbb{R}$, uniquely determines the (perpendicular) plane

$$a^{\perp} = \Pi_a = \{\, x = (x_1, x_2, x_3) \mid a_1x_1 + a_2x_2 + a_3x_3 = 0 \,\}$$

in $\mathbb{R}^3$. This plane Π_a is nothing but a line in the projective plane $\mathbb{R}P^2$; we denote it by the same letter a.

On the other hand, the straight line generated by the vector a determines a point A in $\mathbb{R}P^2$. We have obtained a correspondence $A \longleftrightarrow a$; let us denote it by D. Thus we write $a = DA$ to say that the line a is dual to the point A, and $A = Da$ to say that the point A is dual to the line a; these relations are equivalent because $D^2 = \mathrm{Id}$. In what follows, we denote the line passing through points A and B by (AB) (or merely AB), and the intersection point of lines a and b by (ab). We say that the line (AB) is incident to the points A and B, and that the point (ab) is incident to the lines a and b.

It is easy to see that a line (AB) is incident to a point C if and only if the point (ab) is incident to the line c.

The correspondence $A \longleftrightarrow a$ makes the following assertion obvious.

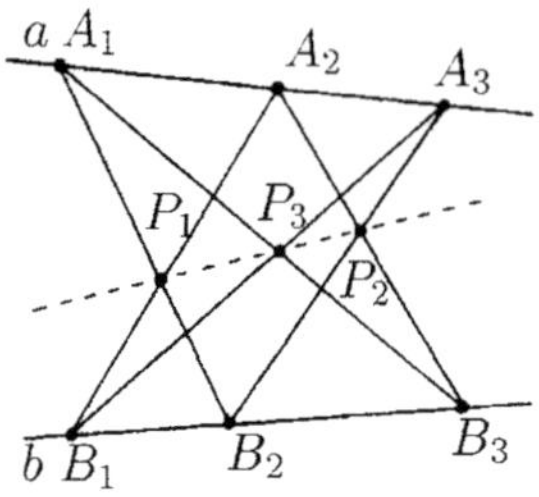

FIGURE 3.6

THEOREM 4 (on dual correspondence). *The interchange of words 'line' and 'point' in any statement about a configuration of points and lines related by incidence does not affect the validity of the statement.*

Before proceeding to the main theorems, we state the following assertion.

LEMMA 2. *Let* A_1, A_2, A_3, A_4 *and* B_1, B_2, B_3, B_4 *be two sets of points in general position in* $\mathbb{R}P^2$. *Then there exists a unique projective transformation* $f\colon \mathbb{R}P^2 \to \mathbb{R}P^2$ *such that* $f(A_i) = B_i$ *for* $1 \leqslant i \leqslant 4$.

Points $\{C_i\}_{i=1}^4$ are said to be *in general position* if the corresponding vectors $\{c_i\}_{i=1}^4$ have the property that any three of them form a basis. The proof of Lemma 2 is based on elementary facts from linear algebra; it is presented in the general case of an n-dimensional space in the next section (see Theorem 2 on p. 59).

PAPPUS' THEOREM. *Suppose we are given two lines* a *and* b *and three points on each,* A_1, A_2, A_3 *on* a, *and* B_1, B_2, B_3 *on* b. *Let* P_1, P_2, P_3 *be the intersection points of* (B_1A_2) *with* (A_1B_2), (B_2A_3) *with* (A_2B_3), (B_1A_3) *with* (A_1B_3), *respectively* (see Figure 3.6). *Then the points* P_1, P_2, *and* P_3 *belong to one line.*

Pappus of Alexandria, ancient Greek mathematician of the third century A.D.

Like any meaningful result, Pappus' theorem can be proved in different ways, and this book contains several proofs. Here is a proof in the spirit of those presented in Chapter 1 to advertise the coordinate method.

First proof. By Lemma 2, there exists a projective transformation that maps A_1, A_2, B_1, and B_2 to the points $A_1' = (0,0)$, $A_2' = (1,0)$, $B_1' = (0,1)$, and $B_2' = (1,1)$ (which are the vertices of a square). Let the images of A_3 and B_3 under this transformation be $A_3' = (a,0)$ and $B_3' = (1,b)$. It is easy to find the coordinates of the points P_i' for $i = 1,2,3$; it is also easy to see that all three points lie on one line. □

Another proof of Pappus' theorem can be obtained by using the fact that a projective transformation of a line with three fixed points is the identity map.

Second proof. Let P_3' and C be the intersection points of B_1A_3 with P_1P_2 and A_1B_2, respectively. We must prove that $P_3 = P_3'$.

Consider the composition of projections

$$B_1A_3 \xrightarrow{A_1} b \xrightarrow{A_2} B_2A_3 \xrightarrow{P_1} B_1A_3;$$

the letters above the arrows indicate the centers of the respective projections. It is easy to verify that this projective transformation of the line B_1A_3 does not move

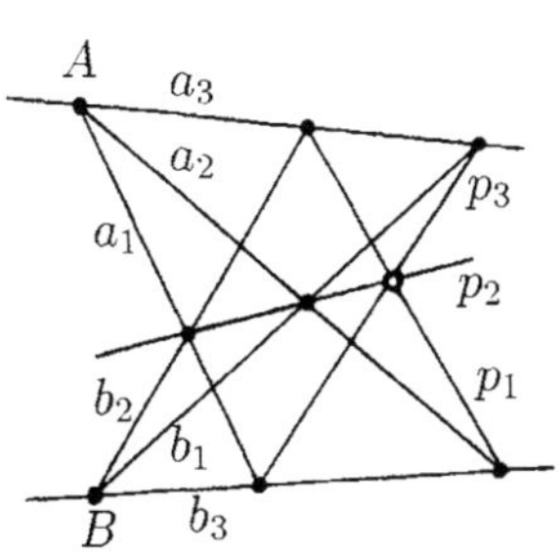

FIGURE 3.7

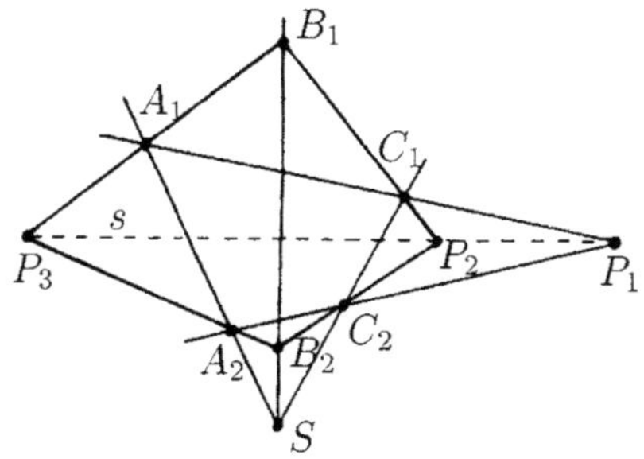

FIGURE 3.8

the points B_1, C, and A_3 and maps P_3 to P_3'. Therefore, this transformation is the identity map, and $P_3 = P_3'$. □

The statement dual to Pappus' theorem is as follows:

Suppose given lines a_1, a_2, a_3 sharing a point A and lines b_1, b_2, b_3 sharing a point B. If p_1 is the line through the intersection points (b_1a_2) and (a_1b_2), p_2 is the line through the points (b_1a_3) and (b_3a_1), and p_3 is the line through the points (b_2a_3) and (b_3a_2), then the lines p_1, p_2, p_3 share a point.

Figure 3.7 convinces us that Pappus' theorem is equivalent to the dual statement. To conclude, note that Pappus' theorem is a special case of Pascal's theorem (see p. 71 below).

DESARGUES' THEOREM. *If the straight lines passing through the respective vertices of triangles A_1, B_1, C_1 and A_2, B_2, C_2 meet at one point S, then the intersection points of the respective sides of these triangles are collinear* (Figure 3.8).

Gérard Desargues (1591–1661), French architect, engineer, and mathematician. Laid the foundations of projective and descriptive geometries. Introduced the notions of elements at infinity and of polarity. Desargues' contemporaries did not comprehend his ideas; projective geometry was revived only in the nineteenth century in works of Poncelet, Steiner, and others.

Desargues' theorem, like Pappus' theorem, can be proved in many different ways. In our first proof, as in the first proof of Pappus' theorem, we apply a projective transformation to attain a simplest arrangement of the points under consideration.

First proof. Let us apply the projective transformation mapping the points A_1, A_2, B_1, and B_2 to the points (vertices of a square)

$$A_1' = (0,0), \quad A_2' = (1,0), \quad B_1' = (0,1), \quad \text{and} \quad B_2' = (1,1),$$

respectively. The lines $A_1'B_1'$ and $A_2'B_2'$ meet at the point at infinity P_3' of the axis Ox_2, and the points C_1' and C_2' (the images of C_1 and C_2) lie on a line parallel to the axis Ox_1; S' is the point at infinity of this line. Let P_1' and P_2' be the intersection points of $C_1'A_1'$ with $C_2'A_2'$ and of $C_1'B_1'$ with $C_2'B_2'$, respectively. It is easy to verify that P_1' and P_2' belong to the same vertical line. Therefore, the point P_3' is the point at infinity of the line $P_1'P_2'$, and hence the preimages of the points P_1', P_2', P_3' belong to one line. □

Another proof of Desargues' theorem can be obtained by using the properties of projective transformations mentioned in Theorem 2 on p. 50.

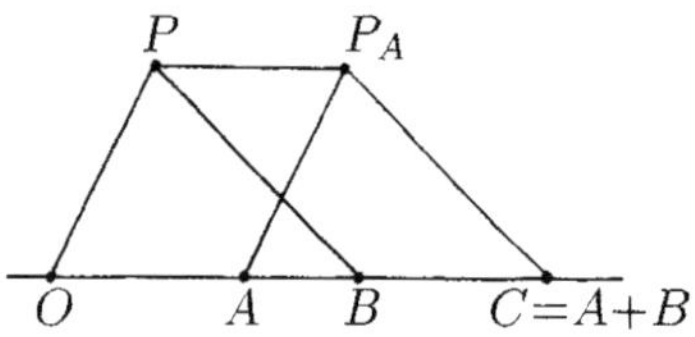

FIGURE 3.9

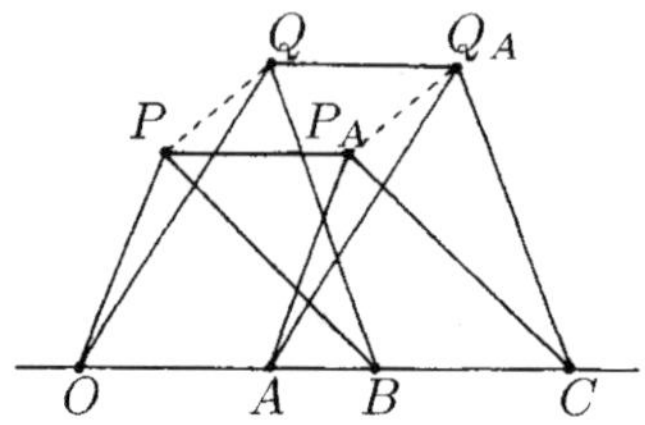

FIGURE 3.10

Second proof. Consider the projection of the line A_1B_1 onto the line A_2B_2 from the point S. For an arbitrary point X_1, let X_2 denote its image under this projection. The correspondence $C_1X_1 \mapsto C_2X_2$ is projective, and the line C_1C_2 corresponds to the line C_2C_1. Therefore, the intersection points of C_1A_1 with C_2A_2, of C_1B_1 with C_2B_2, and of C_1P_3 with C_2P_3 (P_3 is the intersection point of A_1B_1 with A_2B_2) are collinear. □

One more proof follows from the fact that the plane $\mathbb{R}^2$ is embedded in the space $\mathbb{R}^3$.

Third proof. In $\mathbb{R}^3$, the following assertion is obviously valid: *If two planes* Π_1 *and* Π_2 *intersect the edges of a trihedral angle with vertex* S *at points* A_1, B_1, C_1 *and* A_2, B_2, C_2, *respectively, then the intersection points of* A_1B_1 *with* A_2B_2, *of* B_1C_1 *with* B_2C_2, *and of* A_1C_1 *with* A_2C_2 *lie on the common line* s *of the planes* Π_1 *and* Π_2. Projecting this configuration onto some plane, we obtain Desargues' theorem. □

Like Pappus' theorem, Desargues' theorem is equivalent to its dual statement. The Pappus configuration consists of nine points and nine lines; each line contains three points, and through each point, three lines pass. The Desargues configuration consists of ten points and ten lines; each line contains three points, and through each point, three lines pass.

On a straight line in the plane, we can define the algebraic structure of a field by means of only geometric constructions in the projective space. These constructions are closely related to the theorems of Desargues and Pappus; namely, Desargues' theorem ensures that the addition and multiplication operations in the field are well defined, and Pappus' theorem implies their commutativity.

To define addition and multiplication on the line, we choose points O (the zero element) and E (the identity element). For any points A and B on the line, the point $C = A + B$ can be defined by using the construction shown in Figure 3.9, where P is an arbitrary point, $PP_A \parallel OA$, $AP_A \parallel OP$, and $CP_A \parallel BP$. The independence of C from the choice of P is equivalent to the following assertion:

Let three sets of points $\{O, A, B, C\}$, $\{P, P_A\}$, $\{Q, Q_A\}$ *lie on three parallel lines (each set lying on one line),* $OP \parallel AP_A$, $OQ \parallel AQ_A$, $PB \parallel P_AC$ (Figure 3.10). *Then* $QB \parallel Q_AC$.

This assertion is proved by applying Desargues' theorem twice. First, we apply it to the triangles OPQ and AP_AQ_A, obtaining $PQ \parallel P_AQ_A$. Next, we apply it to the triangles PQB and P_AQ_AC, obtaining $QB \parallel Q_AC$.

To define the multiplication on the line, we use in addition to O, the point E. The product C of points A and B is defined by the construction shown in Figure 3.11, where P is an arbitrary point, $BP_B \parallel EP$, and $CP_B \parallel AP$. The independence of C from the choice of P is again implied by Desargues' theorem.

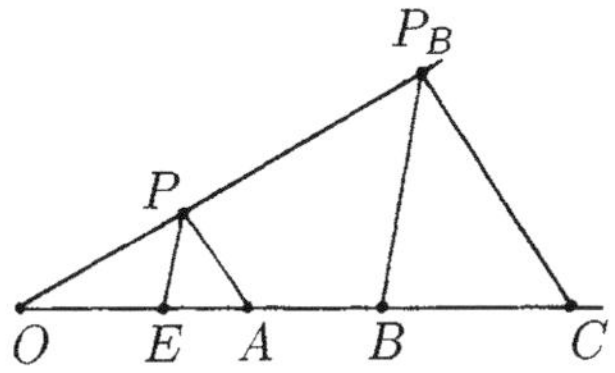

FIGURE 3.11

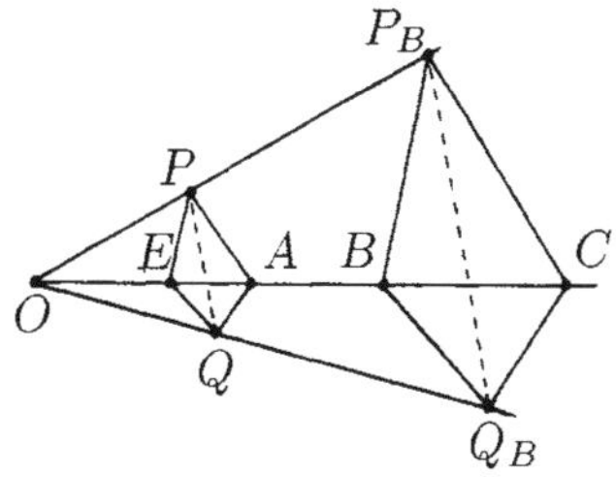

FIGURE 3.12

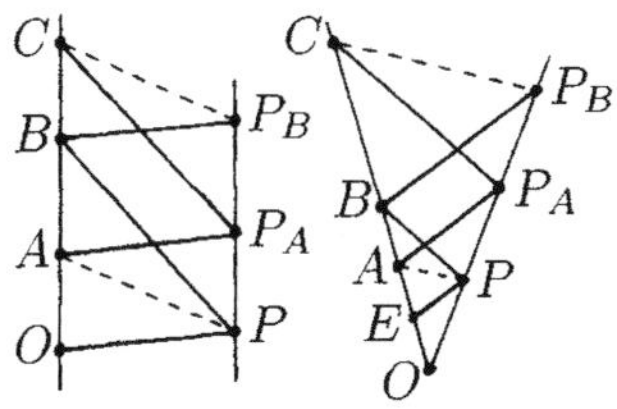

FIGURE 3.13

Indeed, according to this theorem, $QP \parallel Q_B P_B$ (Figure 3.12). Therefore, applying Desargues' theorem once more, we obtain $QA \parallel Q_B C$, as required.

The commutativity of addition and multiplication follows from Pappus' theorem. Indeed, geometrically, the points $A + B$ and $B + A$ are defined differently; the points $A \cdot B$ and $B \cdot A$ are also defined differently (Figure 3.13). But in both cases, Pappus' theorem and the relations $AP_A \parallel BP_B$ and $BP \parallel CP_A$ imply $AP \parallel P_B C$.

3.2. Projective n-space

A direct generalization of the notions of projective line and projective plane is the notion of *projective n-space* $\mathbb{R}P^n$.

The projective space $\mathbb{R}P^n$ has several expressive interpretations. It can be represented as the set of all straight lines containing 0 in the space $\mathbb{R}^{n+1}$, as the sphere

$$\mathbb{S}^n = \{ x \in \mathbb{R}^{n+1} \mid \sum_{i=1}^{n+1} x_i^2 = 1 \}$$

with identified antipodal points, or as the space $\mathbb{R}^n$ completed by a hyperplane at infinity. To obtain the last representation, we can use the same picture as in the one- and two-dimensional cases. Namely, we can consider the hyperplane in $\mathbb{R}^{n+1}$ given by the equation $x_{n+1} = 1$ and the "traces" of the lines passing through the origin on this hyperplane. The lines lying in the parallel hyperplane containing the origin (i.e., the hyperplane $x_{n+1} = 0$) form the $n-1$-dimensional projective plane "at infinity".

Let us give precise definitions (in other words, we construct a model of projective n-space).

The *points* in $\mathbb{R}P^n$ are the $(n+1)$-tuples $(x_1, \dots, x_{n+1})$ of real numbers such that not all $x_1, \dots, x_{n+1}$ are zero. Proportional $(n+1)$-tuples are considered equivalent. (In other words, the points in $\mathbb{R}P^n$ are the straight lines through the origin in $\mathbb{R}^{n+1}$.) The numbers $x_1, \dots, x_{n+1}$ are called the *homogeneous coordinates* of the corresponding point in the projective space, and this point is denoted by $(x_1 : x_2 : \dots : x_{n+1})$.

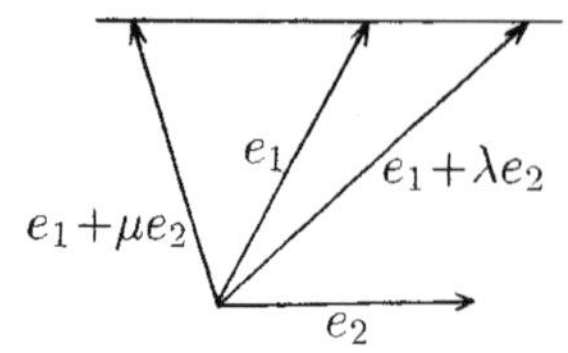

FIGURE 3.14

Consider a $(k+1)$-dimensional plane passing through the origin in $\mathbb{R}^{n+1}$. The points of $\mathbb{R}P^n$ that correspond to the lines lying in this plane form a *k-dimensional subspace* in $\mathbb{R}P^n$. The 1-dimensional subspaces of $\mathbb{R}P^n$ are called *lines*, and the $(n-1)$-dimensional subspaces are called *hyperplanes*.

In homogeneous coordinates, a k-dimensional subspace of the projective n-space is specified by a homogeneous linear system of the form

$$\sum_{j=1}^{n+1} a_{ij}x_j = 0, \quad 1 \leqslant i \leqslant n-k, \quad \operatorname{rank}(a_{ij}) = n-k. \tag{1}$$

We already know that *on the projective plane $\mathbb{R}P^2$, any two distinct lines have exactly one intersection point.* (Indeed, two lines on the projective plane $\mathbb{R}P^2$ correspond to two planes in $\mathbb{R}^3$ containing the origin. These two planes have a common line passing through the origin, as required.) A natural generalization of this property is as follows: *any n hyperplanes in the projective n-space intersect, and if these hyperplanes are "in general position," i.e., the rank of system* (1) *is n, then they intersect at precidely one point.*

The main objects in the projective n-space are its subspaces of various dimensions and *quadrics*, i.e., hypersurfaces specified (in homogeneous coordinates) by

$$\sum_{i,j=1}^{n+1} a_{ij}x_ix_j = 0. \tag{2}$$

Now, let us turn to the group of projective transformations. Just as for the line and plane, the transformations of projective n-space correspond to transformations in the general linear group $GL(n+1,\mathbb{R})$; in other words, a *projective transformation* of the space $\mathbb{R}P^n$ is a transformation induced by a linear transformation of the vector space $\mathbb{R}^{n+1}$.

THEOREM 1. *A projective transformation maps a line to a line and preserves the cross ratios of all quadruples of collinear points.*

Proof. A linear transformation maps 2-dimensional subspaces to 2-dimensional subspaces. Therefore, the induced transformation of the projective space maps lines to lines.

Consider four collinear points in the projective space. They correspond to the vectors e_1, e_2, $e_1+\lambda e_2$, and $e_1+\mu e_2$ (Figure 3.14) in the space $\mathbb{R}^{n+1}$. The cross ratio of the points under consideration is $[0,\infty,\lambda,\mu]$. A linear transformation maps the four corresponding vectors to vectors ε_1, ε_2, $\varepsilon_1+\lambda\varepsilon_2$, and $\varepsilon_1+\mu\varepsilon_2$. Hence the cross ratio of the points corresponding to the transformed vectors is also $[0,\infty,\lambda,\mu]$. □

Now, let us generalize the theorem claiming that a projective transformation of the projective line $\mathbb{R}P^1$ is uniquely determined by the images of three pairwise distinct points to $\mathbb{R}P^n$. We say that points $A_1,\dots,A_{n+2} \in \mathbb{R}P^n$ are *in general position* if the corresponding vectors $a_1, \dots, a_{n+2} \in \mathbb{R}^{n+1}$ have the property that every set of $n+1$ of these points forms a basis, or, in other words, if the vectors $a_1, \dots, a_{n+1}$ form a basis and

$$a_{n+2} = \lambda_1 a_1 + \dots + \lambda_{n+1} a_{n+1}, \qquad \text{where} \quad \lambda_1 \cdots \lambda_{n+1} \neq 0.$$

THEOREM 2. *Let $A_1, \dots, A_{n+2}$ and $B_1, \dots, B_{n+2}$ be two sets of points in general position in $\mathbb{R}P^n$. Then there exists a unique projective transformation $f\colon \mathbb{R}P^n \to \mathbb{R}P^n$ such that $f(A_i) = B_i$ for $i = 1, \dots, n+2$.*

Proof. Suppose that the points $A_1, \dots, A_{n+2}$ correspond to vectors

$$a_1, \dots, a_{n+1}, a_{n+2} = \lambda_1 a_1 + \dots + \lambda_{n+1} a_{n+1},$$

and the points $B_1, \dots, B_{n+2}$ correspond to vectors

$$b_1, \dots, b_{n+1}, b_{n+2} = \mu_1 b_1 + \dots + \mu_{n+1} b_{n+1}.$$

We must prove that there exists a linear transformation A such that $Aa_i = x_i b_i$ for $i = 1, \dots, n+2$, and that A is defined uniquely up to proportionality. For $i \neq n+2$, we put $Aa_i = x_i b_i$, where x_i is an arbitrary nonzero number. The numbers $x_1, \dots, x_{n+1}$ uniquely determine the transformation A. We have $Aa_{n+2} = x_{n+2} b_{n+2}$ if and only if

$$x_i = \frac{\mu_i}{\lambda_i} x_{n+2} \qquad \text{for} \quad i = 1, \dots, n+1.$$

This means that the required set of numbers $x_1, \dots, x_{n+1}$ exists, and it is defined uniquely up to proportionality. □

Problems

Cross ratio.

3.1. Given pairwise distinct collinear points A, B, C, D, E, prove that

$$[A, B, C, D] \cdot [A, B, D, E] \cdot [A, B, E, C] = 1.$$

3.2. Given the line $l(t)$ with the equation $a_1 x + b_1 y = t(a_2 x + b_2 y)$, prove that

$$[l(a), l(b), l(c), l(d)] = [a, b, c, d].$$

3.3. Given four lines a, b, c, d in $\mathbb{R}^3$ each intersecting lines l_1, l_2, l_3, prove that the cross ratios of the three quadruples of the intersection points on the lines l_1, l_2, l_3, are equal.

3.4. Prove that the cross ratio of the intersection points of the planes containing the faces of a tetrahedron $ABCD$ with a line l is equal to the cross ratio of the four planes that pass through l and the vertices of the tetrahedron.

3.5. Given four hyperplanes in $\mathbb{R}^n$ with a common subspace of dimension $n-2$, prove that the cross ratio of the points at which a line l intersects these hyperplanes does not depend on the choice of l.

Linear-fractional and projective transformations.

3.6. Given a nonidentity map $x \mapsto (ax+b)/(cx+d)$, prove that its square is the identity map if and only if $a+d=0$. (By the square of a transformation $f(x)$, we mean the transformation $f(f(x))$.)

3.7. Prove that an arbitrary nonidentity linear-fractional transformation with fixed points a and b has the form

$$x \mapsto \frac{\lambda x - ab}{x + \lambda - a - b}.$$

3.8. Given a linear-fractional transformation f such that $f(a) \neq a$ and $f(f(a)) = a$ for some a, prove that $f(f(x)) = x$ for all x.

3.9. Prove that any linear-fractional transformation can be represented as the composition of no more than three linear-fractional transformations whose squares are the identity transformation.

3.10. Let l, l_1, l_2 be pairwise skew lines in $\mathbb{R}^3$. To each point $A_1 \in l_1$, we assign the intersection point A_2 of the line l_2 with the plane containing A_1 and l. Prove that the map $A_1 \mapsto A_2$ of the line l_1 to the line l_2 is projective.

3.11. Given points A_1, A_2, A_3, O, and O' in the plane, draw a straight line through the intersection points of A_iO with A_jO' and of A_iO' with A_jO for each pair of distinct i and j. Prove that the resulting lines meet at one point.

3.12. Lines $l_1, \dots, l_{n-1}$, and l lie in the plane, and points $O_1, \dots, O_n$ belong to the line l. The vertices $A_1, \dots, A_{n-1}$ of an n-gon $A_1 \dots A_n$ move along the lines $l_1, \dots, l_{n-1}$. Given that the lines containing the sides of this n-gon always pass through the points $O_1, \dots, O_n$, prove that the vertex A_n moves along some straight line.

CHAPTER 4

Conics and Quadrics

The preceding chapters were largely concerned with linear objects, such as level lines and surfaces of linear functions. In this chapter, we consider level lines and surfaces of *quadratic functions*. The level lines of such functions on the plane are called *conics*, and the level surfaces in higher-dimensional spaces, *quadrics*.

The theory of plane conics (later, we show that they are sections of the right circular cone; hence the name conics) was constructed by Apollonius (c. 260–c. 190 B.C.). His *Konika* (Conic Sections) was a jewel of ancient mathematics. For scholars of new times, the pioneers of the modern natural sciences (such as Galileo, Kepler, Huygens, and Newton), Apollonius' study of parabola, hyperbola, and ellipse was a point of departure to explore the laws of nature.

That is why we devote our first section to these remarkable curves.

4.1. Plane curves of the second order

Metric, affine, and projective classification of second-order curves.

Metric classification. Let Ox_1x_2 be an orthogonal coordinate system on the plane. This section examines properties (largely metric) of the curves given by the formula

$$Q(x_1, x_2) = a_{11}x_1^2 + 2a_{12}x_1x_2 + a_{22}x_2^2 + 2b_1x_1 + 2b_2x_2 = c \tag{1}$$

(i.e., the level lines of quadratic functions). Without loss of generality, we assume that $a_{11} \geqslant 0$.

THEOREM 1 (Metric classification of conics). (a) *If* $a_{11}a_{22} - a_{12}^2 \neq 0$, *then the curve* (1) *either is isometric to one of the curves* $x_1^2/a_1^2 + x_2^2/a_2^2 = 1$ (*ellipse*), $x_1^2/a_1^2 - x_2^2/a_2^2 = 1$ (*hyperbola*), *and* $x_1^2/a_1^2 = x_2^2/a_2^2$ (*a pair of intersecting straight lines*) *or is a point or the empty set.*

(b) *If* $a_{11}a_{22} - a_{12}^2 = 0$ (*but not all numbers* a_{11}, a_{12}, a_{22} *are equal to* 0), *then the curve* (1) *either is isometric to one of the curves* $x_1^2 = 2px_2$ (*parabola*), $x_2^2 = c^2$ (*a pair of parallel lines*), *and* $x_2^2 = 0$ (*a pair of merged lines*) *or is the empty set.*

Proof. (a) Applying the translation $x_1' = x_1 + a_1$, $x_2' = x_2 + a_2$, we obtain

$$\begin{aligned} c = Q(x_1, x_2) &= a_{11}(x_1' - a_1)^2 + 2a_{12}(x_1' - a_1)(x_2' - a_2) + a_{22}(x_2' - a_2)^2 \\ &\quad + 2b_1(x_1' - a_1) + 2b_2(x_2' - a_2) \\ &= a_{11}x_1'^2 + 2a_{12}x_1'x_2' + a_{22}x_2'^2 + 2(-a_{11}a_1 - a_{12}a_2 + b_1)x_1' \\ &\quad + 2(-a_{12}a_1 - a_{22}a_2 + b_2)x_2' + Q(a_1, a_2) - 2(b_1a_1 + b_2a_2). \end{aligned}$$

Solving the system $a_{11}a_1 + a_{12}a_2 = b_1$, $a_{12}a_1 + a_{22}a_2 = b_2$ and setting

$$c' = c - Q(a_1, a_2) + 2(b_1a_1 + b_2a_2),$$

we reduce (1) to the form

$$\text{(i)} \qquad Q(x_1', x_2') = a_{11}x_1'^2 + 2a_{12}x_1'x_2' + a_{22}x_2'^2 = c'.$$

If $a_{12} = 0$, then there is nothing to prove. Otherwise, we perform a rotation and denote the new variables, similarly to the initial ones, by x_1 and x_2 (rather than by x_1'' and x_2''). We have

$$\text{(ii)} \qquad x_1' = x_1 \cos\varphi + x_2 \sin\varphi, \quad x_2' = -x_1 \sin\varphi + x_2 \cos\varphi.$$

This gives

$$\begin{aligned}
\text{(iii)} \qquad c' = Q(x_1', x_2') &= Q(x_1 \cos\varphi + x_2 \sin\varphi, -x_1 \sin\varphi + x_2 \cos\varphi) \\
&\overset{\text{Id}}{=} x_1^2(a_{11}\cos^2\varphi - 2a_{12}\cos\varphi\sin\varphi + a_{22}\sin^2\varphi) \\
&\quad + 2x_1x_2(a_{11}\sin\varphi\cos\varphi + a_{12}(\cos^2\varphi - \sin^2\varphi) - a_{22}\cos\varphi\sin\varphi) \\
&\quad + x_2^2(a_{11}\sin^2\varphi + 2a_{12}\sin\varphi\cos\varphi + a_{22}\cos^2\varphi).
\end{aligned}$$

Solving the equation $(a_{11} - a_{22})/2a_{12} = -\cot 2\varphi$ and performing necessary computation, we obtain

$$\frac{x_1^2}{a_1^2} \pm \frac{x_2^2}{a_2^2} = \theta,$$

where $\theta = 0$ or 1, which proves (a).

Let us prove (b). By assumption, $a_{11}a_{22} = a_{12}^2$. Suppose that $a_{12} \neq 0$. Then a_{11} and a_{22} are positive. Applying the rotation

$$x_1 = x_1' \cos\varphi - x_2' \sin\varphi, \quad x_2 = x_1' \sin\varphi + x_2' \cos\varphi$$

with $\tan\varphi = \sqrt{a_{11}/a_{22}}$, we kill the coefficient at x_1' and, after simple calculations, obtain the required form. If $a_{12} = 0$, then, without loss of generality, we can assume that $a_{11} = 0$ and hence the coefficient at x_1 equals zero; thus the curve (1) itself has the desired form. □

REMARK. Theorem 1 implies that the conics split into two classes, nondegenerate (ellipse, hyperbola, and parabola) and degenerate (pairs of lines, sometimes merged). Let us demonstrate that the conic (1) is degenerate if and only if the *principal determinant*

$$\det Q = \begin{vmatrix} a_{11} & a_{12} & b_1 \\ a_{21} & a_{22} & b_2 \\ b_1 & b_2 & -c \end{vmatrix}$$

vanishes. Indeed, our calculations show that, in case (a), a pair of intersecting lines is obtained if and only if the system

$$\begin{aligned}
a_{11}a_1 + a_{12}a_2 &= b_1, \\
a_{21}a_1 + a_{22}a_2 &= b_2, \\
a_1b_1 + a_2b_2 &= -c
\end{aligned}$$

has a nonzero solution, i.e., if and only if $\det Q = 0$. In case (b), such a pair is obtained if and only if $b_1/\sqrt{a_{11}} = b_2/\sqrt{a_{22}}$, and this is exactly what we get when equate the principal determinant to zero under the condition $a_{21} = \sqrt{a_{11}a_{22}}$.

Affine and projective classification. Suppose again that

$$Q(x_1, x_2) = a_{11}x_1^2 + 2a_{12}x_1x_2 + a_{22}x_2^2 + 2b_1x_1 + 2b_2x_2 = c \tag{1}$$

is a second-order curve on the affine plane; as we remember (see Chapter 2), the motions of the affine plane are arbitrary affine transformations

$$(x_1, x_2) \mapsto (\alpha_{11}x_1 + \alpha_{12}x_2 + \beta_1, \ \alpha_{21}x_1 + \alpha_{22}x_2 + \beta_2).$$

THEOREM 2 (Affine classification of plane conics). (a) *If* $a_{11}a_{22} - a_{12}^2 \neq 0$ *in* (1), *then the curve* (1) *either is an affine equivalent of the circle* $x_1^2 + x_2^2 = 1$, *of the equilateral hyperbola* $x_1^2 - x_2^2 = 1$, *or of the pair of intersecting lines* $x_1^2 - x_2^2 = 0$, *or it is a point or the empty set.*

(b) *If* $a_{11}a_{22} - a_{12}^2 = 0$, *then the curve* (1) *either is an affine equivalent of the canonical parabola* $x_1^2 = x_2$, *of the pair of parallel lines* $x_2^2 = 1$, *or of the pair of merged lines* $x_2^2 = 0$, *or it is the empty set.*

Proof. The circle $x_1^2 + x_2^2 = 1$ is obtained from the ellipse $x_1^2/a_1^2 + x_2^2/a_2^2 = 1$ by contractions along the axes Ox_1 and Ox_2; the other curves mentioned in the statement of the theorem are obtained similarly. In remains to prove that they are not equivalent to each other. This follows from topological considerations. The circle is compact and the affine maps are continuous; therefore, the circle cannot be equivalent to any other curve under consideration. The hyperbola consists of two pieces, while the other curves (except the parallel lines, which remain parallel lines under affine maps) are connected; thus the hyperbola is also not equivalent to the other curves. The affine transformations map pairs of merged lines to pairs of merged lines and pairs of intersecting lines to pairs of intersecting lines; in particular, such pairs cannot be affinely transformed into parabolas. □

A second-order curve is written in homogeneous coordinates as

$$Q(x_1, x_2, x_3) = a_{11}x_1^2 + 2a_{12}x_1x_2 + a_{22}x_2^2 + a_{33}x_3^2 + 2a_{13}x_1x_3 + 2a_{23}x_2x_3 = 0. \tag{2}$$

Such quadratic forms reduce to one of the following forms:

$$x_1^2 + x_2^2 - x_3^2 = 0, \quad x_1^2 + x_2^2 + x_3^2 = 0, \quad x_1^2 - x_2^2 = 0, \quad x_1^2 + x_2^2 = 0, \quad x_1^2 = 0.$$

This implies the following theorem.

THEOREM 3 (Projective classification of plane conics). *A second-order curve on the projective plane is projectively equivalent to the circle* $x_1^2 + x_2^2 = 1$, *to the pair of intersecting lines* $x_1^2 - x_2^2 = 0$, *or to the pair of merged lines, or it is a point or the empty set.*

The ellipse, hyperbola, and parabola. As already mentioned, a second-order curve isometric to the curve $x_1^2/a_1^2 + x_2^2/a_2^2 = 1$ is called an *ellipse* (we assume that $a_1 \geqslant a_2 > 0$), a curve isometric to the curve $x_1^2/a_1^2 - x_2^2/a_2^2 = 1$ is called a *hyperbola*, and a curve isometric to $x_2^2 = 2px_1$ is called a *parabola*.

The equations written above are called the *canonical equations* of ellipse, hyperbola, and parabola. The points with coordinates $(\pm c, 0) = (\pm\sqrt{a_1^2 - a_2^2}, 0)$ are called the *foci* of the ellipse, the points $(\pm c, 0) = (\pm\sqrt{a_1^2 + a_2^2}, 0)$ are the *foci* of the hyperbola, and $(p/2, 0)$ is the *focus* of the parabola; the numbers $e = \sqrt{1 - a_2^2/a_1^2}$, $\sqrt{1 + a_2^2/a_1^2}$, and 1 are said to be the *eccentricities* of the ellipse, hyperbola, and parabola, respectively. The straight lines $x_1 = \pm a_1/e$ are called the *directrices* of the ellipse and hyperbola, and $x_1 = -p/2$ is the *directrix* of the parabola. The line $x_2 = 0$ is the *axis of the parabola*.

THEOREM 4. (a) *The ratio of the distances from a point on an ellipse, hyperbola, or parabola to the (corresponding) focus and to the (corresponding) directrix is a constant equal to* e.

(b) *The sum of the distances from a point on an ellipse to its foci is constant; the difference of the distances from a point on a hyperbola to its foci is constant.*

(c) *For given point* F *and line* l *in the Euclidean plane, the set of all points* X *such that the ratio of the distance between* X *and* F *to the distance between* X *and* l *equals a constant* e *is an ellipse if* $e < 1$, *a hyperbola if* $e > 1$, *and a parabola if* $e = 1$.

Proof. (a) Let us start with an ellipse. We put $d = a_1/e$. Then $de^2 = a_1^2/\sqrt{a_1^2 - a_2^2} \cdot (a_1^2 - a_2^2)/a_1^2 = c$, and d is the distance from the origin to the directrix. The set of points $X = (x_1, x_2)$ for which the ratio of the distances to the focus $(c, 0)$ and to the directrix $x_1 = d$ is e, i.e., the set of points determined by the equation

$$\frac{(x_1 - c)^2 + x_2^2}{(x_1 - d)^2} = e^2, \tag{3}$$

is the ellipse specified by the canonical equation. Indeed, under the condition $c = de^2$, (3) is equivalent to

$$\frac{x_1^2}{d^2e^2} + \frac{x_2^2}{d^2e^2(1 - e^2)} = 1,$$

which coincides with the canonical equation of an ellipse in which $a_1^2 = d^2e^2$ and $a_2^2 = d^2e^2(1 - e^2)$.

For a hyperbola and for a parabola assertion (a) is proved similarly.

Let us prove (b) for the ellipse. Let F_1 and F_2 be the foci of the ellipse. Consider the projections H_1 and H_2 of a point X of the ellipse to the directrices. According to (a), we have

$$|F_1X| + |F_2X| = e(|XH_1| + |XH_2|) = e|H_1H_2| = \text{const}.$$

For a hyperbola, (b) is proved similarly.

Let us prove (c) for the ellipse. Suppose that the line l is defined by the equation $x_1 = d$ and the point F has coordinates $(c, 0)$. Then the set under consideration contains two points with coordinates $(x, 0)$ satisfying the equation $(x - c)/(x - d) = \pm e$. Let us take the midpoint between these two points as the origin. Then

$$\frac{de + c}{1 + e} + \frac{-de + c}{1 - e} = 0,$$

i.e., $c = de^2$. Under this condition, (3) is equivalent to the canonical equation of the ellipse. □

THEOREM 5. *The equations of ellipse, hyperbola, and parabola in polar coordinates have the form*

$$r = \frac{pe}{1 + e\cos\varphi}, \tag{4}$$

where e *is the eccentricity and* p *is the distance between a focus and the corresponding directrix.*

Proof. The definitions and Theorem 4 imply (see Figure 4.1)

$$\frac{r}{p - r\cos\varphi} = e,$$

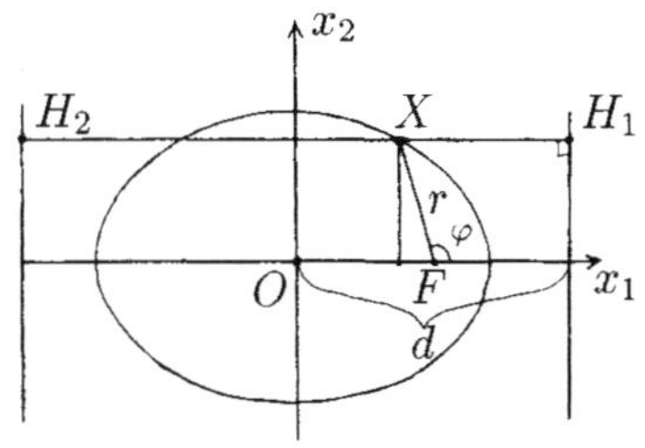

FIGURE 4.1

which yields (4). The cases of hyperbolas and parabolas are similar. □

THEOREM 6. *The set of points with constant sum (difference) of distances to two given points F_1 and F_2 is an ellipse (a hyperbola).*

Proof (for an ellipse). Consider the coordinate system with origin at the midpoint between F_1 and F_2, with the Ox_1 axis directed along $[F_1, F_2]$, and Ox_2 axis perpendicular to Ox_1. Let the coordinates of F_1 and F_2 be $(c, 0)$ and $(-c, 0)$, respectively. By $2a_1$, we denote the sum of distances from a point $X = (x_1, x_2)$ on the hypothetical ellipse to F_1 and F_2. We have

$$\begin{aligned}
&\sqrt{(x_1 - c)^2 + x_2^2} = 2a_1 - \sqrt{(x_1 + c)^2 + x_2^2} \\
&\quad\Longrightarrow (x_1 - c)^2 + x_2^2 = 4a_1^2 - 4a_1\sqrt{(x_1 + c)^2 + x_2^2} + (x_1 + c)^2 + x_2^2 \\
&\quad\Longrightarrow a_1\sqrt{(x_1 + c)^2 + x_2^2} = a_1^2 + x_1 c \\
&\quad\Longrightarrow a_1^2(x_1^2 + 2x_1 c + c^2) + a_1^2 x_2^2 = a_1^4 + 2a_1^2 x_1 c + x_1^2 c^2 \\
&\quad\Longrightarrow (a_1^2 - c^2)x_1^2 + a_1^2 x_2^2 = a_1^2(a_1^2 - c^2) \\
&\quad\Longrightarrow x_1^2/a_1^2 + x_2^2/a_2^2 = 1,
\end{aligned}$$

where $a_2^2 = a_1^2 - c^2$; this is the equation of the ellipse. The proof for a hyperbola is similar. □

THEOREM 7. *The midpoints of parallel chords of an ellipse, a hyperbola, or a parabola lie on one line (for a parabola, this line is parallel to the axis of the parabola).*

Proof. As before, we give the proof for an ellipse. The intersection points (x_1', x_2') and (x_1'', x_2'') of the ellipse (we assume that it is specified by the canonical equation) with the line $x_2 = px_1 + q$ are found by solving the quadratic equation

$$\frac{x_1^2}{a_1^2} + \frac{(px_1 + q)^2}{a_2^2} = 1.$$

By Viète's theorem, $(x_1' + x_1'')/2 = -a_1^2 pq/(a_2^2 + a_1^2 p^2)$, and therefore,

$$\frac{x_2' + x_2''}{2} = p\frac{x_1' + x_1''}{2} + q = \frac{a_2^2 q}{a_2^2 + a_1^2 p^2}.$$

Thus the midpoints of those chords of the ellipse that are parallel to the line $x_2 = px_1$ lie on the line $x_2 = (-a_2^2/a_1^2)px_1$.

For hyperbolas and parabolas, the arguments are similar. □

We know from the calculus course that the equation of a tangent line to an ellipse, hyperbola, or parabola (specified by the canonical equation) at a point $X = (\xi_1, \xi_2)$ has the form

$$\frac{\xi_1 x_1}{a_1^2} + \frac{\xi_2 x_2}{a_2^2} = 1 \qquad \text{(for an ellipse)},$$

$$\frac{\xi_1 x_1}{a_1^2} - \frac{\xi_2 x_2}{a_2^2} = 1 \qquad \text{(for a hyperbola)},$$

$$p(x_1 - \xi_1) = \xi_2(x_2 - \xi_2) \qquad \text{(for a parabola)}.$$

For an ellipse, the equation of a tangent line can be obtained without using calculus, from the observation that a tangent line to an ellipse is a line having exactly one common point with the ellipse. Indeed, if

$$\frac{\xi_1^2}{a_1^2} + \frac{\xi_2^2}{a_2^2} = 1, \quad \frac{x_1^2}{a_1^2} + \frac{x_2^2}{a_2^2} = 1, \quad \frac{\xi_1 x_1}{a_1^2} + \frac{\xi_2 x_2}{a_2^2} = 1,$$

then

$$\frac{(\xi_1 - x_1)^2}{a_1^2} + \frac{(\xi_2 - x_2)^2}{a_2^2} = 0;$$

therefore, $(\xi_1, \xi_2) = (x_1, x_2)$.

The equation of the tangent line to a parabola at a point X implies that the vertex of the parabola is the midpoint of $[A, B]$, where A is the projection of X to the axis and B is the intersection point of the axis and the tangent line to the parabola at X.

THEOREM 8. (a) *An elliptic mirror has the property that light beams emitted from one focus converge at the other focus.*

(b) *Incident light beams parallel to the axis of a parabola, being reflected from the parabola, converge at the focus of the parabola.*

Proof. (a) Let us draw the tangent line to the ellipse at a point X and the bisector of the exterior angle of the vertex X of the triangle F_1XF_2. If the bisector is not tangent to the ellipse, then it has a second intersection point $M \neq X$ with the ellipse. Reflecting F_2 about the bisector, we obtain a point F_2'. We have

$$|F_1M| + |F_2M| = |F_1M| + |F_2'M| > |F_1F_2'| = |F_1X| + |F_2X|,$$

i.e., M lies outside the ellipse. We have arrived at a contradiction. Therefore, the bisector coincides with the tangent line, and the angle of incidence (i.e., the angle between the line F_1X and the tangent line) equals the angle of reflection (i.e., the angle between the line F_2X and the tangent line). Thus we have proved that the beams emitted from F_1 converge in F_2.

(b) Let X be a point on the parabola. If H is its projection to the directrix, then $|FX| = |XH|$ by Theorem 4. Let us draw the bisector of the angle FXH. If it is not a tangent line, then it intersects the parabola at some point $M \neq X$, and we have $|FM| = |MH| \neq |MH'|$, where H' is the projection of M to the directrix. This contradiction implies that the angle of incidence of any ray parallel to the axis equals the angle between the tangent line and FX; i.e., all beams parallel to the axis converge at the focus, as required. □

Now, let us discuss one special feature of the family of all ellipses. The ellipses form a five-parameter family on the plane because each ellipse is determined by the lengths a and b of its principal semiaxes, the coordinates x_0 and y_0 of its

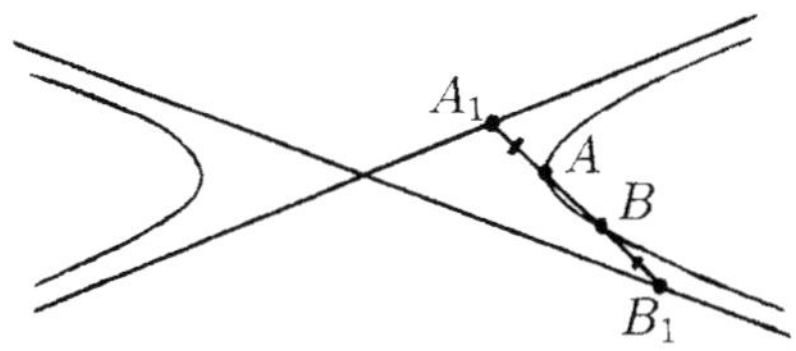

FIGURE 4.2

center, and the angle φ between one of its principal axis and a fixed line. Imposing one condition $a = b$ on these five parameters, we seemingly should obtain a four-parameter family. But in reality, we obtain a three-parameter family (a circle is determined by its radius and the coordinates of its center). How can we explain this paradox?

The very superficial explanation is that two of the three parameters a, b, and φ become redundant under the specified condition. To gain a deeper insight, let us write out the equation of the ellipse with parameters a, b, x_0, y_0, and φ:

$$\left(\frac{x\cos\varphi + y\sin\varphi - x_0}{a}\right)^2 + \left(\frac{-x\sin\varphi + y\cos\varphi - y_0}{b}\right)^2 = 1.$$

The quadratic part of this equation has the form

$$px^2 + 2qxy + ry^2,$$

where

$$p = b^2\cos^2\varphi + a^2\sin^2\varphi, \qquad r = b^2\sin^2\varphi + a^2\cos^2\varphi,$$
$$q = 2\cos\varphi\sin\varphi(a^2 - b^2) = (a^2 - b^2)\sin 2\varphi.$$

It is easy to verify that $p - r = (b^2 - a^2)\cos 2\varphi$; hence $(p-r)^2 + q^2 = (a^2 - b^2)^2$. Thus the single constraint $a = b$ implies two conditions, $p = r$ and $q = 0$.

An important distinction between a hyperbola and an ellipse is that the hyperbola $x^2/a^2 - y^2/b^2 = 1$ has the asymptotes $y = (\pm b/a)x$ (the asymptotes do not have common points with the hyperbola but they do have points arbitrarily close to the hyperbola).

Let us discuss some properties of the hyperbola that are related to asymptotes.

THEOREM 9. *If a line intersects a hyperbola at points A and B and its asymptotes at points A_1 and B_1, then $|AA_1| = |BB_1|$* (Figure 4.2).

Proof. Consider the hyperbolas

$$\frac{x^2}{a^2} - \frac{y^2}{b^2} = \lambda, \quad 0 < \lambda \leqslant 1.$$

The midpoints of their chords parallel to the line $y = kx$ lie on the line $y = (b^2k/a^2)\,x$ (see the proof of Theorem 7). Therefore, for all hyperbolas from this family, the chords lying on a certain line have the same midpoints. Passing to the limit as $\lambda \to 0$, we obtain the coincidence of the midpoints of $[A, B]$ and $[A_1, B_1]$, which implies $|AA_1| = |BB_1|$. □

COROLLARY. *If a line segment $[A_1, B_1]$ with endpoints on the asymptotes of a hyperbola is tangent to the hyperbola at a point X, then X is the midpoint of $[A_1, B_1]$.*

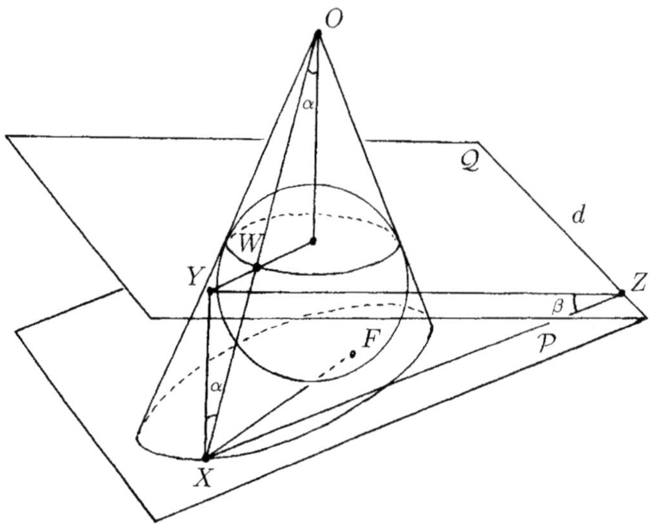

FIGURE 4.3

THEOREM 10. *The area of the triangle formed by the asymptotes of a hyperbola and a tangent line to the hyperbola does not depend on the tangent line.*

Proof. An arbitrary tangent line to the hyperbola has the form $x_0x/a^2 - y_0y/b^2 = 1$ and intersects the asymptotes $y = (\pm b/a)x$ at the points with coordinates $x_{1,2} = a\,(x_0/a \pm y_0/b)^{-1}$. Therefore, $x_1x_2 = a^2$. On the other hand, the area of the triangle under consideration is obviously proportional to x_1x_2. □

The ellipse, hyperbola, and parabola as conic sections.

THEOREM 11. *Consider the right circular cone $x_3^2 = x_1^2 + x_2^2$ in $\mathbb{R}^3$ intersected by a plane not containing its vertex. If the plane intersects all generatrices of one sheet of the cone, then the section is an ellipse; if it intersects all generatrices of one sheet except one, then the section is a parabola; and if it intersects both sheets of the cone, then the section is a hyperbola.*

Proof. Let $\mathcal{P}$ be the secant plane. Consider the sphere tangent to the generatrices of the cone (at the points of some circle) and to $\mathcal{P}$ (at a point F). Let $\mathcal{Q}$ be the plane of the circle of tangency. We denote the intersection of $\mathcal{P}$ with $\mathcal{Q}$ by d. Choose a point X in the intersection of the cone's surface with the plane $\mathcal{P}$. Let Y be the projection of X to $\mathcal{Q}$, and let Z be the projection of Y to d (Figure 4.3).

The angle YZX (we denote it by β) is the angle between $\mathcal{Q}$ and $\mathcal{P}$. Let O be the vertex of the cone. Then XO is a generatrix of the cone, and the angle YXO is equal to the angle α between the generatrix and the axis of the cone. Let W be the intersection point of OX with $\mathcal{Q}$.

Considering the triangles XYW and XYZ, we obtain $|XY|/|XW| = \cos\alpha$ and $|XY|/|XZ| = \sin\beta = \cos(\pi/2 - \beta)$. Clearly, the segments $[X, W]$ and $[X, F]$ have equal lengths, because they are both tangent to a sphere and share an endpoint. This allows us to conclude that the ratio of the distances from the point X (which was chosen arbitrarily on the curve that is a plane section of the cone) to the fixed point F and to the fixed line d is constant. Therefore, this curve is an ellipse, a hyperbola, or a parabola (see Theorem 4). It remains to note that if the plane intersects all the generatrices of one sheet of the cone, then the obtained curve is bounded; if it intersects all rays except one, then the curve has one connected component; and if it intersects both sheets of the cone, then the curve has two connected components. □

THEOREM 12. *If a plane $\mathcal{P}$ intersects all generatrices of one sheet of a straight circular cone K, does not contain the vertex of K, and is tangent to two spheres inscribed in the cone at points F_1 and F_2, then the sum of distances from the intersection point X of $\mathcal{P}$ with K to F_1 and from X to F_2 is constant.*

Proof. Let us connect X to the vertex O of the cone. Consider points W_1 and W_2 on the line XO that belong to the first and second spheres, respectively. We have $|XF_1| = |XW_1|$ (because these are the lengths of line segments sharing the endpoint X and tangent to the first sphere) and $|XF_2| = |XW_2|$. Thus $|XF_1| + |XF_2| = |W_1W_2|$, which is a constant value. □

Historical comments. The most important properties of the ellipse, hyperbola, and parabola stated in Theorems 4, 6, and 9 were known to Apollonius (and even earlier). Theorem 12 and the proof of Theorem 11 are due to Belgian mathematicians Quetelet and Dandelin.

4.2. Additional remarks

We start with corollaries to the theorem on the metric classification of second-order curves (Theorem 1 on p. 61).

Fourth-degree equations. Everybody who knows how to solve the third-degree equations can easily solve a fourth-degree equation using the remark on p. 62.

Indeed, suppose that we need to solve the equation $x^4 + ax^3 + bx^2 + cx + d = 0$. For this purpose, it suffices to find the intersection points of the conics $f = y - x^2 = 0$ and $g = y^2 + axy + by + cx + d = 0$. The intersection points do not change if we replace the curve $g = 0$ with a curve of the form $\lambda f + g = 0$. If the latter curve is degenerate, i.e., its equation is the product of two linear factors, then the intersection points are easy to find. The degeneracy condition for the conic

$$g + \lambda f = y^2 - \lambda x^2 + axy + cx + (b + \lambda)y + d = 0,$$

is that the principal determinant

$$\begin{vmatrix} -\lambda & a/2 & c/2 \\ a/2 & 1 & (b+\lambda)/2 \\ c/2 & (b+\lambda)/2 & d \end{vmatrix}$$

equals zero, i.e., we must solve an equation in λ of only the third degree!

The theorem about the conic passing through five points. Let us prove that *through five given points on the plane, there passes exactly one conic if and only if no four of the five points are collinear.*

Indeed, the condition for a conic to pass through five points can be written as a system of five linear homogeneous equations in six variables. Such a system always has a nonzero solution.

If certain four points are collinear, then, multiplying a linear function vanishing at these four points by any linear function vanishing at the fifth one, we obtain a family of conics passing through the given five points.

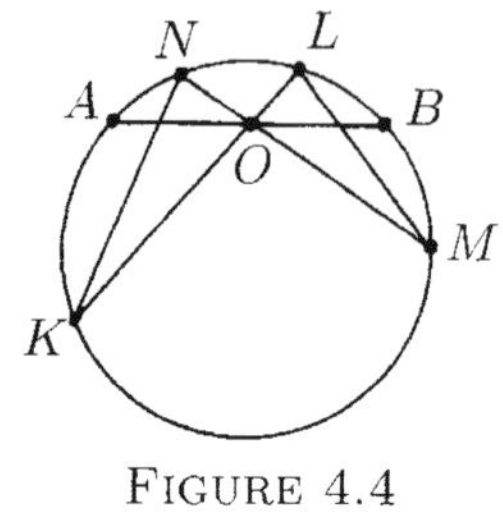

FIGURE 4.4

Suppose that no four points are collinear. Let us prove the uniqueness of the conic. If certain three points lie on one line, then the conic must be degenerate, because a nondegenerate conic cannot intersect a straight line at three points. Two lines can be drawn through the system of points under consideration in only one way, namely, one line must pass through the three collinear points and the other one, through the remaining two points.

Finally, suppose that no three points are collinear. Let there exist two different (certainly, nondegenerate) conics $f = 0$ and $g = 0$ that pass through the five given points. Consider the family of conics $f + \lambda g = 0$. The principal determinant of conics from this family is a third-degree polynomial in λ; therefore, it has a real root, and the conic corresponding to this root is degenerate, which is impossible. This completes the proof of the theorem. □

The theorem about the pencil of conics passing through four points. *If no three of points A_1, A_2, A_3, and A_4 are collinear and $f = 0$ is the equation of a conic passing through these points, then there exist numbers λ and μ such that $f = \lambda g + \mu h := \lambda l_{A_1A_2}l_{A_3A_4} + \mu l_{A_2A_3}l_{A_1A_4}$, where each $l_{A_iA_j}$ is a first-degree polynomial vanishing at A_i and A_j.*

Indeed, choose an arbitrary point A_5 different from A_i $(i = 1, 2, 3, 4)$ for which $f(A_5) \neq 0$. Find λ_1 and μ_1 such that $f(A_5) - \lambda_1 g(A_5) - \mu_1 h(A_5) = 0$. By the uniqueness theorem for a conic passing through five points, the function f is proportional to $\lambda_1 g + \mu_1 h$. □

The butterfly problem. *Let two chords KL and MN of a circle pass through the midpoint O of a third chord AB. Then the distances from O to the intersection points of the line AB with the lines KN and ML are equal* (Figure 4.4).

We have already mentioned this problem (see p. 16). Below, we give yet another solution.

The following three conics pass through the points K, L, M, N (we use the notation from the preceding theorem): the circle $f = 0$, $g = l_{KL}l_{MN} = 0$, and $h = l_{KN}l_{ML}$; by the preceding theorem, $h = \lambda f + \mu g$. This equality also holds for the restrictions of the functions to the line AB. Let us introduce a coordinate x on AB; for the origin, we take O. We can assume that $f = x^2 - a$ and $g = x^2$; then $h = bx^2 - c$. Therefore, the roots of the equation $h = 0$ are equidistant from O. □

Hyperbolas with perpendicular asymptotes. Let us prove the following theorem: *Any hyperbola passing through the vertices of a triangle ABC and the intersection point of its altitudes has perpendicular asymptotes* (Figure 4.5).

The proof of Theorem 1 readily implies that the conic (1) is a hyperbola with perpendicular asymptotes if and only if $a_{11} + a_{22} = 0$; therefore, any linear combination of the equations of hyperbolas with perpendicular asymptotes is also a

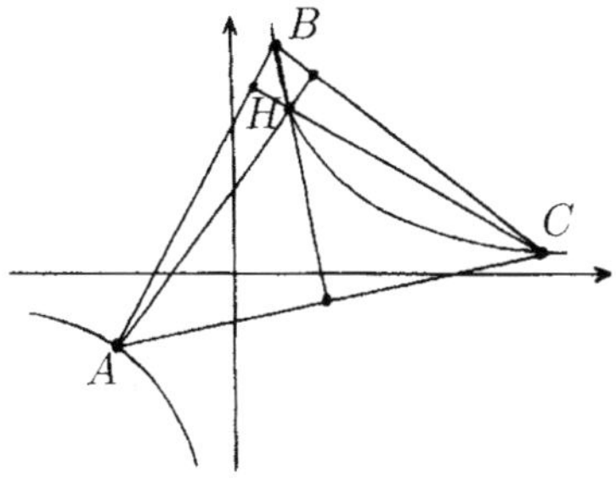

FIGURE 4.5

hyperbola with perpendicular asymptotes. The pencil of conics through A, B, C, H contains two degenerate conics with perpendicular asymptotes $l_{AB}l_{CH}$ and $l_{BC}l_{AH}$. By the theorem about the pencil of conics, all conics of the pencil under consideration are hyperbolas with perpendicular asymptotes.

Pascal's theorem. Consider a hexagon $ABCDEF$ with vertices on a conic $f = 0$. The quadrilaterals $ABCD$, $AFED$, and $BEFC$ are inscribed in this conic; therefore, f can be represented in any of the following forms:

(1) $$f = \lambda_1 l_{AB} \cdot l_{CD} + \mu_1 l_{AD} \cdot l_{BC},$$

(2) $$f = \lambda_2 l_{AF} \cdot l_{ED} + \mu_2 l_{AD} \cdot l_{EF},$$

(3) $$f = \lambda_3 l_{BE} \cdot l_{CF} + \mu_3 l_{BC} \cdot l_{EF}.$$

Equating (1) to (2), we obtain

$$\lambda_1 l_{AB} \cdot l_{CD} - \lambda_2 l_{AF} \cdot l_{ED} = (\mu_1 l_{BC} - \mu_2 l_{EF}) l_{AD}.$$

Let X be the intersection point of the lines AB and ED. Then the function given by $\mu_1 l_{BC} - \mu_2 l_{EF}$ vanishes at the point X, while the function l_{AD} does not. Hence X lies on the line $\mu_1 l_{BC} = \mu_2 l_{EF}$. Similarly, the line $\mu_1 l_{BC} = \mu_2 l_{EF}$ contains the intersection point of the lines CD and AF. Obviously, the intersection point of BC and EF lies on the line $\mu_1 l_{BC} = \mu_2 l_{EF}$. As a result, we obtain the following assertion.[1]

PASCAL'S THEOREM. *If points A, B, C, D, E, F lie on one conic, then the intersection points of the line AB with the line DE, of BC with EF, and of CD with FA are collinear.*

Let us continue our arguments. Equating (2) to (3), we see that the intersection points of AF with BE, of ED with CF, and of AD with BC lie on the line $\mu_2 l_{AD} = \mu_3 l_{BC}$. Finally, equating (1) to (3), we see that the intersection points of AB with CF, of CD with BE, and of AD with EF lie on the line $\mu_1 l_{AD} = \mu_3 l_{EF}$. It is easy to verify that the lines $\mu_1 l_{BC} = \mu_2 l_{EF}$, $\mu_2 l_{AD} = \mu_3 l_{BC}$, and $\mu_1 l_{AD} = \mu_3 l_{EF}$ meet at one point. Indeed, if X is the intersection point of the first two lines, then

$$\mu_1 \mu_2 l_{BC}(X) l_{AD}(X) = \mu_2 \mu_3 l_{EF}(X) l_{BC}(X).$$

Reducing this by the factor $\mu_2 l_{BC}(X)$, we obtain $\mu_1 l_{AD}(X) = \mu_3 l_{EF}(X)$ (we do not consider the degenerate case $\mu_2 l_{BC}(X) = 0$).

The *Pascal line* of a hexagon inscribed in a conic is the straight line containing the intersection points of the pairs of opposite sides of the hexagon. The hexagon

[1]This is our second encounter with Pascal's name; we first mentioned it on p. 15.

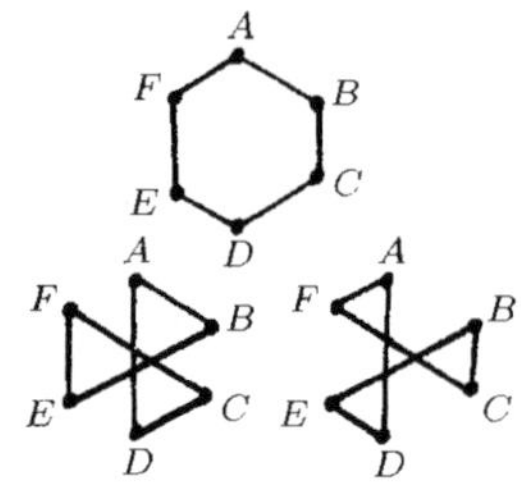

FIGURE 4.6

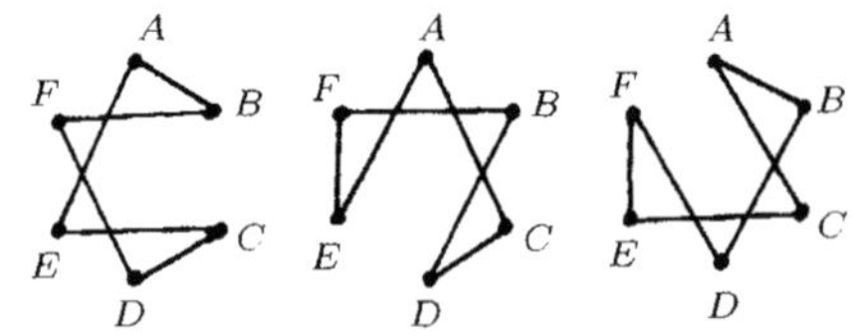

FIGURE 4.7

may be a polygon with self-intersections. The assertion proved above can be stated as follows.

STEINER'S THEOREM. *If points A, B, C, D, E, F lie on one conic, then the Pascal lines of the hexagons $ABCDEF$, $ADCFEB$, $ADEBCF$* (see Figure 4.6) *intersect at one point.*

Recall that the proof of this theorem began by considering the quadrilaterals $ABCD$, $AFED$, $BEFC$. Departing from the quadrilaterals $ABFE$, $ABDC$, $CDFE$, we obtain the following assertion.

KIRKMAN'S THEOREM. *The Pascal lines of the hexagons*

$$ABFDCE, \quad AEFBDC, \quad ABDFEC$$

(Figure 4.7) *intersect at one point.*

It is easy to see that six points on a conic determine sixty Pascal lines, and each Pascal line is contained in exactly one Steiner triple and in three Kirkman triples.

Common chords of two conics inscribed in the same conic. The theorem about the pencil of conics passing through four points remains valid when some pairs of points merge, i.e., when the conics not only pass through a given point but also are tangent to each other at this point. Using this observation, we can prove the following assertion.

THEOREM ON COMMON CHORDS OF CONICS. *Suppose that two conics Γ and Γ_1 are tangent to each other at points A and B, conics Γ and Γ_2 are tangent to each other at points C and D, and Γ_1 has four common points with Γ_2. Then the conics Γ_1 and Γ_2 have a pair of common chords passing through the intersection point of the lines AB and CD* (Figure 4.8).

Proof. Let $p_1 = 0$ and $p_2 = 0$ be the equations of the common tangent lines to the conics Γ and Γ_1 at the points A and B, and let $q = 0$ be the equation of the line AB. Then the equations of the conics Γ and Γ_1 can be written as

$$f = \lambda p_1 p_2 + \mu q^2 = 0 \quad \text{and} \quad f_1 = \lambda_1 p_1 p_2 + \mu_1 q^2 = 0,$$

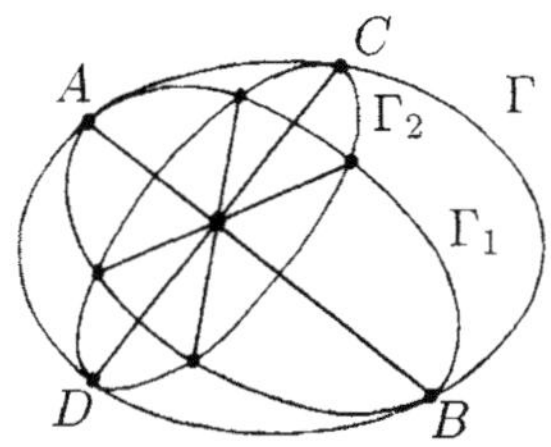

FIGURE 4.8

respectively. We assume that $\lambda = \lambda_1$ (otherwise, we multiply f_1 by λ/λ_1), and hence $f_1 = f + \alpha q^2$. Similarly, $f_2 = f + \beta r^2$, where $r = 0$ is the equation of the line CD. Consider the equation $f_1 - f_2 = 0$, which is equivalent to $\alpha q^2 - \beta r^2 = 0$. It is satisfied by the four common points of Γ_1 and Γ_2. On the other hand, this equation decomposes into the product of the linear equations $\sqrt{\alpha}q + \sqrt{\beta}r = 0$ and $\sqrt{\alpha}q - \sqrt{\beta}r = 0$. Therefore, the lines $\sqrt{\alpha}q \pm \sqrt{\beta}r = 0$ contain common chords of the conics Γ_1 and Γ_2.

Clearly, the intersection point of these lines coincides with the intersection point of the lines $q = 0$ and $r = 0$. □

4.3. Some properties of quadrics

Quadrics are the surfaces in $\mathbb{R}^n$ given by $\sum a_{ij}x_ix_j + \sum b_ix_i + c = 0$. We do not give their metric, affine, and projective classifications (these descriptions are usually included in linear algebra courses); rather, we shall discuss some geometric properties of the quadrics.

A quadric in $\mathbb{R}^3$ is called a *cone* if in some Cartesian coordinate system it has the equation $ax^2 + by^2 = z^2$ with $a, b > 0$. If $a = b$, the cone is called a *cone of rotation*, or a *right circular cone*. The point $(0, 0, 0)$ is the *vertex* of the cone, and the axis Oz is its *axis*.

A quadric in $\mathbb{R}^n$ is called an *ellipsoid* if it is given by $(x_1/a_1)^2 + \dots + (x_n/a_n)^2 = 1$ in some Cartesian coordinate system. A hyperplane intersecting an ellipsoid at exactly one point is called a *tangent hyperplane* to the ellipsoid (of course, this definition coincides with the notion of tangent hyperplane to an ellipsoid in the sense of classical analysis).

THEOREM 1. *The equation of the hyperplane tangent to the ellipsoid given by the equation* $\sum(x_i/a_i)^2 = 1$ *at a point* $(\xi_1, \dots, \xi_n)$ *is*

$$\frac{\xi_1 x_1}{a_1^2} + \dots + \frac{\xi_n x_n}{a_n^2} = 1. \tag{1}$$

Proof. Suppose that a point $(x_1, \dots, x_n)$ belongs to both the ellipsoid and the hyperplane (1), i.e., $\sum(x_i/a_i)^2 = 1$ and $\sum(\xi_ix_i/a_i^2) = 1$. Then

$$\sum\left(\frac{x_i - \xi_i}{a_i}\right)^2 = \sum\left(\frac{x_i}{a_i}\right)^2 - 2\sum\frac{x_i\xi_i}{a_i^2} + \sum\left(\frac{\xi_i}{a_i}\right)^2 = 0,$$

and therefore $x_i = \xi_i$. □

For a quadric $\sum a_{ij}x_ix_j = 0$, the tangent plane at a point $(\xi_1, \dots, \xi_n)$ is given by $\sum a_{ij}\xi_ix_j = 0$.

To an ellipsoid $\sum(x_i/a_i)^2 = 1$ and an orthonormal basis $\varepsilon_1, \dots, \varepsilon_n$, we can assign the system of hyperplanes Π_i and the system of line segments $[O, A_i]$ such that each Π_i is tangent to the ellipsoid and orthogonal to the corresponding vector ε_i, O is the origin, all A_i lie on the ellipsoid, and each $[O, A_i]$ is parallel to ε_i.

THEOREM 2. (a) *Let d_i be the distance from the origin to the hyperplane Π_i. Then*

$$d_1^2 + \cdots + d_n^2 = a_1^2 + \cdots + a_n^2.$$

(b) *Let r_i be the length of $[O, A_i]$. Then*

$$\frac{1}{r_1^2} + \cdots + \frac{1}{r_n^2} = \frac{1}{a_1^2} + \cdots + \frac{1}{a_n^2}.$$

Proof. Let $e_1, \ldots, e_n$ be an orthonormal basis in which the ellipsoid has the equation $\sum (x_i/a_i)^2 = 1$. Then $\varepsilon_j = \sum_i a_{ij} e_i$ for some orthogonal matrix $A = (a_{ij})$.

(a) Let ξ be an intersection point of the hyperplanes $\Pi_1, \ldots, \Pi_n$. Then we have $d_1^2 + \cdots + d_n^2 = (\xi, \xi)$; thus we must prove that $\sum \xi_i^2 = \sum a_i^2$.

The equation of the plane Π_j has the form $(\varepsilon_j, x) = (\varepsilon_j, \xi)$, i.e.,

$$\sum_i \frac{a_{ij} x_i}{(\varepsilon_j, \xi)} = 1.$$

On the other hand, it has the form $\sum_i (\eta_i^{(j)} x_i / a_i^2) = 1$. Therefore,

$$\eta_i^{(j)} = \frac{a_i^2 a_{ij}}{(\varepsilon_j, \xi)}, \quad \text{and} \quad \sum_i \left(\frac{\eta_i^{(j)}}{a_i} \right)^2 = 1, \quad \text{i.e.,} \quad \sum_i a_i^2 a_{ij}^2 = (\varepsilon_j, \xi)^2.$$

Hence

$$\sum a_i^2 = \sum_i \Big(a_i^2 \sum_j a_{ij}^2\Big) = \sum (\varepsilon_j, \xi)^2 = \sum \xi_i^2.$$

(b) One endpoint of the vector $r_j \varepsilon_j = \sum_i a_{ij} r_j e_i$ lies on the ellipsoid determined by $\sum (x_i/a_i)^2 = 1$; hence $\sum_i (a_{ij} r_j / a_i)^2 = 1$, i.e.,

$$\frac{1}{r_j^2} = \sum_i \frac{a_{ij}^2}{a_i^2}.$$

Therefore,

$$\sum_j \frac{1}{r_j^2} = \sum_{i,j} \frac{a_{ij}^2}{a_i^2} = \sum_i \frac{1}{a_i^2}. \quad \square$$

REMARK. Theorem 2 (a) can be reformulated as follows: *The length of the diagonal of a rectangular parallelepiped circumscribed about an ellipsoid does not depend on the parallelepiped; i.e., the vertices of all such parallelepipeds lie on one sphere.*

THEOREM 3. *All sections of an ellipsoid in $\mathbb{R}^3$ by parallel planes are similar to each other.*

Proof. We can assume that the equations of the secant planes have the form $z = \lambda$. Each section of the quadric

$$ax^2 + bxy + cy^2 + d + (a_1 x + b_1 y + c_1) z + a_2 z^2 = 0$$

by such a plane is the conic

$$ax^2 + bxy + cy^2 + \lambda a_1 x + \lambda b_1 y + a_2 \lambda^2 + c_1 \lambda + d = 0.$$

For an ellipsoid, this conic is an ellipse (if it is nonempty and does not degenerate to a point). The directions of the axes of the ellipse and the ratio of their lengths are completely determined by the quadratic part $ax^2 + bxy + cy^2$. □

THEOREM 4. *A plane section of an ellipsoid*

$$\frac{x^2}{a^2} + \frac{y^2}{b^2} + \frac{z^2}{c^2} = 1$$

with $0 < a < b < c$ *is a circle if and only if the secant plane has the equation* $x\sqrt{\alpha} \pm z\sqrt{\beta} = \lambda$, *where* $\alpha = 1/a^2 - 1/b^2$ *and* $\beta = 1/b^2 - 1/c^2$.

Proof. According to Theorem 3, it is sufficient to consider the case in which the secant plane passes through the origin. Suppose that such a section is a circle of radius r. Then the points of the circle lie on the sphere $x^2 + y^2 + z^2 = r^2$. Since these points also lie on the ellipsoid, they belong to the surface

$$x^2\left(\frac{1}{a^2} + \frac{1}{r^2}\right) + y^2\left(\frac{1}{b^2} - \frac{1}{r^2}\right) + z^2\left(\frac{1}{c^2} - \frac{1}{r^2}\right) = 0. \tag{2}$$

If the coefficients of x^2, y^2, and z^2 are nonzero, the surface (2) is a cone (or a point). Therefore, a circle centered at the origin can only lie on the surface (2) if one of the coefficients $1/a^2 - 1/r^2$, $1/b^2 - 1/r^2$, and $1/c^2 - 1/r^2$ is zero and the other two coefficients have different signs. Taking into account that $0 < a < b < c$, we obtain $r = b$. □

Two families of straight lines on a quadric. The equation of a quadric in $\mathbb{C}P^3$ can be reduced to the form $x_1^2+x_2^2 = x_3^2+x_4^2$. After the change $y_{1,2} = x_1 \pm ix_2$, $y_{3,4} = x_3 \pm ix_4$, this equation takes the form $y_1y_2 = y_3y_4$.

The surface $y_1y_2 = y_3y_4$ contains two families of straight lines, namely,

$$y_1 = \lambda y_3, \quad \lambda y_2 = y_4 \qquad \text{and} \qquad y_1 = \mu y_4, \quad \mu y_2 = y_3$$

(we do not exclude the values $\lambda, \mu = \infty$, which correspond to the lines $y_3 = 0$, $y_2 = 0$ and $y_4 = 0$, $y_2 = 0$). It is easy to verify that this surface contains no other lines. Indeed, consider the line passing through the points with coordinates (a_i) and (b_i). It consists of points of the form $sa_i + tb_i$. We can assume that $b_4 = 0$. If the line under consideration lies on the surface $y_1y_2 = y_3y_4$, then

$$(sa_1 + tb_1)(sa_2 + tb_2) = (sa_3 + tb_3)sa_4$$

for all s and t. Therefore, $b_1b_2 = 0$, i.e., $b_1 = 0$ or $b_2 = 0$. If $b_1 = 0$, then $y_1 = \mu y_4$, and if $b_2 = 0$, then $\lambda y_2 = y_4$.

Through every point (a_1, a_2, a_3, a_4) of the surface $y_1y_2 = y_3y_4$, there passes exactly one line from each family, because the parameters λ and μ are uniquely determined by the point, namely, $\lambda = a_1/a_3 = a_4/a_2$ and $\mu = a_1/a_4 = a_3/a_2$. For a surface in $\mathbb{R}^3$, these lines are not necessarily real. For example, an ellipsoid cannot contain real straight lines.

The line $y_1 = \mu y_4$, $\mu y_2 = y_3$ intersects the lines $y_1 = \lambda y_3$, $\lambda y_2 = y_4$ and $y_1 = \lambda' y_3$, $\lambda' y_2 = y_4$ at the points $A = (\lambda\mu, 1, \mu, \lambda)$ and $A' = (\lambda'\mu, 1, \mu, \lambda')$, respectively. The map $A \mapsto A'$ of the line with parameter λ to the line with parameter λ' is projective, because μ can be treated as a coordinate on both lines.

Through the two lines $y_1 = 0$, $y_4 = 0$ and $y_2 = 0$, $y_3 = 0$, there pass an entire family of quadrics, namely, the quadrics $y_1y_2 = \alpha y_3y_4$. Let us add one more line $y_1 = \lambda y_3$, $\lambda y_2 = y_4$ ($\lambda \neq 0, \infty$). Only one quadric $y_1y_2 = y_3y_4$ passes through the three lines. This is because, if a, b, and c are pairwise skew lines in $\mathbb{R}^3$, then through any point A on the line a, there passes exactly one line meeting the lines b and c (possibly, at points at infinity). To find this line, we must draw a plane containing A and b and find its intersection point with c.

We have seen that no more than one quadric passes through three pairwise skew lines. Let us show that any three straight lines in $\mathbb{R}^3$ are contained in a quadric (possibly degenerate). A quadric in $\mathbb{R}^3$ is determined by the coefficients of x^2, y^2, z^2, xy, yz, xz, x, y, and z and the constant term. Therefore, the condition that a quadric passes through N given points is equivalent to a system of N linear equations in ten variables. For $N \leqslant 9$, this system has a nonzero solution; hence through any nine points in $\mathbb{R}^3$, at least one quadric passes. Let us choose three points on each of the given lines. Some quadric passes through the nine points obtained. Since this quadric intersects each given line at three points, it contains all these lines entirely.

Through a point (a_1, a_2, a_3, a_4) of the quadric $x_1x_2 = x_3x_4$, there pass the lines $a_2x_1 = a_4x_3$, $a_1x_2 = a_3x_4$ and $a_1x_2 = a_4x_3$, $a_2x_1 = a_3x_4$ lying on the quadric. The plane containing these lines is given by the equation $a_2x_1 + a_1x_2 = a_4x_3 + a_3x_4$. It is easy to verify that the tangent plane to the quadric at the point (a_1, a_2, a_3, a_4) is specified by the same equation. Therefore, the tangent plane to a quadric at a given point is the plane that contains two lines passing through the point and lying on the quadric.

More than three lines in $\mathbb{R}^3$ do not necessarily lie on one quadric. There are several theorems describing certain configurations of straight lines lying on one quadric. As an example, we give generalizations of Pascal's and Brianchon's theorems.

DEFINITION. A hexagon $ABCDEF$ in $\mathbb{R}^3$ is called a *Brianchon hexagon* if its diagonals AD, BE, CF meet at one point (possibly at infinity).

THEOREM 5. *A nonplanar hexagon $ABCDEF$ in $\mathbb{R}^3$ is a Brianchon hexagon if and only if all its sides lie on one quadric.*

Proof. Suppose that the diagonals of a nonplanar hexagon $ABCDEF$ in $\mathbb{R}^3$ meet at one point. Let us show that in this case the line AB belongs to a quadric containing BC, DE, and AF. It suffices to prove that the line AB intersects DE (it intersects BC and AF at the points B and A, respectively). By assumption, the lines AD and BE intersect; therefore, the points A, B, D, and E are coplanar. Hence the lines AB and DE intersect. The lines CD and EF also lie on the quadric under consideration; this is proved similarly.

Now, suppose that the lines AB, CD, and EF lie on a quadric containing BC, DE, and AF. Then AB intersects DE, and hence AD intersects BE. Similarly, the lines AD, BE, and CF pairwise intersect. By assumption, these lines do not lie in one plane; therefore, they must all intersect at one point. □

Pascal's theorem can be stated as follows: *Let the intersection points of a conic with the sides of a triangle ABC be denoted as shown in* Figure 4.9. *Then the intersection points of A_bA_c with BC, of B_aB_c with AC, and of C_aC_b with AB are collinear.* Thus the following assertion can be regarded as a generalization of Pascal's theorem.

THEOREM 6. *Let the intersection points of a quadric with the edges of a tetrahedron $ABCD$ be denoted as shown in* Figure 4.10. *Then the intersection lines a, b, c, and d of the pairs of planes $A_bA_cA_d$ and BCD, $B_aB_cB_d$ and ACD, $C_aC_bC_d$ and ABD, and $D_aD_bD_c$ and ABC lie on one quadric.*

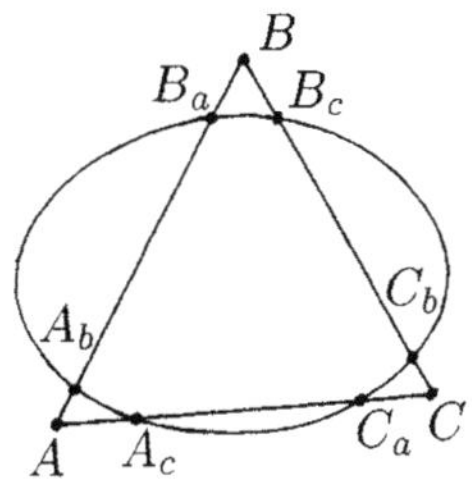

FIGURE 4.9

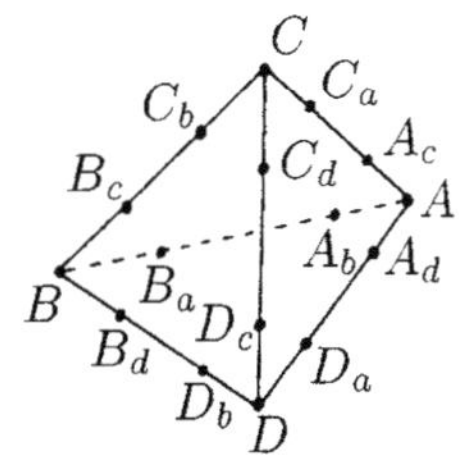

FIGURE 4.10

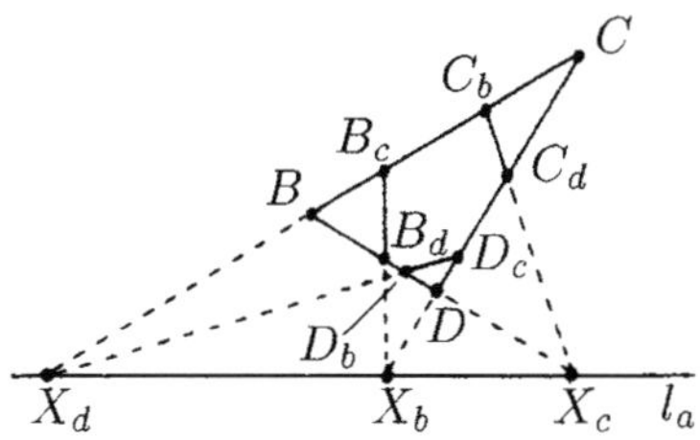

FIGURE 4.11

Proof. Consider the plane BCD. According to Pascal's theorem, the points

$$X_b = B_cB_d \cap C_dD_c, \quad X_c = C_bC_d \cap B_dD_b, \quad X_d = D_bD_c \cap C_bB_c$$

lie on one line l_a (Figure 4.11). The line l_a intersects the line a, because they both lie in the plane BCD. In addition, l_a intersects b, c, and d at the points X_b, X_c, and X_d, respectively. Similarly, lines l_b, l_c, and l_d each intersecting a, b, c, and d are constructed.

Now, it is easy to prove that the lines a, b, c, and d lie on one quadric. Indeed, consider a quadric containing a, b, and c. This quadric also contains l_a, l_b, l_c, and l_d. Clearly, any quadric containing l_a, l_b, and l_c must contain d. □

THEOREM 7. *The altitudes of a tetrahedron $ABCD$ lie on one quadric.*

Proof. The planes passing through the edges DA, DB, and DC and perpendicular to the faces DBC, DAC, and DAB, respectively, have a common line l_d. (This is equivalent to the assertion that the altitudes of a spherical triangle meet at one point; the proof is given in the solution of Problem 5.12 (b).) The line l_d intersects the altitudes from the vertices A, B, and C. In addition, it intersects the altitude from D (at D). Similarly, lines l_a, l_b, and l_c, each intersecting all altitudes of the tetrahedron, are constructed. The conclusion of the proof repeats the conclusion of the proof of Theorem 6. □

Problems

A pencil of conics through four points.

4.1. Given an ellipse tangent to the sides of a quadrilateral $ABCD$ at points A_1, B_1, C_1, D_1, prove that the lines AC, BD, A_1C_1, B_1D_1 meet at one point.

4.2. Given two conics with four intersection points, prove that the common points lie on one circle if and only if the axes of the conics are perpendicular.

4.3. Prove that any pencil of conics passing through four points contains at most two parabolas.

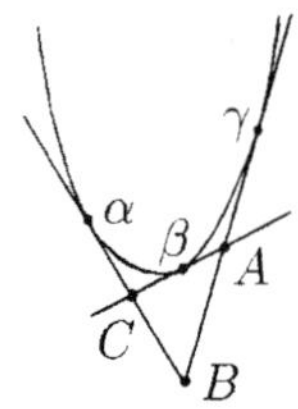

FIGURE 4.12

4.4. Prove that the centers of all conics passing through points A, B, C, D form a conic Γ.

4.5. Prove the following properties of the conic Γ mentioned in the previous problem:

(a) Γ passes through the six midpoints of the line segments joining the pairs of the given points and through the three intersection points of the lines through the pairs of the given points;

(b) the center of Γ coincides with the centroid of the system of points A, B, C, and D;

(c) if D is the intersection point of the altitudes of the triangle ABC, then Γ is the nine-point circle of this triangle;

(d) if the quadrilateral $ABCD$ can be inscribed in a circle, then Γ is a hyperbola with perpendicular asymptotes, and the axes of all conics in the pencil are parallel to the asymptotes of Γ.

The parabola.

4.6. Prove that the parabola $4ay = x^2$ can be transformed into the parabola $y = x^2$ by a homothety with center $(0, 0)$.

4.7. Given a circle intersecting a parabola at four points, prove that the centroid of the points lies on the axis of the parabola.

4.8. Given two parabolas with perpendicular axes and four intersection points, prove that the intersection points lie on one circle.

4.9. Given a parabola with a chord P_1P_2 passing through the focus F, prove that the value $1/|P_1F| + 1/|P_2F|$ does not depend on the choice of the chord.

4.10. Prove that the tangent lines to the parabola $4y = x^2$ at the points $(2t_1, t_1^2)$ and $(2t_2, t_2^2)$ intersect at the point $(t_1 + t_2, t_1 \cdot t_2)$.

4.11. Given tangent lines OA and OB through a point O to a parabola with focus F, prove that $\angle AFB = 2\angle AOB$ and the ray OF is the bisector of the angle AFB.

4.12. Prove that tangent lines OA and OB to a parabola are perpendicular if and only if one of the following equivalent conditions holds:

(a) the segment $[A, B]$ passes through the focus of the parabola;

(b) the point O lies on the directrix of the parabola.

4.13. Prove that the transformation $z \mapsto 2(1 - z)^{-2}$ of the complex plane maps the unit circle $|z| = 1$ into a parabola.

4.14. Does there exist a transformation of the plane that maps the family of parabolas $x = a + by + y^2$ to a family of straight lines?

4.15. The tangent lines to a parabola at points α, β, and γ form a triangle ABC (Figure 4.12). Prove that

(a) the circle circumscribed about the triangle ABC passes through the focus of the parabola;

(b) the intersection point of the altitudes of the triangle ABC lies on the directrix of the parabola;

(c) $S_{\alpha\beta\gamma} = 2S_{ABC}$;

(d) $\sqrt[3]{S_{\alpha\beta C}} + \sqrt[3]{S_{\beta\gamma A}} = \sqrt[3]{S_{\alpha\gamma B}}$.

4.16. Given a line l obtained from the directrix of a parabola by homothety with factor 2 centered at the focus of the parabola and tangent lines OA and OB through a point O on l to a parabola, prove that the orthocenter of the triangle AOB is the vertex of the parabola.

4.17. Parallel incident beams of light, reflected from a curve C, converge in a point F. Prove that C is a parabola with focus F and axis parallel to the light beams.

The ellipse.

4.18. (a) Prove that for any parallelogram, there exists an ellipse tangent to the sides of the parallelogram at their midpoints.

(b) Prove that for any triangle, there exists an ellipse tangent to the sides of the triangle at their midpoints.

4.19. Given conjugate diameters AA' and BB' of an ellipse centered at O, prove that

(a) the area of the triangle AOB does not depend on the choice of the conjugate diameters;

(b) the value OA^2+OB^2 does not depend on the choice of the conjugate diameters.

4.20. (a) Prove that the projections of the foci of an ellipse to all tangent lines lie on one circle.

(b) Given the distances d_1 and d_2 between the foci of an ellipse and a tangent line to the ellipse, prove that the value $d_1 d_2$ does not depend on the choice of the tangent line.

4.21. Given tangent lines OA and OB to an ellipse with foci F_1 and F_2 passing through a point O, prove that $\angle AOF_1 = \angle BOF_2$ and $\angle AF_1O = \angle BF_1O$.

4.22. Given a parallelogram circumscribed about an ellipse, prove that its diagonals contain conjugate diameters of the ellipse.

4.23. Given complex numbers a and b, prove that as φ varies from 0 to 2π, the points of the form $ae^{i\varphi} + be^{-i\varphi}$ sweep out an ellipse or a line segment.

4.24. Prove that the length of the diagonal of a rectangle circumscribed about an ellipse does not depend on the rectangle.

4.25. In the circle $x^2+y^2 = a^2+b^2$ centered at O, a chord PQ tangent to the ellipse $x^2/a^2 + y^2/b^2 = 1$ is given. Prove that the lines PO and QO contain conjugate diameters of the ellipse.

4.26. (a) In an ellipse centered at O, conjugate diameters AA' and BB' are given; from the point B, the perpendicular to OA is drawn; $[B,P]$ and $[B,Q]$ are its segments whose lengths equal $|OA|$. Prove that the principal axes of the ellipse are the bisectors of the angles made by the lines OP and OQ.

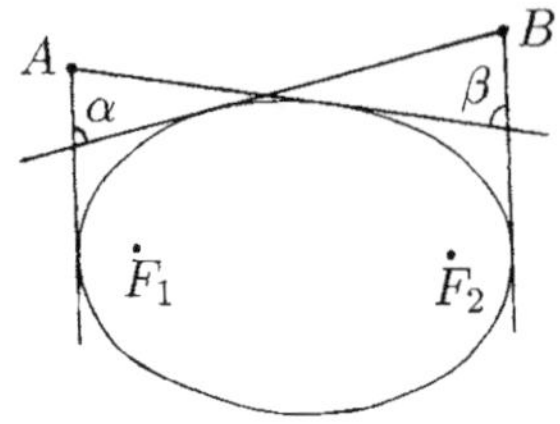

FIGURE 4.13

(b) Given a pair of conjugate diameters of an ellipse drawn on the plane, construct the principal axes of the ellipse with a compass and a straightedge.

4.27. Given a normal to an ellipse at a point A intersecting the minor semiaxis at a point Q and the projection P of the center of the ellipse to the normal, prove that $|AP| \cdot |AQ| = a^2$, where a is the major semiaxis.

4.28. Prove that all rhombi inscribed in an ellipse are circumscribed about one circle.

4.29. Given a circle centered on an ellipse and tangent to two conjugate diameters of the ellipse, prove that the radius of the circle does not depend on the choice of the conjugate diameters.

4.30. (a) Given tangent lines OP and OQ through a point O to an ellipse with foci F_1 and F_2, prove that

$$\angle POQ = \pi - \frac{1}{2}(\angle PF_1O + \angle PF_2O).$$

(b) Given view angles φ_1 and φ_2 of a line segment $[A, B]$ from the foci F_1 and F_2 of an ellipse, prove that $\varphi_1 + \varphi_2 = \alpha + \beta$ (see Figure 4.13).

4.31. Given an ellipse centered at O, two parallel tangent lines l_1 and l_2 to the ellipse, and a circle centered at O_1 and tangent to the ellipse (externally) and to the lines l_1 and l_2, prove that the length of $[O, O_1]$ is equal to the sum of the semiaxes of the ellipse.

4.32. Given an ellipse with semiaxes a and b centered at O and a circle of radius r centered at a point C on the major semiaxis of the ellipse and tangent to the ellipse at two points, prove that

$$|OC|^2 = \frac{(a^2 - b^2)(b^2 - r^2)}{b^2}.$$

4.33. Given three circles of radii r_1, r_2, and r_3 centered on the major semiaxis of an ellipse and tangent to the ellipse, the circle of radius r_2 being externally tangent to the circles of radii r_1 and r_3, prove that

$$r_1 + r_3 = \frac{2a^2(a^2 - 2b^2)}{a^4} r_2.$$

4.34. Given N circles of radii r_i ($1 \leqslant i \leqslant N$) centered on the major semiaxis of an ellipse and tangent to the ellipse, the circle of radius r_i being tangent to the circles of radii r_{i-1} and r_{i+1} for each $i = 2, \ldots, N-1$, prove that, if $3n - 2 < N$, then

$$r_{2n-1}(r_1 + r_{2n-1}) = r_n(r_n + r_{3n-2}).$$

4.35. Given a triangle ABC in the complex plane and an ellipse with foci F_1 and F_2 tangent to the sides of the triangle at their midpoints, prove that F_1 and F_2 are the roots of the derivative of the third-degree polynomial with roots A, B, and C.

The hyperbola.

4.36. Given a hyperbola, consider the parallelograms with sides on the asymptotes of the hyperbola such that one vertex of each parallelogram lies on the hyperbola and the opposite vertex is the intersection point of the asymptotes. Prove that all the parallelograms have equal areas.

4.37. Prove that the asymptotes of the hyperbola

$$ax^2 + 2bxy + cy^2 + dx + cy + f = 0$$

are orthogonal if and only if $a + c = 0$.

4.38. Given a triangle with vertices on the hyperbola $xy = 1$, prove that the orthocenter of the triangle lies on the same hyperbola.

4.39. Find the set of intersection points of all pairs of perpendicular tangent lines to a hyperbola.

4.40. Given the intersection points $(-x_0, -x_0^{-1})$, A, B, and C of the circle of radius $2\sqrt{x_0^2 + x_0^{-2}}$ centered at (x_0, x_0^{-1}) with the hyperbola $xy = 1$, prove that the triangle ABC is equilateral.

Conics as point sets.

4.41. Prove that the set of points equidistant from a given point and a given circle is an ellipse, a hyperbola, or a ray.

4.42. Prove that the set of all centers of the circles passing through a given point and tangent to a given circle (or a straight line) not containing the point is an ellipse or a hyperbola (or a parabola).

4.43. Given points $A_t = (1+t, 1+t)$ and $B_t = (-1+t, 1-t)$ in the plane, describe the set swept out by all lines A_tB_t with $t \in \mathbb{R}$.

4.44. Given a point O, a line l and a point X moving along l, describe the set swept out by the perpendiculars to XO from X.

4.45. Points X and X' move along lines l and l' at constant velocities $v \neq v'$. What set is swept out by the lines XX'?

4.46. A circle S and a point O outside S are given. Through each point X inside S, the line l orthogonal to XO is drawn. Describe the set swept out by all lines l.

4.47. Prove that the centers of all regular triangles inscribed in a conic lie on one conic.

The basic properties of quadrics.

4.48. Prove that the projection of any plane section of the quadric $z = x^2 + y^2$ to the plane $z = 0$ is a circle or a line.

4.49. Prove that any cone in $\mathbb{R}^3$ has a circular section.

4.50. (a) Prove that, if two quadrics in $\mathbb{R}^3$ have a plane section, then they have at least two plane sections.

(b) Prove that, if three quadrics in $\mathbb{R}^3$ have a plane section, then the three planes of the other pairwise common sections whose existence is asserted by (a), have a common line.

4.51. (a) Given an ellipsoid in $\mathbb{R}^3$ with pairwise distinct lengths of the principal axes centered at O, prove that if a point A of the ellipsoid does not belong to the principal axes, then there exists one and only one plane section for which OA is a principal semiaxis.

(b) Given the plane $x\cos\alpha + y\cos\beta + z\cos\gamma = 0$, where $\cos^2\alpha + \cos^2\beta + \cos^2\gamma = 1$, whose intersection with the ellipsoid $x^2/a^2 + y^2/b^2 + z^2/c^2 = 1$ is an ellipse with principal semiaxes r_1 and r_2, prove that the numbers r_1 and r_2 satisfy the equation

$$\frac{a^2\cos^2\alpha}{a^2 - r^2} + \frac{b^2\cos^2\beta}{b^2 - r^2} + \frac{c^2\cos^2\gamma}{c^2 - r^2} = 0.$$

4.52. An ellipsoid in $\mathbb{R}^3$ is obtained by rotating an ellipse with foci F_1 and F_2 about the line F_1F_2.

(a) Prove that F_1 is a focus for the section of the ellipsoid by any plane containing F_1.

(b) Let C be a plane section of the ellipsoid. Prove that the lines F_1X, where $X \in C$, form a right circular cone.

CHAPTER 5

The World of Non-Euclidean Geometries

In this chapter, we discuss two non-Euclidean geometries, Lobachevsky geometry (known also as *hyperbolic*) and Riemannian (*elliptic*) geometry.[1] We start with Riemannian geometry, which is easier to understand, because locally (in the small) it coincides with the geometry of the sphere in ordinary three-dimensional space, and theorems of spherical geometry can be interpreted as ordinary space geometry theorems.

In Lobachevsky geometry, through any point not belonging to a line, one can draw many lines disjoint with the given one. For many centuries, people could not believe that this was possible. Euclid's fifth postulate about parallelism seemed to be a theorem following from the other axioms; in addition, Euclid stated it in a cumbersome and awkward way.[2] This belief proved to be false. The basic theorems of hyperbolic geometry were proved by Gauss, Lobachevsky, and Bolyai; then, Beltrami, Cayley, Klein, and Poincaré constructed models of this geometry, and the reality of this non-Euclidean geometry, known as Lobachevsky geometry, became evident. All of this is the subject of this chapter.

5.1. The circle and the two-dimensional sphere: one- and two-dimensional Riemannian geometries

The circle and the sphere. Like the straight line, the circle is one of the most familiar objects of Euclidean geometry. We imagine it as a wheel rim without thickness, which can slide along itself or be mapped onto itself by a "turnover", when the endpoints of some diameter are fixed.

In the Cartesian model of the Euclidean plane $\mathbb{R}^2$, which consists of points X represented by pairs (x_1, x_2) of real numbers, the circle of unit radius centered at the origin is the set of points X whose coordinates satisfy the relation $x_1^2 + x_2^2 = 1$. This set is denoted by $\mathbb{S}^1$. It can be made into a metric space by choosing the length of the shorter arc joining points X and Y as the distance $d(X, Y)$ between these points. The motions of the circle can be defined as one-to-one transformations of the circle into itself that do not change the distances between points. Such transformations include rotations

$$(x_1, x_2) \mapsto (x_1 \cos\varphi + x_2 \sin\varphi,\ -x_1 \sin\varphi + x_2 \cos\varphi),$$

reflections $(x_1, x_2) \mapsto (x_1, -x_2)$, and compositions of these maps. It is easy to prove that all motions of the circle can be described in this manner.

Let us describe the two-dimensional sphere as a geometric object. We imagine the sphere as the surface (without thickness) of a globe or of a ball. It can slide along

[1]These names were suggested by Klein in the Erlangen Program.

[2]*If a straight line intersects two straight lines so as to make the interior angles on one side of it together less than two right angles, the two straight lines will intersect, if indefinitely produced, on the side on which are the angles which are together less than two right angles.*

itself. This surface transforms into itself under symmetry maps (central symmetry and symmetry about any plane through the center of the sphere).

In the Cartesian model of the Euclidean 3-space $\mathbb{R}^3$, which consists of points X represented by triples (x_1, x_2, x_3) of real numbers, the sphere of unit radius centered at the origin is the set of points X whose coordinates satisfy the relation $x_1^2 + x_2^2 + x_3^2 = 1$. This set is denoted by $\mathbb{S}^2$.

The mathematical discipline that studies geometric images on the sphere is called *spherical geometry.*

The circle in which the sphere intersects a plane passing through the center of the sphere is called a *great circle*. The sphere $\mathbb{S}^2$ can be made into a metric space by defining the distance between two distinct antipodal points X and Y on the sphere as the length of the shorter arc cut out by the points X and Y on the unique great circle passing through X and Y. The distance between two antipodal points on $\mathbb{S}^2$ equals π.

The motions of the sphere $\mathbb{S}^2$ are its one-to-one transformations that do not change the distances between points. Motions include rotations through arbitrary angles about lines containing the center of the sphere and symmetries about planes containing the center of the sphere. We shall show later that any motion of the sphere is a composition of symmetries about planes through the center of the sphere.

Spherical geometry and plane geometry have much in common, but there are also many differences. Spherical geometry is considered in a separate chapter.

Spherical geometry has been known from ancient times. The properties of right spherical triangles were known to Ptolemy. Even earlier, spherical geometry was developed (also for astronomy purposes) by Menelaus. His treatise *Sphaerica* has survived; the first two books of it contain various theorems on spherical triangles (including theorems on the equality of spherical triangles).

Claudius Ptolemy (c. 100–178 A.D.), ancient Greek scientist, author of the famous *Mathematical Collection in XIII Books*, known under the Arabic name *Almagest.* It described the motion of planets. Ptolemy obtained many geometric results, such as the famous Ptolemy theorem about inscribed quadrilaterals.

One of the characterizations of a straight line in Euclidean space is that each segment of the line with endpoints A and B is the shortest curve with endpoints A and B. Precisely this characterization serves as a basis for defining a straight line (geodesic) in non-Euclidean geometries. A curve with endpoints A and B is said to be a *geodesic* if its length is less than the length of any other curve with the same endpoints. On a nonclosed curve, two points determine only one segment, while two points on a closed curve determine two segments. For this reason, a nonclosed curve is said to be *geodesic* if each segment of this curve is geodesic, and a closed curve is said to be *geodesic* if at least one of the two segments into which the curve is divided by two arbitrary points is always geodesic.

Let us examine the structure of geodesics on the 2-sphere $\mathbb{S}^2$. First, we must explain what the length of a curve on $\mathbb{S}^2$ is. The sphere $\mathbb{S}^2$ is embedded in the Euclidean space $\mathbb{R}^3$, so any curve on $\mathbb{S}^2$ is also a curve in $\mathbb{R}^3$. The *length of a curve on the sphere* $\mathbb{S}^2$ is the length of this curve in $\mathbb{R}^3$ (the length of a curve in $\mathbb{R}^3$ is defined as the least upper bound of the lengths of the polygonal lines inscribed in this curve).

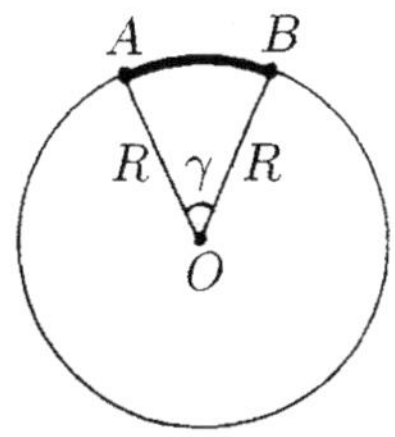

FIGURE 5.1

In Euclidean geometry, we can determine whether or not a curve is geodesic using the triangle inequality $|AC| \leqslant |AB| + |BC|$; if a point B does not lie on the line segment $[A, C]$, then the inequality is strict. Indeed, if $AA_1A_2 \ldots A_nB$ is an arbitrary polygonal line with endpoints A and B, then the triangle inequality implies

$$\begin{aligned}|AB| &\leqslant |AA_1| + |A_1B| \leqslant |AA_1| + |A_1A_2| + |A_2B| \leqslant \cdots \\ &\leqslant |AA_1| + |A_1A_2| + \cdots + |A_{n-1}A_n| + |A_nB|,\end{aligned}$$

and if the polygonal line does not coincide with $[A, B]$, then at least one of these inequalities is strict. Thus the length of any polygonal line with endpoints A and B is larger than the length of $[A, B]$. Since a curve can be approximated by a polygonal line to arbitrary accuracy, the length of a curve with endpoints A and B is larger than the length of the straight line segment $[A, B]$.

As we shall soon see, the geodesics on the sphere $\mathbb{S}^2$ are the great circles. Through any two nonantipodal points on the sphere, there passes exactly one great circle, and through a pair of antipodal points, there pass infinitely many great circles. In order that great circles be indeed geodesics, the triangle inequality should read as follows: *The sum of lengths of the shorter arcs of the great circles joining A to B and B to C is not smaller than the length of the arc of the great circle joining C to A.* Since the length of the arc AB equals $R\gamma$ (Figure 5.1), an equivalent statement is: The sum of two plane angles in a trihedral angle is no less than the third plane angle. Let us prove this assertion for a nondegenerate trihedral angle.

THEOREM 1. *For any trihedral angle $OABC$ with vertex O, we have the inequality $\angle AOC < \angle AOB + \angle BOC$.*

Proof. It is sufficient to consider the case in which the angle AOC is the largest. We can then choose a point B_1 inside the angle AOC so that $\angle AOB_1 = \angle AOB$ and $|OB_1| = |OB|$. Let us draw a line segment with endpoints on the rays OA and OC through B_1; we can assume that A and C are the endpoints of this segment. The equality of the triangles AOB and AOB_1 implies that $|AB| = |AB_1|$. In addition, $|AB_1| + |B_1C| = |AC| < |AB| + |BC|$; hence $|B_1C| < |BC|$. The law of cosines implies that the length of the side BC monotonically increases with the angle BOC if the lengths of BO and OC are constant. Therefore, $\angle B_1OC < \angle BOC$, and

$$\angle AOC = \angle AOB_1 + \angle B_1OC < \angle AOB + \angle BOC. \quad \square$$

The spherical triangle inequality makes it possible to prove that the great circles coincide with the geodesics on the sphere. (The proof is a word-for-word repetition of the arguments used above to show that the geodesics in Euclidean space are straight lines.) For this reason, the great circles are sometimes called spherical straight lines.

The distance between points A and B on the sphere $\mathbb{S}^2$ is defined as the length of the shorter arc of the great circle joining A and B. Note that the spherical distance between points A and B monotonically increases with the Euclidean distance between A and B. In particular, spherical distances are equal if and only if the corresponding Euclidean distances are equal.

Elementary spherical geometry. In spherical geometry, we can define the following correspondence called the *polar correspondence*: to each great circle S we assign the pair of endpoints of the diameter perpendicular to S, and each pair of antipodal points A and B is assigned the great circle lying in the plane perpendicular to AB. The two points assigned to a great circle are called its *poles*, and the great circle assigned to a pair of antipodal points is their *polar*. It is easy to verify that if a great circle S passes through a point A, then the polar of the point A passes through a pole of the great circle S. Thus the polar correspondence transforms points into lines and lines into points, and the statement "a line l contains a point A" transforms into the statement "the point $l^{\perp}$ lies on the line $A^{\perp}$."

To a spherical triangle ABC we can assign the polar triangle $A'B'C'$ as follows: A' is the pole of the spherical line BC that lies on the same side of this line as A; the points B' and C' are defined similarly. The following properties of a polar triangle are easy to verify:

(i) if a triangle $A'B'C'$ is polar to a triangle ABC, then the triangle ABC is polar to the triangle $A'B'C'$;

(ii) if α, β, γ are the angles of a triangle ABC, and aR, bR, cR are its side lengths (R is the radius of the sphere), then the polar triangle $A'B'C'$ has angles $\pi - a$, $\pi - b$, $\pi - c$ and side lengths $(\pi - \alpha)R$, $(\pi - \beta)R$, $(\pi - \gamma)R$.

To prove property (i), consider the center O of the sphere. Since $OA' \perp OC$ and $OB' \perp OC$, we have $OC \perp OA'B'$.

Property (ii) follows from the observation that the inward normals to the planes forming a dihedral angle α form an angle of $\pi - \alpha$.

In spherical geometry, as opposed to Euclidean geometry, triangles with equal respective angles are necessarily equal. Indeed, the equality of respective angles in two triangles implies the equality of respective sides in their polar triangles, and the equality of triangles with equal respective sides is proved in exactly the same way as in the Euclidean case. In particular, in spherical geometry, we can compute the area of a triangle if we know its angles.

THEOREM 2. *The area of a spherical triangle ABC with angles α, β, and γ equals $(\alpha + \beta + \gamma - \pi)R^2$, where R is the radius of the sphere.*

Proof. First, consider a *spherical digon*, i.e., one of the four pieces into which the sphere is partitioned by two spherical lines. Let $S(\alpha)$ be the area of a spherical digon with angle α. Clearly, $S(\alpha)$ is proportional to α, and $S(\pi) = 2\pi R^2$ (the area of the hemisphere). Therefore, $S(\alpha) = 2\alpha R^2$.

The pairs of the spherical lines AB, BC, and CA determine twelve spherical digons. Let us choose the six spherical digons each containing either the triangle ABC or the triangle $A_1B_1C_1$ symmetric to ABC with respect to the center of the sphere. Each point in ABC and in $A_1B_1C_1$ is covered by exactly three such digons, and any other point of the sphere is covered by exactly one digon (we do not consider the points on the spherical lines AB, BC, and CA). Thus

$$4(\alpha+\beta+\gamma)R^2 = 4\pi R^2 + 2S_{ABC} + 2S_{A_1B_1C_1},$$

and since $S_{ABC} = S_{A_1B_1C_1}$, we have $S_{ABC} = (\alpha+\beta+\gamma-\pi)R^2$. □

COROLLARY. *The sum of angles of a spherical triangle is greater than π.*

The sides a, b, and c of a spherical triangle ABC satisfy relations similar to those given in the law of sines and law of cosines for triangles in the Euclidean plane. For convenience, we assume that the sphere has radius 1. Let α, β, γ be the angles with vertices A, B, C, respectively.

THE LAW OF SINES.

$$\frac{\sin a}{\sin\alpha} = \frac{\sin b}{\sin\beta} = \frac{\sin c}{\sin\gamma}.$$

Proof. Let O be the center of the sphere. If H is the projection of the point A to the plane OBC, and A_b and A_c are the projections of A to the lines OB and OC, respectively, then A_b and A_c coincide with the projections of H to the lines OB and OC. Hence

$$AH = A_bA\sin\beta = \sin c\sin\beta, \qquad AH = A_cA\sin\gamma = \sin b\sin\gamma.$$

Therefore, $\sin b : \sin\beta = \sin c : \sin\gamma$. Similarly, $\sin b : \sin\beta = \sin a : \sin\alpha$. □

If a is small, then $\sin a \approx a$; so the spherical law of sines transforms into the Euclidean law of sines (see p. 12) as $a, b, c \to 0$.

The spherical law of sines implies the spherical form of Thales' theorem: *The base angles in an isosceles triangle are equal.*

THE FIRST LAW OF COSINES.

$$\cos a = \cos b\cos c + \sin b\sin c\cos\alpha.$$

Proof. Let O be the center of the sphere. Denote $e_a = \overrightarrow{OA}$, $e_b = \overrightarrow{OB}$, and $e_c = \overrightarrow{OC}$. The projections of the vectors e_b and e_c to e_a equal $e_a\cos c$ and $e_a\cos b$, respectively. Therefore, the vectors $u = e_b - e_a\cos c$ and $v = e_c - e_a\cos b$ are orthogonal to e_a, their lengths are $\sin c$ and $\sin b$, and the angle between them is α. Hence

$$\begin{aligned}\cos a &= (e_b, e_c) = (u + e_a\cos c, v + e_a\cos b)\\ &= (u, v) + \cos c\cos b = \sin b\sin c\cos\alpha + \cos b\cos c.\end{aligned}$$ □

If a is small, then $\cos a \approx 1 - a^2/2$. So, as $a, b, c \to 0$, the first spherical law of cosines transforms into the Euclidean law of cosines (see p. 12).

THE SECOND LAW OF COSINES.

$$\cos\alpha = -\cos\beta\cos\gamma + \sin\beta\sin\gamma\cos a.$$

Proof. The second law of cosines for a triangle ABC is equivalent to the first law of cosines for the polar triangle $A'B'C'$. □

As $a \to 0$, the second law of cosines becomes the following equality:

$$\cos\alpha = -\cos(\beta+\gamma), \qquad \text{i.e.,} \quad \alpha+\beta+\gamma = \pi.$$

Geometry of the n-sphere. For the n-sphere $\mathbb{S}^n$ in the space $\mathbb{R}^{n+1}$, the great circles are defined as sections by planes passing through the center of the sphere. It is easy to see that the great circles are geodesic. The distance between points A and B on the sphere is defined as the length of the shorter arc of the great circle joining these points. Similarlly to the case of the 2-sphere, the equality of the spherical distances between points on the n-sphere is equivalent to the equality of the Euclidean distances between these points.

In the Euclidean n-space, the pairwise distances between $n + 2$ points $A_1, \dots, A_{n+2}$ are related as follows. The distances from A_{n+2} to $A_1, \dots, A_n$ determine the position of the point A_{n+2} up to symmetry about the plane $A_1 \dots A_n$. Thus the length of $[A_{n+1}, A_{n+2}]$ can take only two values, the lengths of all other segments being set. For $n + 2$ points on the n-sphere $\mathbb{S}^n$, a similar relation holds.

THEOREM 3. *Let points $A_1, \dots, A_{n+2}$ lie on the n-sphere of radius* 1, *and let d_{ij} be the (spherical) distance between A_i and A_j. Then* $\det(\cos d_{ij}) = 0$.

Proof. Let O be the center of the sphere. Put $e_i = \overrightarrow{OA_i}$. Consider the $(n+2)$-parallelepiped spanned by $e_1, \dots, e_{n+2}$. The squared volume of this parallelepiped is equal to $\det((e_i, e_j)) = \det(\cos d_{ij})$.

On the other hand, the vectors $e_1, \dots, e_{n+2}$ lie in a space of dimension $n + 1$, hence the volume of the $(n + 2)$-parallelepiped spanned by these vectors is 0. □

Riemannian, or elliptic, geometry. Departing from the sphere, Riemann suggested to consider a slightly different geometric object in which the straight lines have the basic property that through any two points, exactly one line can be drawn. He suggested to "glue together" the antipodal points of the circle $\mathbb{S}^1$ or of the sphere $\mathbb{S}^2$. We can imagine this as follows: The object consists of all straight lines through the origin (in $\mathbb{R}^2$ or in $\mathbb{R}^3$, respectively). These are points. The distance between two points is the angle between the lines. The motions are the maps of the lines induced by the orthogonal transformations of $\mathbb{R}^2$ and of $\mathbb{R}^3$. In $\mathbb{R}^3$, a line in the new geometry is the family of all lines in $\mathbb{R}^3$ that lie in one plane passing through the origin. Thus through any two distinct points, there passes exactly one line. But unlike ordinary plane geometry, Riemannian geometry does not have any parallelism, because any two lines intersect.

This resembles the projective line and plane, which also can be treated as families of lines through the origin in $\mathbb{R}^2$ and in $\mathbb{R}^3$. But in projective geometry, the traces of these lines on the Euclidean coordinate line and plane were considered, while in Riemannian geometry, they leave traces on the circle $\mathbb{S}^1$ and on the sphere $\mathbb{S}^2$.

Locally, elliptic geometry and spherical geometry are isometric. But the global properties of elliptic geometry significantly differ from the properties of all other geometries. Its main distinguishing feature is due to the existence of noncontractible curves, in particular, triangles, on the projective plane $\mathbb{R}P^2$.

One of the most striking differences between elliptic geometry and the other geometries is that in elliptic geometry, triangles with equal respective sides are not necessarily equal. This property can be stated differently as follows: *An isometry defined on a finite set of points cannot always be extended to an isometry of the entire space.*

To construct an example of two nonequal triangles with equal sides, consider the spherical triangle with sides $\pi/3$, $\pi/3$, and α on the sphere of radius 1. Such

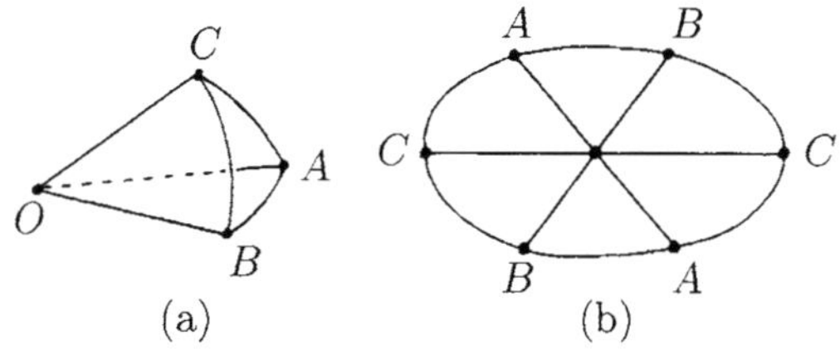

FIGURE 5.2

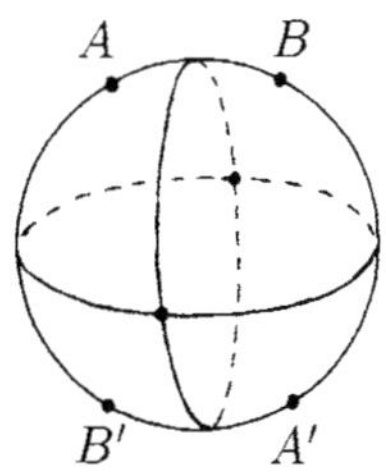

FIGURE 5.3

a triangle exists if $0 < \alpha < 2\pi/3$; hence triangles with sides $(\pi/3, \pi/3, \alpha)$ and $(\pi/3, \pi/3, \pi - \alpha)$ exist if $\pi/3 < \alpha < 2\pi/3$. If $\alpha < \pi/2$, then these triangles correspond to elliptic triangles with sides $(\pi/3, \pi/3, \alpha)$. The simplest way to see that the obtained triangles are not equal (i.e., their corresponding angles are not equal) is to consider the degenerate case $\alpha = \pi/3$ (Figure 5.2). It is also evident that in this case, the difference between the angles of the triangles is too large, and a small change of α cannot make them equal.

Figure 5.2 shows that there are two types of elliptic triangles, namely, those obtained from the spherical triangles symmetric with respect to the center of the sphere (Figure 5.2 (a)) and those obtained from one closed six-sided polygon symmetric with respect to the center of the sphere (Figure 5.2 (b)). The first type corresponds to contractible curves on the projective plane and the second to non-contractible curves.

Another distinguishing feature of elliptic geometry is that the set of points equidistant from two given distinct points consists of two rather than one, lines (or hyperplanes, in multidimensional spaces). This is because the set of points equidistant from points $\{A, A'\}$ and from points $\{B, B'\}$ on the elliptic plane corresponds to the union of the set of points equidistant from A and B with the set of points equidistant from A and B' on the sphere (Figure 5.3).

5.2. Lobachevsky geometry

The fact that both Lobachevsky and Euclid geometries can be equally valid became quite clear in the 1870s after the work of Cayley and Klein. In 1871 Klein constructed the projective model of Lobachevsky geometry.

The Klein model of Lobachevsky geometry. In the Klein model, the Lobachevsky plane is represented as the interior of the unit disk. The points of the Lobachevsky plane are the points of this disk, the straight lines are the chords, and the distance between points X and Y is given by

$$d(X, Y) = \frac{c}{2} \left| \ln \frac{|AX|}{|BX|} : \frac{|AY|}{|BY|} \right|,$$

where A and B are the endpoints of the chord through X and Y. The motions of the Lobachevsky plane are the projective transformations that map the disk onto itself (they also form the group of isometries).

A rigorous description of the model must include the following assertions: the function d introduced above is well defined, it satisfies the axioms of distance, and the group of motions is transitive.

We begin by proving the first assertion; as a byproduct, we shall construct the Lobachevsky line.

Define the distance $d(a, b)$ between points a and b on an interval (x, y) by setting

$$d(a,b) = \left|\ln[a,b,x,y]\right| = \left|\ln\left(\frac{x-a}{x-b} : \frac{y-a}{y-b}\right)\right|.$$

It is easy to see that such a definition makes sense, i.e.,

$$\frac{x-a}{x-b} : \frac{y-a}{y-b} > 0.$$

Indeed, $x - a < 0$, $x - b < 0$, $y - a > 0$, and $y - b > 0$. Clearly, $d(a, a) = 0$ and $d(a, b) \to \infty$ as $b \to x$ and $b \to y$. In addition, $d(a, b) = d(b, a)$, because

$$\frac{x-a}{x-b} : \frac{y-a}{y-b} = \left(\frac{x-b}{x-a} : \frac{y-b}{y-a}\right)^{-1}.$$

Note that $\ln[a, b, x, y] = -\ln[a, b, y, x]$, so there is no need to distinguish the order of the points x and y, i.e., to specify an orientation of the interval (x, y).

The identity

$$\left(\frac{x-a}{x-b} \cdot \frac{y-b}{y-a}\right)\left(\frac{x-b}{x-c} \cdot \frac{y-c}{y-b}\right)\left(\frac{x-c}{x-a} \cdot \frac{y-a}{y-c}\right) = 1$$

implies $\pm d(a, b) \pm d(b, c) \pm d(c, a) = 0$. A more careful examination shows that if c lies between a and b, then $d(a, c) + d(c, b) = d(a, b)$.

The distance $d(a, b)$ does not change under the projective transformations of the line that preserve the interval (x, y).

The interval (x, y) with the distance $d(a, b)$ defined above is called the *Lobachevsky line in the Klein model.* The Lobachevsky line differs from the Euclidean line only slightly because these lines are isometric. But still they have some distinctions. Below we give an example showing that the motion groups of the Lobachevsky and Euclidean lines act differently, even though they are isomorphic.

The map $x \mapsto (x+v)/(xv+1)$ preserves the interval $(-1, 1)$ and takes the point 0 to v. This map can be called the *shift* of the Lobachevsky line by the vector v. It is easy to verify that the composition of shifts by the vectors v_1 and v_2 is the shift by the vector

$$v = \frac{v_1 + v_2}{1 + v_1 v_2}. \tag{1}$$

Formula (1) coincides with the formula for adding velocities in special relativity theory (in the system of units in which the speed of light c equals 1); see pp. 137–139 for more details.

As already mentioned, the Lobachevsky line is isometric to the Euclidean line. However, the Lobachevsky plane is not isometric to the Euclidean plane. Now, we are prepared to describe the Lobachevsky plane; more precisely, we shall describe the *Klein model* of the Lobachevsky plane. The points in the Klein model are the interior points of some disk; the boundary circle of this disk is called the *absolute.*

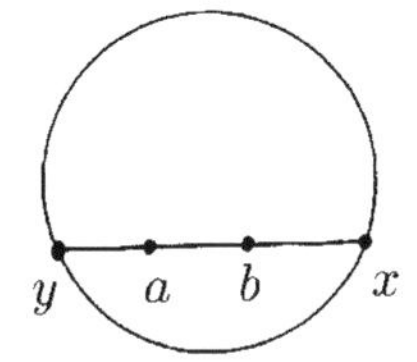

FIGURE 5.4

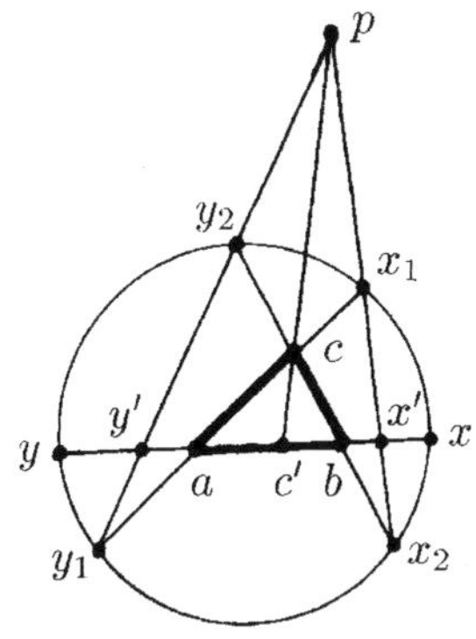

FIGURE 5.5

The distance between points a and b is defined as $d(a,b)$ for the interval (x,y) whose endpoints x and y are the intersection points of the line ab with the absolute. At this point, we must slightly alter the definition; the meaning of the alteration will be explained later. We have defined $d(a,b) = \big|\ln[a,b,x,y]\big|$; instead, we now define

$$d(a,b) = \frac{c}{2}\big|\ln[a,b,x,y]\big|.$$

The parameter c plays the same role in Lobachevsky geometry as the radius R of the sphere plays in spherical geometry. Many formulas of spherical geometry take the simplest form at $R = 1$. Similarly, many formulas of Lobachevsky geometry take their simplest forms at $c = 1$. In what follows, we shall often assume that $c = 1$.

If the points a and b are arranged in the same order as in Figure 5.4, then $\ln[a,b,x,y] > 0$, and the absolute value sign in the definition of distance can be removed: $d(a,b) = \frac{1}{2}\ln[a,b,x,y]$ (here $c = 1$).

THEOREM 1. *In the Klein model, the function d satisfies the axioms of distance.*

Proof. We have already seen that d is nonnegative and symmetric. It remains to prove that $d(a,c) + d(c,b) \geqslant d(a,b)$, and if c does not lie on the segment (a,b), then $d(a,c) + d(c,b) > d(a,b)$. Suppose that the absolute intersects the rays ab and ba at points x and y, respectively, the rays ac and ca at points x_1 and y_1, and the rays cb and bc at points x_2 and y_2 (Figure 5.5). Then the intersection point x' of the chords x_1x_2 and xy lies on the segment $[x,b]$, and the intersection point y' of the chords y_1y_2 and xy lies on the segment $[a,y]$. Let p be the intersection point of the lines x_1x_2 and y_1y_2, and let c' be the intersection point of the lines pc and xy. Then c' lies on the segment $[a,b]$.

The cross ratios are preserved by a projection of one line to another. Therefore, $[a,c,x_1,y_1] = [a,c',x',y']$ and $[c,b,x_2,y_2] = [c',b,x',y']$ (we consider projections from the point p to the line xy).

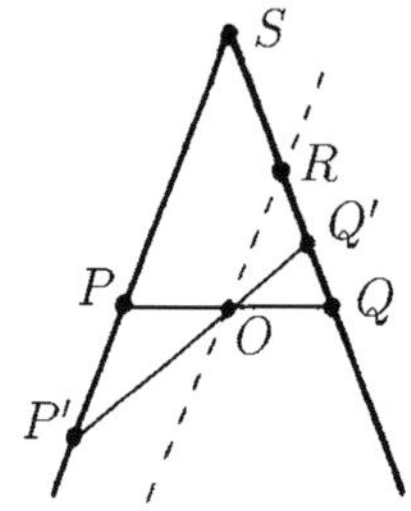

FIGURE 5.6

Let us show that $[a,c',x',y'] > [a,c',x,y]$ and $[c',b,x',y'] > [c',b,x,y]$, or, in other words, that if the points a, b, x, y are arranged as in Figure 5.4, then an increase of the segment $[x,y]$ decreases the cross ratio $[a,b,x,y]$. For the positive direction, we take the direction of the ray yx; thus, to increase the segment $[x,y]$, we must add a positive number ε to the coordinate of the point x. The second endpoint of the segment is left unchanged for a while. The cross ratio becomes smaller, because

$$\frac{x-a}{x-b} - \frac{x+\varepsilon-a}{x+\varepsilon-b} = \frac{\varepsilon(b-a)}{(x-b)(x+\varepsilon-b)} > 0.$$

The second endpoint is handled similarly.

As a result, we obtain $[a,c,x_1,y_1] > [a,c',x,y]$ and $[c,b,x_2,y_2] > [c',b,x,y]$; these inequalities imply

$$[a,c,x_1,y_1][c,b,x_2,y_2] > [a,c',x,y][c',b,x,y] = [a,b,x,y],$$

i.e., $d(a,c)+d(c,b) > d(a,b)$. □

Lobachevsky geometry, as well as spherical and plane geometries, has a fairly large group of isometries: given two arbitrary points A and B, a line through A, and a line through B, we can transform A into B in such a way that the given line through A will transform into the given line through B. To prove this, it is sufficient to verify that there exists a transformation of the plane that preserves the cross ratios, transforms a given disk into itself, and maps the center of the disk to a given interior point B. Indeed, such a transformation is an isometry; to map any point A to any other point B, we can map first A to the center O of the disk and then O to B; an arbitrary line through O can simultaneously be transformed into any other line through O by rotation.

THEOREM 2. *There exists a transformation of the Lobachevsky plane that preserves the cross ratios, transforms a given disk into itself, and maps its center to an arbitrary given interior point; i.e., the group of motions of the Lobachevsky plane is transitive.*

Proof. Consider the right circular cone with vertex S. The section of this cone by the plane passing through O and perpendicular to its axis is a circle with diameter PQ centered at O. Consider also the section of the cone by the plane passing through O and perpendicular to the plane SPQ (the cone is assumed to be infinite in one direction). If the point Q' belongs to the interval QR (Figure 5.6), then the section under consideration is an ellipse.

We can introduce affine coordinates on the plane Π' that contains this ellipse and on the plane Π that contains the circle with diameter PQ so that the circle

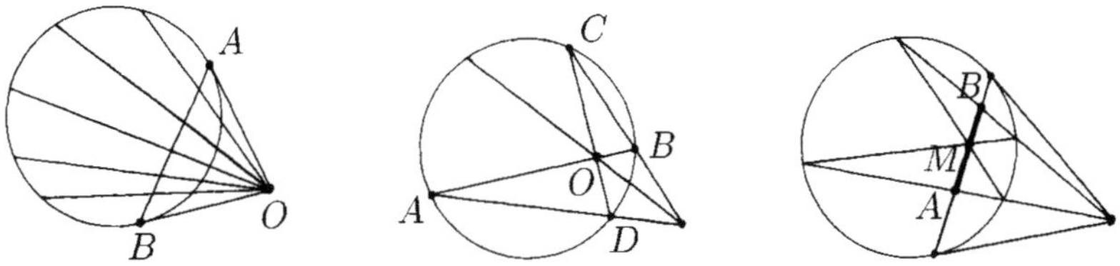

FIGURE 5.7 FIGURE 5.8 FIGURE 5.9

and the ellipse coincide under identification of points with equal coordinates. As the origins, we take the centers of the ellipse and the circle, and as the Ox axes, the lines $P'Q'$ and PQ. The point O inside the ellipse is then identified with the point O_1 of the disk such that $|P'O| : |OQ'| = |PO_1| : |O_1Q|$.

As the point Q' moves along the segment $[Q, R]$, the ratio $|P'O| : |OQ'|$ varies from 1 to ∞. Hence O_1 can be any point inside the segment $[O, Q]$.

The required transformation is the composition of the maps $f\colon \Pi \to \Pi_1$ and $g\colon \Pi \to \Pi_1$, where f is the projection from the point S and g is the identification of points with equal coordinates. □

The projective transformations of the plane that preserve a given disk are motions of the Lobachevsky plane in the Klein model. This observation makes it possible to find out what perpendiculars to a given line, bisectors of angles between lines, etc. look like in the Klein model. First, we define perpendicular lines. Two intersecting lines partition the plane into four angles. If there exist motions that transform one of these angles into each of the other three angles, then the lines are called *perpendicular*. The family of lines perpendicular to a line AB is shown in Figure 5.7. To prove this, it suffices to apply a projective transformation that maps the point O to a point at infinity. Figure 5.8 demonstrates a method for constructing the bisector of the angle between lines AB and CD (this can be proved by transforming the point O into the center of the disk), and Figure 5.9 demonstrates a method for constructing the midpoint M of a segment $[A, B]$ (this can be proved by transforming M into the center of the disk).

Linear-fractional transformations and stereographic projections. To get acquainted with two other important models of Lobachevsky geometry, we need to learn some properties of linear-fractional transformations and stereographic projections.

A map $f\colon \mathbb{C} \to \mathbb{C}$ given by

$$f(z) = \frac{az+b}{cz+d}, \qquad \text{where} \quad a, b, c, d \in \mathbb{C} \quad \text{and} \quad ad - bc \neq 0,$$

is called a *linear-fractional transformation*. Actually, this is a transformation of $\mathbb{C} \cup \{\infty\}$ rather than $\mathbb{C}$, but we usually disregard the point ∞ and its preimage.

The formula

$$\frac{az+b}{cz+d} = \frac{a}{c} + \frac{bc-ad}{c(cz+d)}$$

shows that any linear-fractional transformation can be represented as the composition of a transformation of the form $z \mapsto az + b$ and the transformation $z \mapsto z^{-1}$. The first transformation is the composition of a rotation, a dilation or contraction (homothety), and a shift. The second coincides with the inversion with respect to the unit circle up to complex conjugation. Recall that an *inversion* with respect to a circle of radius R centered at O is the transformation of the plane that takes each point A to the point A^* lying on the ray OA and satisfying the relation $|OA| \cdot |OA^*| = R^2$. The inversion with respect to the unit circle is given by $z \mapsto (\bar{z})^{-1}$. (The *unit circle* has the equation $|z| = 1$, where $z \in \mathbb{C}$.)

It is proved in the same way as for linear-fractional transformations of the real line that the linear-fractional transformations of the complex plane preserve the cross ratio

$$[z_1, z_2, z_3, z_4] = \frac{z_3 - z_1}{z_3 - z_2} : \frac{z_4 - z_1}{z_4 - z_2}$$

of four points.

THEOREM 3. *The linear-fractional transformations*

(a) *transform a circle or a line into a circle or a line;*

(b) *do not change the angles.*

Proof. (a) It is easy to verify that points $z_1, z_2, z_3 \in \mathbb{C}$ are collinear if and only if $(z_1 - z_2)/(z_1 - z_3) \in \mathbb{R}$. In addition, any three points can be mapped to three collinear points by a linear-fractional transformation. Using these two observations, it is easy to prove that four points $z_1, z_2, z_3, z_4 \in \mathbb{C}$ lie on one circle or on one line if and only if $[z_1, z_2, z_3, z_4] \in \mathbb{R}$. This property is preserved under linear-fractional transformations.

(b) It suffices to prove that the angles do not change under inversions. An inversion with center O transforms a line l into a circle passing through O and such that the tangent line to the circle at the point O is parallel to l. Therefore, an inversion with center O transforms two lines intersecting at A into two circles forming an angle at the point O equal to the angle between the lines; but the angle between the circles at their intersection point O equals the angle between them at the other intersection point A^* (the image of A under the inversion).

Thus inversions preserve the angles between lines. Instead of an angle between two circles, we can consider the angle between the tangent lines to the circles at their intersection point; therefore, the angles between circles are also preserved under inversions. □

Similarly to inversions in the plane, we can define inversions in space. The inversion with respect to a sphere of radius R centered at O is the transformation of the space that maps each point A to the point A^* lying on the ray OA and satisfying the relation $|OA| \cdot |OA^*| = R^2$. The properties of inversions in space are largely similar to the properties of inversions on the plane.

1. *An inversion transforms spheres and planes into spheres and planes.* To prove this, it suffices to consider the family of planes passing through the center of the inversion and the center of the sphere under consideration (for a plane, the projection of the center of the inversion on this plane should be taken).

2. *An inversion transforms lines and circles into lines and circles.* Indeed, a line or a circle can be represented as an intersection of two spheres or planes, and spheres or planes transform into spheres or planes.

3. *An inversion in space does not change the angles between circles.* As we did for the plane, we can prove that the angles between intersecting lines do not change and conclude that neither do the angles between circles.

4. *An inversion in space does not change the cross ratio of four arbitrary points,* which is defined as the ratio $|CA|/|CB| : |DA|/|DB|$ of the lengths of the corresponding segments. First, note that if an inversion with center O maps the points A

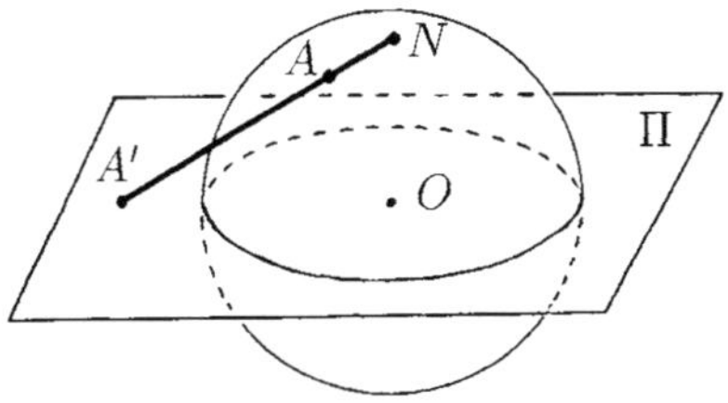

FIGURE 5.10

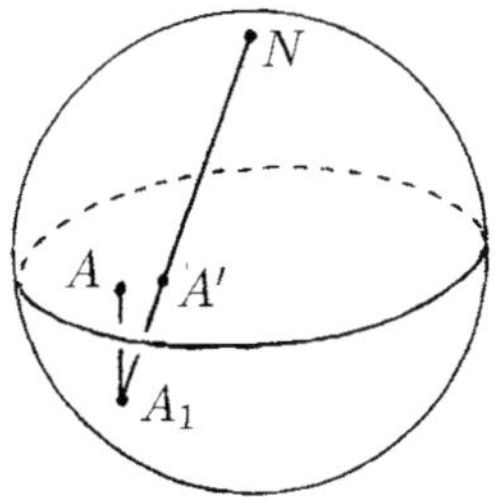

FIGURE 5.11

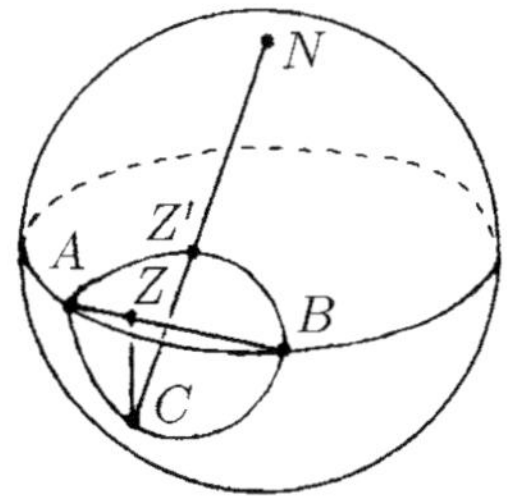

FIGURE 5.12

and B to A^* and B^*, then the triangles OAB and OB^*A^* are similar, and therefore

$$\frac{|AB|}{|A^*B^*|} = \frac{|OB|}{|OA^*|} = \frac{|OB|\cdot|OA|}{|OA^*|\cdot|OA|} = \frac{|OA|\cdot|OB|}{R^2}.$$

This formula readily implies that inversion preserves the cross ratio.

Inversions in space are closely related to the maps of a punctured sphere to a plane called stereographic projections and defined as follows. Consider a sphere S centered at O. Let N be a point on the sphere, and let Π be the plane through O perpendicular to the line ON (Figure 5.10); N can be called the north pole and Π, the equatorial plane. The *stereographic projection* from the point N to the plane Π is the map that takes each point A (different from N) on the sphere to the point A' at which the ray NA intersects Π.

Consider the sphere S_1 with radius $\sqrt{2}|NO|$ centered at N. It is easy to verify that the inversion with respect to S_1 transforms the sphere S into the plane Π, and the stereographic projection is the restriction of this inversion to S. Therefore, the stereographic projections also have properties 2–4.

Other models of Lobachevsky geometry. From the Klein model, we pass to another important model of Lobachevsky geometry. The new model is obtained as follows. Consider the sphere whose equator coincides with the absolute. For a point A of the Klein model, we denote by A_1 the point of the southern hemisphere projected onto A, and by A' the intersection point of the equatorial plane with the line A_1N, where N is the north pole (Figure 5.11). Assigning to each point A the corresponding point A', we obtain a transformation of the equatorial disk. For this transformation to be an isometry, the distance between two points A' and B' in the new model must equal the distance between the points A and B in the initial model. The model of Lobachevsky geometry thus obtained is called the *Poincaré disk model.* As in the Klein model, the boundary of the disk is called the *absolute.*

Henri Poincaré (1859–1912), one of the greatest scientists in history. Made outstanding contributions to the majority of fields of mathematics.

Let us describe the straight lines in the Poincaré model. A chord AB corresponds to a section of the southern hemisphere by a plane perpendicular to the equator. This section is a semicircle perpendicular to the absolute (Figure 5.12). The projection from the pole onto the equatorial plane maps this semicircle onto an arc of a circle perpendicular to the absolute. Thus the lines in the Poincaré disk model are arcs of the circles perpendicular to the absolute.

For the Poincaré model, it is convenient to assume that the disk in which the model is constructed is the unit disk in the complex plane.

It is easy to see that if points Z and W lie on a chord AB and Z' and W' are the corresponding points of the Poincaré model, then

$$\big|[A,B,Z,W]\big| = \big|[A,B,Z',W']\big|^2.$$

Indeed, the stereographic projection is a restriction of a spatial inversion and, therefore, preserves the cross ratios. In addition, in the notation of Figure 5.12, we have

$$|AZ| : |ZB| = \frac{|AC^2|}{|AB|} : \frac{|BC^2|}{|AB|} = |AC^2| : |BC^2|.$$

Thus $\big|\ln[A,B,Z,W]\big| = 2\big|\ln|[A,B,Z',W']|\big|$.

Recall that $d(Z,W) = \frac{1}{2}\big|\ln[A,B,Z,W]\big|$. Therefore,

$$d(Z',W') = \big|\ln|[A,B,Z',W']|\big|.$$

By analogy with the infinite family of different spherical geometries (corresponding to different radii R), we can obtain an infinite family of Lobachevsky geometries by setting $d(Z,W) = \frac{c}{2}\big|\ln[A,B,Z,W]\big| = c\big|\ln|[A,B,Z',W']|\big|$.

Yet another model of Lobachevsky geometry can be obtained by means of a linear-fractional map from the unit disk onto the upper half-plane $H = \{\, x+iy \in \mathbb{C} \mid y > 0 \,\}$. The map that can be used for this purpose is, for example, $z \mapsto w = i(1+z)/(1-z)$. Indeed,

$$\operatorname{Im} w = \operatorname{Re}\left(\frac{1+z}{1-z}\right) = \frac{1}{2}\left(\frac{1+z}{1-z} + \frac{1+\bar{z}}{1-\bar{z}}\right) = \frac{1-|z|^2}{|1-z|^2}.$$

Hence $\operatorname{Im} w > 0 \iff |z| < 1$.

This model of Lobachevsky geometry is called the *Poincaré upper half-plane model*; in this model, the *absolute* is the line $\operatorname{Im} w = 0$.

Lobachevsky geometry is often called *hyperbolic* geometry. For this reason, we call lines, circles, and other objects of Lobachevsky geometry *hyperbolic* to distinguish them from the Euclidean lines and circles.

The linear-fractional transformations map lines and circles to lines and circles, and they preserve the angles. Hence the hyperbolic lines in the upper half-plane H are vertical rays and semicircles centered on the absolute.

The linear-fractional transformations preserve cross ratios; hence the distance between points in the Poincaré upper half-plane model is defined as follows. Let a hyperbolic line AB approach the absolute at points X and Y (Figure 5.13). Then $d(A,B) = c\big|\ln[A,B,X,Y]\big|$. (For brevity, we omit the second absolute value sign; we can do this if the points are arranged as in Figure 5.13.) If the hyperbolic line is a Euclidean ray, we set $Y = \infty$, i.e., $(y-a)/(y-b) = 1$. For the positive part of the imaginary axis, the formula for evaluating the hyperbolic distance acquires the especially simple form $d(ia,ib) = c|\ln(a/b)|$.

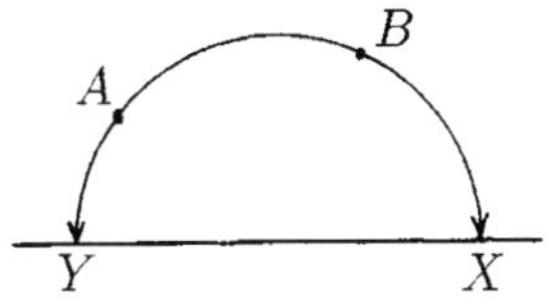

FIGURE 5.13

Next, let us find out what are the motions of the Lobachevsky plane. Any linear-fractional transformation that maps the upper half-plane H onto itself is a motion of the Lobachevsky plane. Let $a, b, c, d \in \mathbb{R}$. It is easy to verify that upper half-plane is invariant under the maps

$$z \mapsto \frac{az+b}{cz+d} \quad \text{with} \quad ad - bc > 0 \qquad \text{and} \qquad z \mapsto \frac{a\bar{z}+b}{c\bar{z}+d} \quad \text{with} \quad ad - bc < 0.$$

Indeed,

$$\operatorname{Im} \frac{az+b}{cz+d} = \operatorname{Im} \frac{(c\bar{z}+d)(az+b)}{|cz+d|^2} = \operatorname{Im} \frac{bc\bar{z}+adz}{|cz+d|^2} = (ad-bc)\frac{\operatorname{Im} z}{|cz+d|^2},$$

$$\operatorname{Im} \frac{a\bar{z}+b}{c\bar{z}+d} = (bc-ad)\frac{\operatorname{Im} z}{|cz+d|^2}.$$

THEOREM 4. *An arbitrary (hyperbolic) motion of the upper half-plane H has the form*

$$z \mapsto \frac{az+b}{cz+d} \quad \text{or} \quad z \mapsto \frac{a(-\bar{z})+b}{c(-\bar{z})+d},$$

where $a, b, c, d \in \mathbb{R}$ and $ad - bc > 0$.

Proof. In Lobachevsky geometry, a point C lies on a line segment $[A, B]$ if and only if $d(A, C) + d(C, B) = d(A, B)$; therefore, any isometry maps hyperbolic lines to hyperbolic lines.

Suppose that an isometry φ maps the positive part L of the imaginary axis to a hyperbolic line $\varphi(L)$ approaching the absolute at points x and y. Then one of the two transformations $z \mapsto \pm(z-x)/(z-y)$ is a hyperbolic motion g^{-1}, and this motion transforms the hyperbolic line $\varphi(L)$ into L (if $y = \infty$, then $g(z) = z - x$.) Thus the isometry $g^{-1}\varphi$ transforms the hyperbolic line L into itself. We can assume that the isometry $g^{-1}\varphi$ does not shift the point i (otherwise, we can apply the isometry $z \mapsto kz$ with $k > 0$). Therefore, every point ia ($a \in \mathbb{R}$) is mapped to a point ib such that $d(i, ia) = d(i, ib)$, i.e., $|\ln a| = |\ln b|$. If the isometry $g^{-1}\varphi$ interchanges the hyperbolic rays (i, ∞) and $(i, 0)$, we additionally apply the isometry $z \mapsto -z^{-1}$. For the obtained isometry $g^{-1}\varphi$, b cannot equal a^{-1}; therefore, the isometry leaves all points of the hyperbolic line L fixed.

The hyperbolic distance d between points $z, w \in H$ can be calculated by the formula

$$\cosh\left(\frac{d}{c}\right) = 1 + \frac{|z-w|^2}{2\operatorname{Im}(z)\operatorname{Im}(w)},$$

where $\cosh t = \frac{1}{2}(e^t + e^{-t})$ is the *hyperbolic cosine*. This formula can be obtained by finding the coordinates of the points at which the hyperbolic line zw approaches the absolute and calculating the corresponding cross ratio. This method, however, requires cumbersome calculations. The formula can be verified in the following simpler way. Both sides of this formula are invariant under real linear-fractional transformations with positive determinants (for the right-hand side, it suffices to check its invariance under the transformations $z \mapsto z + a$, $a \in \mathbb{R}$, and $z \mapsto -1/z$). By such a transformation, the line zw can be mapped to the positive ray L of the imaginary axis; so it remains to show that both sides of the formula give the same result (namely, $x/y + y/x$) for points $ix, iy \in L$, which is easy.

For any positive number t, the isometry $g^{-1}\varphi$ does not move the point it; therefore, $d(it, z) = d(it, g^{-1}\varphi(z))$ for all $z \in H$. Suppose that $z = x + iy$ and $g^{-1}\varphi(z) = u + iv$. The distance formula implies

$$\frac{|it - z|^2}{t\,\mathrm{Im}(z)} = \frac{|it - g^{-1}\varphi(z)|^2}{t\,\mathrm{Im}(g^{-1}\varphi(z))}, \qquad \text{i.e.,} \quad [x^2 + (t - y)^2]v = [u^2 + (t - v)^2]y.$$

This equality holds for all positive t; hence $y = v$ and $x^2 = u^2$, i.e., $g^{-1}\varphi(z) = z$ or $-\bar{z}$. Since any isometry is a continuous map, one of these equalities holds for all points simultaneously. Therefore, $\varphi(z) = g(z)$ or $\varphi(z) = g(-\bar{z})$. □

REMARK 1. The motion $z \mapsto (az + b)/(cz + d)$ is said to be *proper*, and the motion $z \mapsto (a(-\bar{z}) + b)/(c(-\bar{z}) + d)$ is said to be *improper*.

REMARK 2. We can assume that $ad - bc = 1$ in the statement of Theorem 4. Indeed, suppose that $ad - bc = t > 0$. Let us divide the numbers a, b, c, d by $\sqrt{t}$. As a result, we obtain a transformation of the same form with new coefficients a, b, c, d, and this transformation satisfies the required condition.

Spherical and Lobachevsky geometries can be constructed in a unified way. To be more precise, the geometry constructed will be *elliptic* rather than spherical; it is obtained from spherical geometry by identifying antipodal points.

Uniform construction of elliptic and hyperbolic geometries involves the *complex projective plane* $\mathbb{C}P^2$. It is defined similarly to the real projective plane $\mathbb{R}P^2$, using triples of complex rather than real numbers; two triples are considered equivalent if they can be obtained from each other by multiplication by a complex number.

An important role will also be played by the curve $x^2 + y^2 + z^2 = 0$ in $\mathbb{C}P^2$.

Consider two points $P_1 = (x_1, y_1, z_1)$ and $P_2 = (x_2, y_2, z_2)$ on the sphere $x^2 + y^2 + z^2 = R^2$. These points can also be treated as points of $\mathbb{C}P^2$. The line P_1P_2 in $\mathbb{C}P^2$ meets the curve $x^2 + y^2 + z^2 = 0$ at some points J_1 and J_2. Let us show that

$$e^{\pm 2i\varphi} = [P_1, P_2, J_1, J_2],$$

where φ is the angle between the radii OP_1 and OP_2. Indeed, the line P_1P_2 in $\mathbb{C}P^2$ consists of points of the form $(x_1 + \lambda x_2, y_1 + \lambda y_2, z_1 + \lambda z_2)$, where $\lambda \in \mathbb{C} \cup \infty$ (the point P_1 corresponds to $\lambda = 0$, and P_2 corresponds to $\lambda = \infty$). The values of λ corresponding to J_1 and J_2 satisfy the equation

$$(x_1 + \lambda x_2)^2 + (y_1 + \lambda y_2)^2 + (z_1 + \lambda z_2)^2 = 0,$$

i.e., $1 + (2\cos\varphi)\lambda + \lambda^2 = 0$; here we have used the relations

$$x_1^2 + y_1^2 + z_1^2 = R^2, \qquad x_2^2 + y_2^2 + z_2^2 = R^2, \qquad x_1x_2 + y_1y_2 + z_1z_2 = R^2\cos\varphi.$$

Therefore,

$$\frac{\lambda_1}{\lambda_2} = \frac{\cos\varphi \pm \sqrt{\cos^2\varphi - 1}}{\cos\varphi \mp \sqrt{\cos^2\varphi - 1}} = \frac{\cos\varphi \pm i\sin\varphi}{\cos\varphi \mp i\sin\varphi} = e^{\pm 2i\varphi}.$$

On the other hand,

$$[P_1, P_2, J_1, J_2] = \frac{\lambda_1 - 0}{\lambda_2 - 0} : \frac{\lambda_1 - \infty}{\lambda_2 - \infty} = \frac{\lambda_1}{\lambda_2}.$$

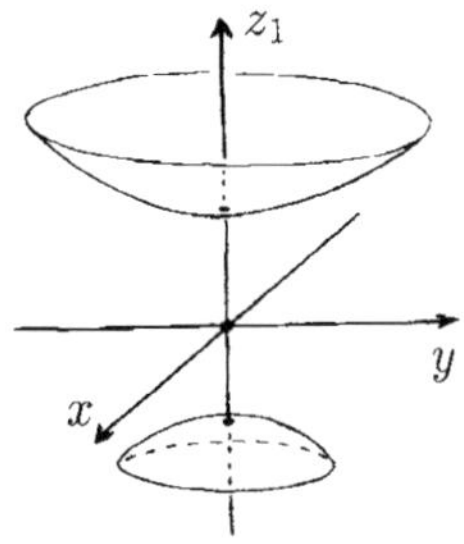

FIGURE 5.14

Thus $[P_1, P_2, J_1, J_2] = e^{\pm 2i\varphi}$. Hence the distance d between points of the sphere of radius R can be defined as

$$d = R\varphi = \pm\frac{R}{2i}\ln[P_1, P_2, J_1, J_2], \tag{2}$$

where ln is the function inverse to the exponential function with base e. As a function of a complex variable, ln is multivalued, so this definition requires additional explanations; but we refrain from giving them because in the case of Lobachevsky geometry, which we are most interested in, a similar formula includes only functions of a real variable.

Recall how formula (2) was obtained. We regarded points $P_1, P_2 \in \mathbb{R}^3$ as points in $\mathbb{C}^3$; then, we assigned the lines OP_1 and OP_2 to these points and regarded them as elements of $\mathbb{C}P^2$. The points P_1 and P_2 lie on the surface $x^2 + y^2 + z^2 = R^2$ in $\mathbb{R}^3$ and in $\mathbb{C}^3$. In $\mathbb{C}^3$, we can consider not only spheres of real radii, but also the sphere $x^2 + y^2 + z^2 = -c^2$ ($c \in \mathbb{R}$) of imaginary radius. Essentially, we replace the radius R by ic.

On the sphere of imaginary radius, as on the ordinary one, it is more convenient to deal only with real points. Setting $z_1 = iz$, we obtain the surface $x^2 + y^2 - z_1^2 = -c^2$. For real x, y, and z_1, this is a two-sheeted hyperboloid (Figure 5.14). Note also that under the transformation described above, the surface $x^2 + y^2 + z^2 = 0$ becomes $x^2 + y^2 - z_1^2 = 0$, which is a cone for real x, y, z_1. Let P_1 and P_2 be points in the upper sheet of the two-sheeted hyperboloid $x^2 + y^2 - z_1^2 = -c^2$. Consider them as points in $\mathbb{C}P^2$. The line P_1P_2 intersects the curve $x^2 + y^2 - z_1^2 = 0$ at some points J_1 and J_2 in $\mathbb{C}P^2$. Similarly to elliptic geometry, we define the distance between P_1 and P_2 by setting

$$d = \pm\frac{ic}{2i}\ln[P_1, P_2, J_1, J_2] = \pm\frac{c}{2}\ln[P_1, P_2, J_1, J_2].$$

We have obtained Lobachevsky geometry! Indeed, consider the section D^2 of the cone $x^2 + y^2 - z_1^2 \leqslant 0$ by a plane perpendicular to the axis of the cone. The rays OP_i and OJ_i intersect D^2 at points P_i' and J_i', respectively, and we clearly have $[P_1, P_2, J_1, J_2] = [P_1', P_2', J_1', J_2']$. Thus, projecting the upper sheet of the two-sheeted hyperboloid to D^2 from the origin, we obtain the Klein model of Lobachevsky geometry.

By analogy with the formula

$$\cos^2\left(\frac{d}{R}\right) = \frac{(u,v)^2}{(u,u)(v,v)},$$

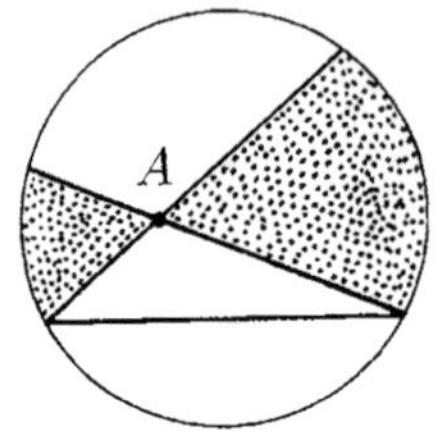

FIGURE 5.15

where $u = \overrightarrow{OP}_1$ and $v = \overrightarrow{OP}_2$, we can write

$$\cos^2\left(\frac{id}{c}\right) = \frac{[u,v]^2}{[u,u][v,v]},$$

where $[u,v] = u_1v_1 + u_2v_2 + (iu_3)(iv_3) = u_1v_1 + u_2v_2 - u_3v_3$. But

$$\cos t = \frac{e^{it} + e^{-it}}{2}, \quad \text{and} \quad \cos(it) = \frac{e^{-t} + e^{t}}{2} = \cosh t.$$

Therefore, $\cos(id/c) = \cosh(d/c)$. The formula

$$\cosh^2\left(\frac{d}{c}\right) = \frac{[u,v]^2}{[u,u][v,v]}$$

is indeed valid; it can be verified directly.

Elementary hyperbolic geometry. In hyperbolic geometry, the *value of an angle* can be defined as follows. We say that the value of an angle equals $2\pi/n$ if n angles congruent to the given one and having a common vertex cover the entire Lobachevsky plane without overlaps. Now, it is easy to define an angle of $2\pi m/n$, and then, by continuity, an arbitrary angle.

In the Klein model and the Poincaré disk model, the rotations about the center of the disk are motions. Therefore, in both models, the hyperbolic angles at the center of the disk equal the corresponding Euclidean angles. The center of the disk can be transferred to any other point by a hyperbolic motion. In the Poincaré model, the motions preserve the Euclidean angles; thus in this model, the Euclidean and hyperbolic angles coincide. (The angle between two intersecting circles is defined as the angle between the tangent lines at their intersection point.) In the Klein model, a motion does not necessarily preserve Euclidean angles, and the hyperbolic angles do not necessarily coincide with the Euclidean ones.

On the Euclidean plane, only one line parallel to a given one can be drawn through a point A. On the Lobachevsky plane, we can draw a family of lines disjoint with a given line l through any point A; these lines fill a pair of vertical angles (Figure 5.15 shows them in the Klein model). One of these four angles with vertex A contains the line l. Its sides are called *rays parallel to the line* l. The lines that contain these rays are called *lines parallel to* l. In Lobachevsky geometry, we must consider parallel rays rather than lines, because the two rays into which a line parallel to l is divided by the point A behave differently with respect to l: one ray approaches l, while the other recedes from it. In addition, only rays have the transitivity property $a \parallel b$, $b \parallel c \Rightarrow a \parallel c$; for lines, this property does not hold.

The *angle of parallelism* for a point A and a line l is the half-angle between the rays with vertex A parallel to l. The angle of parallelism α has the following property. Let us draw a perpendicular AH from A to l and consider the ray AB

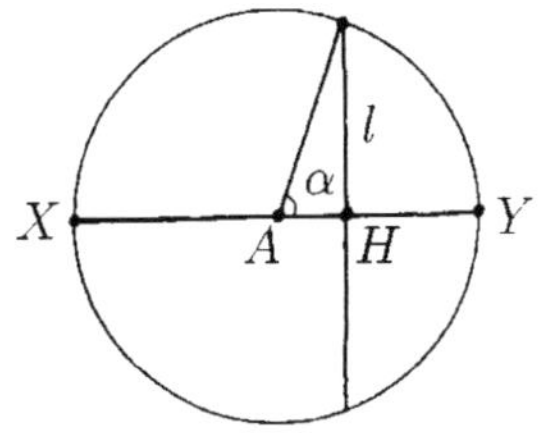

FIGURE 5.16

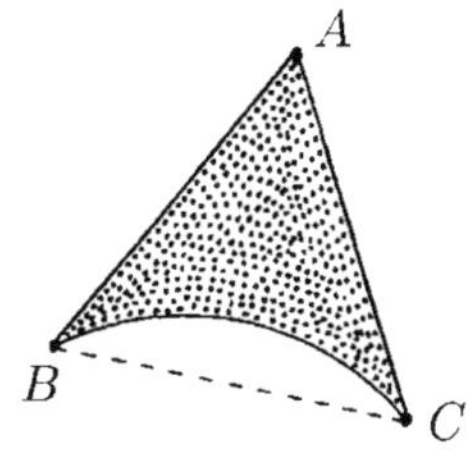

FIGURE 5.17

from A that makes an angle β with AH. The ray AB intersects the line l if and only if $\beta < \alpha$. This property can be taken as another definition of the angle of parallelism.

It is easy to verify that the angle of parallelism α depends only on the distance a from the point to the line. Let us find out exactly how a and α are related. This can easily be done in any model. Consider, for example, the Klein model; we assume that the point A coincides with the center of the disk (Figure 5.16). By definition,

$$a = \frac{c}{2}\left|\ln\left(\frac{XH}{XA}\cdot\frac{YA}{YH}\right)\right| = \frac{c}{2}\ln\left(\frac{1+\cos\alpha}{1-\cos\alpha}\right) = c\ln(\cot\tfrac{\alpha}{2}).$$

Therefore, $e^{-a/c} = \tan(\alpha/2)$. For $a/c \to 0$, we have $\alpha \to \pi/2$; thus Lobachevsky geometry can be transformed into Euclidean geometry by decreasing the distance a (or by increasing the "imaginary radius" c).

In what follows, for simplicity we assume that $c = 1$.

In hyperbolic geometry, the sum of angles of a triangle is less than π. This assertion can easily be proved in the Poincaré disk model by placing one vertex of the triangle at the center of the disk (Figure 5.17). Indeed, the angle A of a hyperbolic triangle ABC equals the angle A of the Euclidean triangle, while the hyperbolic triangle has smaller angles at the vertices B and C.

In Euclidean geometry, the sides and angles of a triangle obey certain relations. Lobachevsky geometry also imposes constraints on elements of triangles. The simplest relations are those for a right triangle, so we start with them.

We denote the angles of a triangle ABC by α, β, γ, and the opposite sides by a, b, c, respectively.

THEOREM 5. *In a triangle with right angle γ, the following relations hold:*
(a) $\cosh c = \cosh a \cosh b$; (b) $\tanh a = \sinh b \tan\alpha$.

Proof. (a) We can assume that $A = ki$ $(k > 1)$, $B = \cos\varphi + i\sin\varphi$, and $C = i$. Then

$$\cosh a = \frac{1}{\sin\varphi}, \qquad \cosh b = \frac{1+k^2}{2k}, \qquad \cosh c = \frac{1+k^2}{2k\sin\varphi}.$$

Therefore, $\cosh a \cosh b = \cosh c$.

For small a, b, and c, the relation $\cosh a \cosh b = \cosh c$ becomes $a^2 + b^2 = c^2$. Thus the formula $\cosh a \cosh b = \cosh c$ can be called the *hyperbolic Pythagorean theorem*.

(b) Again, we assume that $A = ki$ $(k > 1)$, $B = \cos\varphi + i\sin\varphi$, and $C = i$. Let x_0 be the center of the Euclidean circle containing the hyperbolic line AB. Then

$$x_0^2 + k^2 = (\cos\varphi - x_0)^2 + \sin^2\varphi, \qquad \text{i.e.,} \quad k^2 = 1 - 2x_0\cos\varphi.$$

It is easy to verify that $\angle Ax_0O = \alpha$. Therefore,

$$\tan\alpha = \frac{k}{|x_0|} = \frac{2k\cos\varphi}{k^2-1}.$$

Since $\sinh^2 t = \cosh^2 t - 1$, we have

$$\sinh^2 a = \frac{1}{\sin^2\varphi} - 1 = \left(\frac{\cos\varphi}{\sin\varphi}\right)^2, \quad \sinh^2 b = \left(\frac{1+k^2}{2k}\right)^2 - 1 = \left(\frac{k^2-1}{2k}\right)^2.$$

Hence $\cos\varphi = \tanh a$ and $(k^2-1)/2k = \sinh b$, and therefore, $\tan\alpha = \tanh a/\sinh b$.

For small a and b, this relation takes the form $\tan\alpha = a/b$. □

Applying trigonometric and hyperbolic identities to the relations

$$\cosh c = \cosh a\cosh b, \qquad \tanh a = \sinh b\tan\alpha,$$

we can obtain other relations, such as

$$\sinh a = \sinh c\sin\alpha, \qquad \tanh b = \tanh c\cos\alpha,$$
$$\cot\alpha\cot\beta = \cosh c, \qquad \cos\alpha = \cosh a\sin\beta.$$

For small a, b, and c, these relations become

$$a = c\sin\alpha, \qquad b = c\cos\alpha,$$
$$\cot\alpha\cot\beta = 1, \qquad \cos\alpha = \sin\beta,$$

respectively.

In hyperbolic geometry, theorems similar to the law of sines and law of cosines in Euclidean and spherical plane geometry are valid.

THEOREM 6. *For an arbitrary triangle, the following relations hold*:

(a) the law of sines :

$$\frac{\sinh a}{\sin\alpha} = \frac{\sinh b}{\sin\beta} = \frac{\sinh c}{\sin\gamma};$$

(b) the law of cosines :

$$\cosh a = \cosh b\cosh c - \sinh b\sinh c\cos\alpha.$$

Proof. Let us draw a perpendicular $[C,H]$ from the vertex C to the line AB. To be definite, we assume that the point H lies on the line segment $[A,B]$ (the case in which H lies outside AB is considered similarly). Let us denote the lengths of the segments $[C,H]$, $[A,H]$, and $[B,H]$ by h, x, and $c-x$, respectively.

(a) Writing the relations of the form $\sinh a = \sinh c\sin\alpha$ for the right triangles ACH and CBH, we obtain $\sinh b\sin\alpha = \sinh h = \sinh a\sin\beta$. Therefore,

$$\frac{\sinh a}{\sin\alpha} = \frac{\sinh b}{\sin\beta}.$$

(b) The relations of the form $\cosh a\cosh b = \cosh c$ for the triangles ACH and CBH are $\cosh b = \cosh x\cosh h$ and $\cosh a = \cosh(c-x)\cosh h$. Therefore,

$$\begin{aligned}\cosh a &= (\cosh c\cosh x - \sinh c\sinh x)\cosh h \\ &= \cosh b\cosh c - \cosh b\sinh c\tanh x.\end{aligned}$$

In addition, $\tanh x = \tanh b\cos\alpha$. □

Relation (a) implies the hyperbolic form of Thales' theorem: the base angles in an isosceles triangle are equal.

Now, let us describe a *circle* in the Poincaré model, i.e., the set of points equidistant from a given point. For the Poincaré upper half-plane model, the hyperbolic circle of radius r centered at (a, b) is determined by the equation

$$\cosh r = 1 + \frac{(a-x)^2 + (b-y)^2}{2by},$$

i.e., $(x-a)^2+(y-b\cosh r)^2 = b^2(\cosh^2 r - 1)$. Thus in the Poincaré upper half-plane model, the hyperbolic circle of radius r centered at (a, b) is the Euclidean circle of radius $b\sqrt{\cosh^2 r - 1}$ centered at $(a, b\cosh r)$.

For the Poincaré model in a disk Δ, a hyperbolic circle is also a Euclidean circle because linear-fractional transformations map circles into circles and lines. If the center of the hyperbolic circle coincides with the center of the disk Δ, then the Euclidean center of this circle also coincides with the center of the disk Δ. Let us show that in this situation, the Euclidean radius R is related to the hyperbolic radius r as

$$e^r = \frac{1+R}{1-R}, \qquad \text{i.e.,} \quad R = \frac{e^r - 1}{e^r + 1} = \tanh\left(\frac{r}{2}\right).$$

Indeed, if O is the center of the disk Δ, X is a point on the circle under consideration, and A and B are the points at which the line OX intersects the absolute, then

$$r = \left|\ln\left(\frac{|AX|}{|AO|} \cdot \frac{|BO|}{|BX|}\right)\right| = \ln\left(\frac{1+R}{1-R}\right).$$

Our next task is to measure the length of a curve in hyperbolic geometry. Evaluating the length of a curve by integration requires specifying a *length element*, i.e., the distance between infinitely close points. If Δs is the hyperbolic distance between points z and $z + \Delta z$ in the upper half-plane, then

$$\cosh(\Delta s) \approx 1 + \frac{|\Delta z|^2}{2(\operatorname{Im} z)^2} \qquad \text{and} \qquad \cosh(\Delta s) \approx 1 + \frac{(\Delta s)^2}{2}.$$

Therefore,

$$\Delta s \approx \frac{|\Delta z|}{\operatorname{Im} z}, \qquad \text{i.e.,} \quad ds = \frac{|dz|}{\operatorname{Im} z} = \frac{\sqrt{dx^2 + dy^2}}{y}.$$

To pass from the Poincaré model in the upper half-plane H to the Poincaré model in the unit disk Δ, we can use the map $z \mapsto w = (z-i)/(z+i)$. It transforms the upper half-plane H into the disk Δ. Simple calculations show that

$$\frac{|dz|}{\operatorname{Im} z} = \frac{2|dw|}{1 - |w|^2}.$$

Thus the hyperbolic length element in the Poincaré model in the disk Δ is obtained from the Euclidean length element by multiplying by $2/(1 - |w|^2)$. Therefore, the hyperbolic length of the Euclidean circle of radius R centered at the center of the disk Δ is $2\pi R\big(2/(1 - R^2)\big)$. Recall that this curve is a hyperbolic circle of radius r, where r and R are related by $R = \tanh(r/2)$. Hence the length of a hyperbolic circle of radius r equals

$$\frac{4\pi \tanh(r/2)}{1 - \tanh^2(r/2)} = 2\pi \sinh r.$$

REMARK 3. The length of a hyperbolic circle is greater and the length of a spherical circle is less than that of a Euclidean circle of the same radius.

REMARK 4. The law of sines can be stated uniformly in all the three geometries, namely,

$$\frac{l(a)}{\sin\alpha} = \frac{l(b)}{\sin\beta} = \frac{l(c)}{\sin\gamma},$$

where $l(r)$ is the length of a circle of radius r.

The area of a spherical triangle is given by the formula $S = R^2(\alpha+\beta+\gamma-\pi)$. Replacing R by ic, we obtain a formula for the area of a triangle in hyperbolic geometry, namely, $S = c^2(\pi - \alpha - \beta - \gamma)$. This argument can be made into a complete rigorous proof by using the theorem that two analytic functions coinciding at all real values must coincide at all purely imaginary values.

Another approach to the notion of area in hyperbolic geometry involves the so-called *defect* $\delta = \pi - (\alpha + \beta + \gamma)$ of a hyperbolic triangle. A convex n-gon can be cut by the diagonals into $n-2$ triangles. It is easy to see that the sum of their defects is $(n-2)\pi - \sum \alpha_i$, where $\alpha_1, \dots, \alpha_n$ are the angles of the n-gon. The value $\delta = (n-2)\pi - \sum \alpha_i$ is called the *defect* of a hyperbolic n-gon with angles $\alpha_1, \dots, \alpha_n$; the n-gon is not required to be convex. Instead of the interior angles α_i, it is more convenient to consider the exterior angles $\alpha_i' = \pi - \alpha_i$, because the defect of a polygon with exterior angles α_i' equals $(\sum \alpha_i') - 2\pi$.

The following properties of the defect are easy to verify:

(i) The defect of any polygon is positive.
(ii) If polygons M_1 and M_2 are congruent, then $\delta(M_1) = \delta(M_2)$.
(iii) If a polygon M is cut into polygons M_1 and M_2, then $\delta(M) = \delta(M_1)+\delta(M_2)$.

It is natural to define the area of a polygon as a function defined on the set of all polygons and possessing properties (i)–(iii). It can be proved that properties (i)–(iii) determine a function on the set of polygons uniquely up to proportionality (we do not give the proof). Therefore, the defect of a polygon equals its area up to proportionality. To find the proportionality factor k, it suffices to recall that for small a and b, the area of a right triangle with legs a and b must be close to $ab/2$. For such a triangle, we have

$$\begin{aligned}\sin\delta = \cos(\alpha+\beta) &= \frac{\tanh a \tanh b}{\tanh^2 c} - \frac{\sinh a \sinh b}{\sinh^2 c} \\ &= \frac{\sinh a \sinh b}{\sinh^2 c}(\cosh a \cosh b - 1) \approx \frac{ab}{c^2}\cdot\frac{a^2+b^2}{2} = \frac{ab}{2},\end{aligned}$$

where c is the hypotenuse. Hence $k = 1$. Recall that the parameter c of hyperbolic geometry is assumed to be 1. To avoid confusion with the hypotenuse, we temporarily denote the parameter by r. We have $\sin\delta = ab/2r^2$. Therefore, in the general case, $k = r^2$.

One more approach to the definition of the area of a figure in hyperbolic geometry consists in partitioning the figure into infinitesimal "rectangles" (see Problem 5.40).

5.3. Isometries in the three geometries

Euclidean space $\mathbb{R}^n$, the sphere $\mathbb{S}^n$, and the Lobachevsky space H^n are similar in that they all have very large groups of isometries (motions). Namely, any system

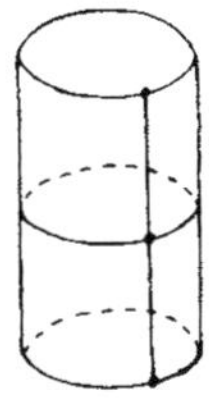

FIGURE 5.18

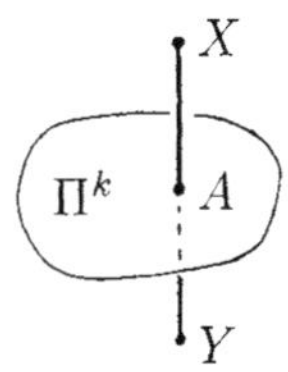

FIGURE 5.19

of n pairwise orthogonal lines meeting at one point can be isometrically transformed into any other system of n pairwise orthogonal lines meeting at one point. This means that the dimension of the group of motions is $n(n+1)/2$. It can be proved that the only spaces with isometry groups of dimension that large are $\mathbb{R}^n$, $\mathbb{S}^n$, $\mathbb{S}^n/\pm I$, and H^n; here $\mathbb{S}^n/\pm I$ is the sphere with antipodal points pairwise identified, i.e., $\mathbb{R}P^n$.

Geometrically, the requirement of maximality of the isometry group is quite natural; it makes all points and all directions equivalent. This means that stating a theorem, we do not have to specify points or directions. If points and directions were nonequivalent, we would have to state special theorems for each point and each direction.

It is easy to give an example of a space in which all points are equivalent (i.e., any point can be mapped to any other point by an isometry), but directions are not. The example is the infinite cylinder (Figure 5.18).

Isometries of Euclidean space. A very important example of an isometry of Euclidean space is a symmetry about a hyperplane Π. The definition of this transformation takes the simplest form when the coordinate system is selected so that the hyperplane Π is specified by the equation $x_1 = 0$. Then the symmetry with respect to Π maps a point $(x_1, \dots, x_n)$ to the point $(-x_1, x_2, \dots, x_n)$. The symmetry about the $(n-k)$-dimensional plane $x_1 = 0, \dots, x_k = 0$ is defined similarly. This transformation maps a point $(x_1, \dots, x_n)$ to $(-x_1, \dots, -x_k, x_{k+1}, \dots, x_n)$.

The symmetry about a k-dimensional plane Π^k can be given an invariant definition. Let us draw a perpendicular $[X, A]$ from a point X to Π^k and consider the point Y such that $|XY| = 2|XA|$ (Figure 5.19). Then Y is the image of X under the symmetry about the k-dimensional plane Π^k.

Any plane Π^k consists of precisely those points that remain fixed under the symmetry about this plane. In addition, if points X and Y are symmetric with respect to Π^k, then they are equidistant from each point of Π^k. This property completely describes Π^k only if $X \neq Y$ and Π^k is a hyperplane.

Next, let us find out what fixed point sets of isometries can look like. First, note that if A and B are distinct fixed points of an isometry, then the entire line AB consists of fixed points of this isometry. Indeed, let X be a point on the line AB. Then either $|AB| = |AX| + |BX|$, or $|AB| = |AX| - |BX|$, or $|AB| = |BX| - |AX|$. In any case, the spheres of radii $|AX|$ and $|BX|$ centered at A and B have exactly one common point X. It is also clear that the image of X under the isometry must belong to both spheres. Therefore, X is a fixed point of the isometry.

Suppose that the isometry has a fixed point C not belonging to AB. Then each point of the line CX, where X is a point on AB, is fixed. Thus all points of the plane ABC are fixed. A similar argument shows that the fixed point set of an isometry of an n-space is either the empty set, or a one-point set, or a k-dimensional plane, where $1 \leqslant k \leqslant n$. In particular, if an isometry of an n-dimensional space has $n+1$ fixed points not lying in one hyperplane, then all points in the space are fixed, i.e., the isometry is the identity transformation.

THEOREM 1. *Any isometry of the affine Euclidean space $\mathbb{R}^n$ can be represented as the composition of at most $n+1$ symmetries about hyperplanes.*

Proof. If an isometry leaves all points fixed, then it is the identity transformation and can be represented as a composition of two symmetries about the same hyperplane. Suppose that an isometry f maps some point A to a point $B \neq A$. Let s be the symmetry with respect to the hyperplane Π consisting of all points equidistant from A and B. The isometry sf that maps each point X to the point $s(f(X))$ has the following properties:

(i) $sf(A) = A$;

(ii) if $f(X) = X$, then $sf(X) = X$.

Property (i) is obvious, and (ii) can be proved as follows. If $f(X) = X$, then f maps the segment $[A, X]$ to $[B, X]$. Therefore, X is equidistant from A and B, and hence $s(X) = X$.

For an isometry f of $\mathbb{R}^n$, we choose points $A_1, \dots, A_{n+1}$ not lying in one hyperplane and construct an isometry $s_k \cdots s_1 f$ for which all these points are fixed. The construction involves $n+1$ steps (one step was described above). Here $k \leqslant n+1$, because multiplication by a symmetry s is not needed if the point A_{i+1} is fixed for the isometry constructed at the preceding step.

Since $s_k \cdots s_1 f = \mathrm{id}$ (the identity isometry), we have $f = s_1 \cdots s_k$ because of the property $s_i^2 = \mathrm{id}$. □

Now, let us discuss one fairly exotic property related to the definition of isometry in $\mathbb{R}^n$ for $n \geqslant 2$. The point is that for a transformation to be an isometry, one does not have to require that it preserves all the distances between points. It is sufficient to require that it preserves only one distance a (i.e., maps any two points at the distance a to points that are also at the distance a). This property already implies that the transformation preserves all distances. First, we prove this in the most interesting case of $n = 2$.

THEOREM 2. *Suppose that a map $f\colon \mathbb{R}^2 \to \mathbb{R}^2$ (not necessarily one-to-one) preserves a distance $a > 0$, i.e., $|X_1Y_1| = a$ whenever $X_1 = f(X)$ and $Y_1 = f(Y)$ for X and Y with $|XY| = a$. Then f is an isometry.*

Proof [**BQ**]. We assume that $a = 1$ (otherwise, we can apply a homothety). The proof involves several steps. We denote the image of a point by the same letter with subscript 1.

Step 1. If $\sqrt{3} - 1 < XY < \sqrt{3} + 1$, then $X_1 \neq Y_1$.

Consider a rhombus $XACB$ with side 1 and diagonals $|AB| = 1$ and $|XC| = \sqrt{3}$ (Figure 5.20). Let S be the circle of radius 1 centered at C. The distances from X to the points of S range from $\sqrt{3} - 1$ to $\sqrt{3} + 1$; therefore, we can assume that Y lies on S. Suppose that $X_1 = Y_1$. Then the points A_1, B_1, and C_1 lie on the circle of radius 1 centered at $X_1 = Y_1$, and the pairwise distances between these points are 1. We have arrived at a contradiction.

Step 2. The map f preserves the distance $\sqrt{3}$.

Let $|XY| = \sqrt{3}$. Consider a rhombus $XAYB$ composed of two equilateral triangles with side 1 (Figure 5.21). Either the points X_1, A_1, Y_1, and B_1 form a

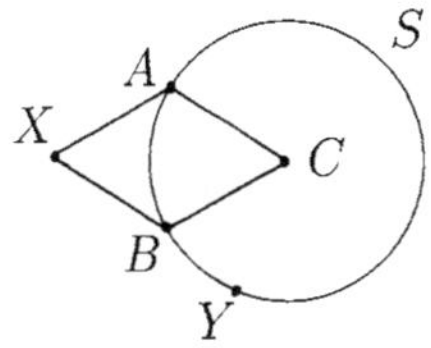

FIGURE 5.20

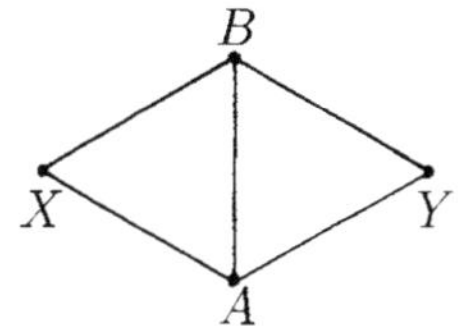

FIGURE 5.21

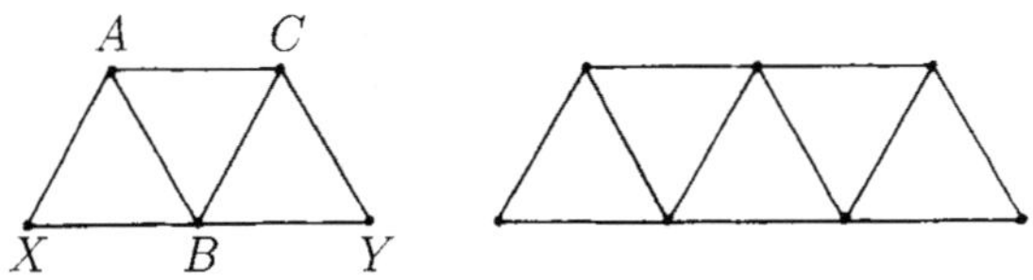

(a) (b)

FIGURE 5.22

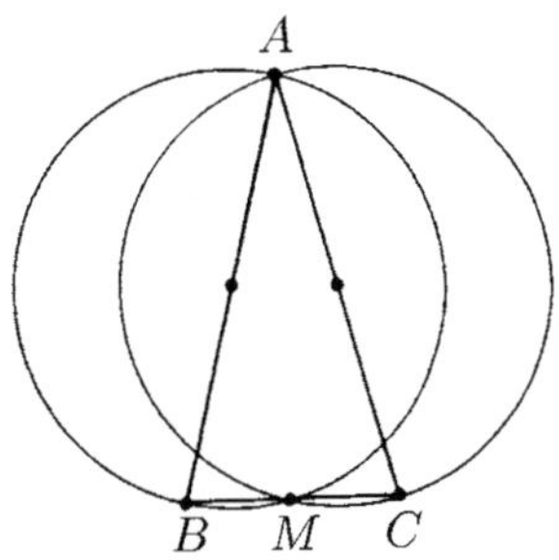

FIGURE 5.23

similar rhombus or $X_1 = Y_1$. The latter cannot happen because $\sqrt{3} - 1 < |XY| < \sqrt{3} + 1$.

Step 3. The map f preserves all positive integer distances n.

Consider rhombi $XACB$ and $ACYB$ composed of pairs of equilateral triangles with side 1 (Figure 5.22 (a)). The points X_1 and Y_1 do not coincide, because $|A_1Y_1| = \sqrt{3} \neq |A_1X_1|$. Therefore, the vertices of the rhombi under consideration are mapped to vertices of like rhombi, and hence $|X_1Y_1| = 2$.

Considering the chain of rhombi shown in Figure 5.22 (b), we see that f preserves the distance 3, etc.

Step 4. The map f preserves the distances $n2^{-m}$, where m and n are positive integers.

Consider an isosceles triangle ABC with sides $2n$, $2n$, and n. Let us construct the circles for which the lateral sides serve as diameters (Figure 5.23). The midpoint M of the shorter side BC coincides with the intersection point of these circles; therefore, the point M_1 coincides with either the point A_1 or the midpoint of $[B_1, C_1]$.

Following the arguments similar to those used at step 1, we find that if

$$n(\sqrt{3} - 1) \leqslant |AM| \leqslant n(\sqrt{3} + 1),$$

then $M_1 \neq A_1$. In the case under consideration, $|AM| = n\sqrt{15}/2$; hence $M_1 \neq A_1$. Therefore, M_1 is the midpoint of $[B_1, C_1]$. Thus f preserves the distance $n/2$. Similarly, it preserves the distances $n/2^m$.

Step 5. The distance between arbitrary points $X, Y \in \mathbb{R}^2$ is preserved.

Let us choose a sequence Y^i of points such that their distances to X are of the form $n/2^m$ and converge to Y. We have $|X_1Y_1^i| = |XY^i|$, and the points Y_1^i converge to Y_1. To prove the last assertion, it suffices to note that if $|YY^i| \leqslant 2^{-k}$, then the points Y and Y^i belong to some circle of radius 2^{-k-1}, and therefore $|Y_1Y_1^i| \leqslant 2^{-k}$. As the result, we obtain

$$|X_1Y_1| = \lim_{i\to\infty} |X_1Y_1^i| = \lim_{i\to\infty} |XY^i| = |XY|.$$

This completes the proof of Theorem 2. □

Let us outline the proof of a similar theorem for a larger dimension $n > 2$. Instead of rhombi composed of two equilateral triangles with side 1, we consider figures composed of two regular n-simplices with edge 1. As in the case of $n = 2$, we obtain the following assertions.

Step 1. If $\sqrt{\frac{2(n+1)}{n}} - 1 < |XY| < \sqrt{\frac{2(n+1)}{n}} + 1$, then $X_1 \neq X_2$.

Step 2. The map f preserves the distance $\sqrt{2(n+1)/n}$.

The next step differs from the corresponding step in Theorem 2.

Step 3. The map f takes a unit circle to a unit circle, and the center of the preimage circle is mapped to the center of the image circle.

For $n = 3$, a unit circle can be represented as the intersection of two spheres of radius $2\sqrt{6}/3$ centered at a distance $2\sqrt{15}/3$ apart. According to the assertion of step 2, these spheres are transformed into spheres (of the same radius), and according to the assertion of step 1, the images of their centers do not coincide. Therefore, the unit circle transforms into a circle. Now, considering the regular hexagon inscribed in the circle, we can easily show that the image circle is of unit radius, and its center is the image of the center of the preimage circle.

For $n > 3$, we take a regular $(n-3)$-simplex with edge s and consider the intersection of the $(n-1)$-spheres of radius s centered at the vertices of this simplex. It is easy to show that the intersection is a 2-sphere of radius $\rho = s\sqrt{(n-1)/2(n-2)}$. Thus, if a map f preserves the distance s, then it transforms a 2-sphere of radius ρ into a 2-sphere of radius ρ. But a unit circle can be represented as the intersection of two 2-spheres of radius

$$\rho = s\sqrt{\frac{n-1}{2(n-2)}}, \qquad \text{where} \quad s = \sqrt{\frac{2(n+1)}{n}}.$$

Step 4. Planes are mapped to planes.

Take the unit circle $\mathbb{S}^1$ centered at O in the plane under consideration. Let $\mathbb{S}^1_A$ be the unit circle centered at $A \in \mathbb{S}^1$ in this plane. Clearly, both these circles are mapped onto two coplanar unit circles. As the point A moves along $\mathbb{S}^1$, the circle $\mathbb{S}^1_A$ sweeps out the disk of radius 2 centered at O. The images of all points of this disk lie in one plane. Performing the same construction for all circles inside the disk, we obtain a disk of radius 3 whose image lies in the same plane, etc.

There exist other sufficient conditions for a map to be an isometry. Say, for spaces of dimension no lower than 3, the preservation of a particular area of triangles is sufficient.

THEOREM 3. (a) *If a map* $f\colon \mathbb{R}^2 \to \mathbb{R}^2$ *takes all triangles of area* 1 *to triangles of area* 1, *then* f *is an affine transformation.*

(b) *If a map* $f\colon \mathbb{R}^n \to \mathbb{R}^n$, *where* $n \geqslant 3$, *takes all triangles of area* 1 *to triangles of area* 1, *then* f *is an isometry.*

Proof. (a) [**L1**] For given points A and B, the set of all points X such that the triangle ABX has area 1 is a pair of parallel lines. The map f takes this pair to a pair of parallel lines. Clearly, any pair of parallel lines can be represented as the set of points X for which the area of the triangle ABX (for some fixed points A and B) is 1.

Next, let us verify that any straight line is mapped to a straight line. Consider a pair of parallel lines and take their segments $[P_1, P_2]$ and $[Q_1, Q_2]$ of equal length a. Let P and Q be the midpoints of these segments. There exist twelve triangles of equal area with vertices at six given points. Varying a, we can make their areas equal 1. Suppose that the images of the points P and Q do not lie on different parallel lines. Then we cannot choose images of the points P_1, P_2, Q_1, Q_2 on these lines so that the images of the twelve triangles defined above are of area 1. Indeed, all these points must lie on one line, and all the four segments $[P_i', Q_j']$ (the prime denotes the image under f) must have equal lengths.

Thus the map f transforms a straight line to a straight line, and hence f is an affine transformation (see p. 39).

(b) [**L2**] First, note that images of distinct points under f are distinct.

The squared area of the triangle whose sides are the vectors a and b is equal to $\frac{1}{4}(|a|^2|b|^2 - (a,b)^2)$; therefore, the area of such a triangle is 1 if and only if

$$|a|^2|b|^2 - (a,b)^2 = 4. \tag{1}$$

Take a three-dimensional subspace in $\mathbb{R}^n$ and consider the points

$$\begin{aligned} O &= (0,0,0), \quad A = (1,1,0), \quad B(1,-1,0), \\ P &= (0,1,\lambda), \quad Q = (1,0,\lambda), \quad R(0,-1,\lambda) \end{aligned}$$

in this subspace. Using (1), it is easy to verify that the areas of the triangles OAP, OAQ, OAR, OBP, OBQ, OBR equal 1 if $2(1+\lambda^2) - 1 = 4$, and the areas of the triangles OPQ, ORQ, APQ, ARQ, BPQ, BRQ equal 1 if $(1+\lambda^2)^2 - \lambda^2 = 4$. Both these equalities hold for $\lambda = \sqrt{3/2}$. The configuration of the points O, A, B, P, Q, R must then transform under f into a configuration of points in which the areas of the images of the twelve triangles and of triangle ABO equal 1.

LEMMA. *Let points A, B, C, D be such that the areas of the triangles ABC, ABD, ACD, BCD are equal to 1. Then either $ABCD$ is a tetrahedron with equal opposite edges or A, B, C, D are vertices of a parallelogram.*

Proof. First, suppose that the points under consideration lie in one plane. To be definite, let A be a vertex of the convex hull of these points, and let AB and AD be the edges incident to A. Then $ABCD$ is a parallelogram, because $DC \parallel AB$ and $BC \parallel AD$.

Now, suppose that the points A, B, C, D do not lie in one plane, i.e., $ABCD$ is a nondegenerate tetrahedron. Through each edge, we draw a plane parallel to the opposite edge. These planes form a parallelepiped $ABCDA_1B_1C_1D_1$. Changing notation, we can say that the vertices of the tetrahedron under consideration coincide with the vertices A, B_1, C, D_1 of the parallelepiped $ABCDA_1B_1C_1D_1$. Consider the projection of this parallelepiped on the plane perpendicular to the line AC. Since the triangles ACB_1 and ACD_1 have equal areas, their altitudes to AC are equal. Therefore, the projection of the tetrahedron is an isosceles triangle, and the point A_1 is projected to the midpoint of its base. Hence $AA_1 \perp B_1D_1$.

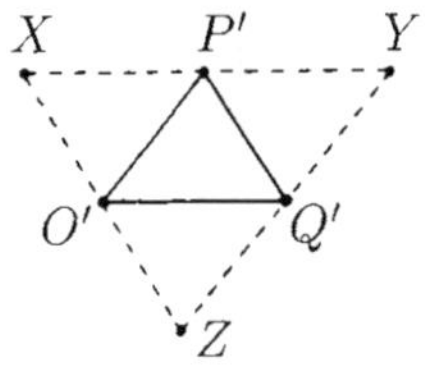

FIGURE 5.24

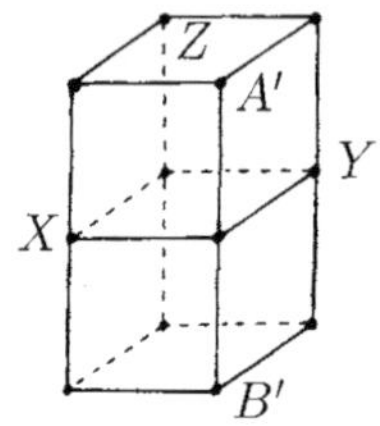

FIGURE 5.25

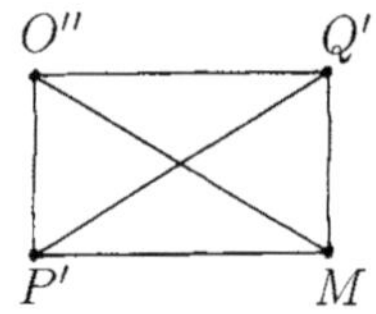

FIGURE 5.26

Similarly, $DD_1 \perp A_1C_1$. Thus the edge AA_1 is perpendicular to the face $ABCD$, i.e., $AA_1 \perp AB$ and $AA_1 \perp AD$. Similarly, $AB \perp AD$.

Thus the parallelepiped corresponding to a tetrahedron with faces of equal areas is rectangular. The equality of the opposite faces in the tetrahedron follows from the equality of the diagonals in a rectangle. □

In the configuration under consideration, the lemma applies to the tetrahedra $OAPQ$ and $OBPQ$ sharing the face OPQ and to the tetrahedra $OARQ$ and $OBRQ$ sharing the face ORQ. If the images of the tetrahedra in one of these pairs are nondegenerate, then either $|O'A'| = |P'Q'| = |O'B'|$ or $|O'A'| = |R'Q'| = |O'B'|$ (here $O' = f(O)$, etc.).

Let us show that $|O'A'|$ equals $|O'B'|$ even if some tetrahedra are degenerate. It is sufficient to consider the case in which both pairs $O'A'P'Q'$, $O'B'P'Q'$ and $O'A'R'Q'$, $O'B'R'Q'$ contain degenerate tetrahedra. Suppose that both tetrahedra $O'A'P'Q'$ and $O'B'P'Q'$ in the first pair are degenerate. According to the lemma, the points A' and B' must coincide with the vertices of a triangle XYZ with sides parallel to the sides of the triangle $O'P'Q'$ (Figure 5.24). But then the area of the triangle $O'A'B'$ equals 0 or 2, while it must be 1. This contradiction shows that each of the pairs $O'A'P'Q'$, $O'B'P'Q'$ and $O'A'R'Q$, $O'B'R'Q'$ contains one degenerate and one nondegenerate tetrahedron. If the tetrahedron $O'A'P'Q'$ is nondegenerate and the tetrahedron $O'B'P'R'$ is degenerate, then the configuration of the points O', A', B', P', Q' is as shown in Figure 5.25. We then have $\{X, Y, Z\} = \{O', P', Q'\}$. If $X = O'$ or $Y = O'$, then $|O'A'| = |O'B'|$. Consider the case $Z = O'$. Let Π be the plane which is perpendicular to the line segment $[A', B']$ and passes through its midpoint M, and let O'' be the projection of O' on the plane Π (Figure 5.26). Considering the pair of tetrahedra $O'A'R'Q'$ and $O'B'R'Q'$, we obtain a similar configuration for the points M, Q', O'', R'. But we then have $R' = P'$, which cannot be. As a result, we obtain $|O'A'| = |O'B'|$.

Thus if $|OA| = |OB| = \sqrt{2}$ and $|AB| = 2$, then $|O'A'| = |O'B'|$. This implies that if $|OA_1| = |OA_2| = \sqrt{2}$, then $|O'A_1'| = |O'A_2'|$. Indeed, let us draw a common perpendicular to the lines OA_1 and OA_2 and take its segment $[O, B]$ of length $\sqrt{2}$. We have $|O'A_1'| = |O'B'| = |O'A_2'|$. Next, let $[A, B]$ and $[C, D]$ be arbitrary segments of length $\sqrt{2}$. The points B and C can be joined by a polygonal line all of whose edges are of length $\sqrt{2}$; therefore, $|A'B'| = |C'D'|$.

Thus the map f takes any segment of length $\sqrt{2}$ to a segment of length $\lambda\sqrt{2}$, where λ is a constant. Hence the map $\lambda^{-1}f$ preserves the distance $\sqrt{2}$ and is therefore an isometry. Since f preserves the unit area of triangles, we have $\lambda = 1$, and f is an isometry itself. □

Isometries of the sphere. Recall that the distance between points A and B on the sphere $\mathbb{S}^2$ is defined as the length of the shorter arc of the great circle that joins A with B. Note that the spherical distance between points A and B is a monotonically increasing function of the Euclidean distance between them. In particular, spherical distances are equal if and only if the corresponding Euclidean distances are equal.

On the n-sphere, the equality of spherical distances between points is also equivalent to the equality of Euclidean distances.

THEOREM 4. *The isometries of the sphere $\mathbb{S}^n$ are in a one-to-one correspondence with the isometries of $\mathbb{R}^{n+1}$ that leave the center of the sphere fixed.*

Proof. Clearly, any isometry of $\mathbb{R}^{n+1}$ leaving the center of the sphere $\mathbb{S}^n$ fixed transforms $\mathbb{S}^n$ into itself and preserves spherical distances. Therefore, we only need to show that any isometry g of the sphere $\mathbb{S}^n$ can be extended to an isometry $\tilde{g}$ of the space $\mathbb{R}^{n+1}$. For convenience, we shall assume that the sphere has radius 1 and is centered at the origin. Any vector in $\mathbb{R}^{n+1}$ can be represented as tv, where $t > 0$ and $|v| = 1$. Let us define $\tilde{g}(tv) = tg(v)$. We must verify that $\tilde{g}$ is an isometry, i.e., $|\tilde{g}(su) - \tilde{g}(tv)| = |su - tv|$. Clearly,

$$|sg(u) - tg(v)|^2 - |su - tv|^2 = s^2 + t^2 - 2st(g(u), g(v)) - s^2 - t^2 + 2st(u, v)$$
$$= 2st((u, v) - (g(u), g(v))) = 0,$$

because $(u, v) = \cos\varphi$, where φ is the angle between u and v. □

THEOREM 5. *Any isometry of the sphere $\mathbb{S}^n$ can be represented as the composition of no more than $n + 1$ symmetries about sections of the sphere by hyperplanes passing through the center of the sphere.*

Proof. An isometry of $\mathbb{S}^n$ corresponds to an isometry of $\mathbb{R}^{n+1}$ with the fixed point O, where O is the center of the sphere. The arguments used in the proof of the theorem about the isometries of Euclidean spaces (see Theorem 1 on p. 106) shows that this isometry can be represented as a composition of no more than $n+1$ symmetries about hyperplanes, and all these hyperplanes pass through O. (The number of symmetries is reduced by 1 because we have one fixed point already.) □

Using Theorem 4, we can easily verify that the sphere $\mathbb{S}^n$ has a large isometry group, i.e., any system of n pairwise disjoint orthogonal spherical lines meeting at one point can be mapped to any other such system of lines by an isometry.

It is easy to show that the sphere $\mathbb{S}^2$ is not isometric to the plane $\mathbb{R}^2$, i.e., spherical geometry is not identical to Euclidean geometry. Indeed, distances between points on the plane can be arbitrarily large, while distances between points on the sphere cannot exceed $2\pi R$, where R is the radius of the sphere.

A less trivial question is whether or not the sphere and the plane are locally isomorphic, i.e., whether a domain on the sphere can be isometric to some domain on the plane. An equivalent question is: Is it possible to draw a plane map of some domain on the sphere without distortions, i.e., so that the distance between any pair of points on the sphere equals the distance between the corresponding points on the map?

THEOREM 6. *No domain on the sphere is isometric to a domain on the plane.*

Proof. We use the term *spherical circle* for the set of points on the sphere at a given distance r (the *radius* of the circle) from a given point on the sphere (the *center* of the circle). If a line segment $[O, X]$ subtends an angle of α at the center of the sphere, then the spherical circle centered at O and passing through X has spherical radius $R\alpha$, while the Euclidean radius of this circle is $R\sin\alpha$. Thus the length of a spherical circle of radius $r = R\alpha$ is

$$2\pi R\sin\alpha = 2\pi R\sin(r/R) < 2\pi R(r/R) = 2\pi r.$$

Any domain on the sphere includes a spherical circle of a sufficiently small radius r. The circumference of this spherical circle is less than that of a Euclidean circle of radius r, while an isometry must transform any circle into a circle of the same radius and circumference. □

REMARK. Instead of the circumference, we could consider the side length of an equilateral triangle inscribed in a circle of radius r.

An isometry $f\colon \mathbb{S}^n \to \mathbb{S}^n$ of the sphere is called a *Clifford translation* if the distance between x and $f(x)$ is constant. (In the Euclidean space, this property is characteristic of parallel translations.) For spheres of arbitrary dimensions, the transformations $f(x) = x$ and $f(x) = -x$ are Clifford translations. Other, nontrivial, Clifford translations only live on spheres of odd dimensions. Recall that the matrix of any orthogonal transformation over the field $\mathbb{C}$ can be reduced to the form $(\lambda_1, \bar\lambda_1, \dots, \lambda_k, \bar\lambda_k, \pm 1, \dots, \pm 1)$, where $|\lambda_i| = 1$ and $\lambda_i \neq \pm 1$, by a change of basis. Over the field $\mathbb{R}$, instead of each pair $\lambda, \bar\lambda$, the matrix

$$\begin{pmatrix} \cos\alpha & \sin\alpha \\ -\sin\alpha & \cos\alpha \end{pmatrix}, \qquad \text{where} \quad \lambda = \cos\alpha + i\sin\alpha,$$

should be taken.

THEOREM 7. *An isometry $f\colon \mathbb{S}^n \to \mathbb{S}^n$ is a nontrivial Clifford translation if and only if the corresponding isometry of $\mathbb{R}^{n+1}$ reduces to the form $(\lambda, \bar\lambda, \lambda, \bar\lambda, \dots, \lambda, \bar\lambda)$, where $|\lambda| = 1$ and $\lambda \neq \pm 1$.*

Proof. Let us choose an orthogonal coordinate system

$$(x_1, y_1, \dots, x_k, y_k, z_1, \dots, z_s)$$

in $\mathbb{R}^{n+1}$ so that $f(x, y, z) = (x', y', z')$, where

$$x_i' = x_i\cos\alpha_i + y_i\sin\alpha_i, \qquad y_i' = -x_i\sin\alpha_i + y_i\cos\alpha_i, \qquad z_i' = \pm z_i.$$

It is easy to verify that the squared Euclidean distance between (x, y, z) and $f(x, y, z)$ equals

$$2\Big(\sum_{i=1}^{k}(x_i^2 + y_i^2)(1 - \cos\alpha_i) + \sum_{j=1}^{s}\varepsilon_j z_j^2\Big),$$

where $\varepsilon_j = 0$ or 1. The squared Euclidean distance must be constant for all points whose coordinates satisfy

$$\sum_{i=1}^{k}(x_i^2 + y_i^2) + \sum_{j=1}^{s} z_j^2 = 1.$$

This is only possible in the following cases:

(i) $k = 0$ and $\varepsilon_1 = \cdots = \varepsilon_s = \pm 1$;

(ii) $s = 0$ and $\cos\alpha_1 = \cdots = \cos\alpha_k$.

Case (i) corresponds to the trivial Clifford translations, and in (ii), the matrix of the transformation has the form indicated in the theorem. □

The sphere $\mathbb{S}^3$ is not locally isometric to the Euclidean space $\mathbb{R}^3$. But the sphere $\mathbb{S}^3$ contains a two-dimensional surface locally isometric to the Euclidean plane $\mathbb{R}^2$. An example of such a surface in the sphere

$$\mathbb{S}^3 = \{ (z, w) \in \mathbb{C}^2 \mid |z|^2 + |w|^2 = 2 \}$$

is the surface T^2 defined by $|z| = |w|$. This surface consists of the points $(e^{i\varphi}, e^{i\psi})$. Let us show that the local coordinates (φ, ψ) on the surface T^2 are Euclidean, i.e., the length of the shortest curve joining points $(e^{i\varphi}, e^{i\psi})$ and $(e^{i(\varphi+\Delta\varphi)}, e^{i(\psi+\Delta\psi)})$ on T^2 is $\sqrt{(\Delta\varphi)^2 + (\Delta\psi)^2}$ (the length of a curve on $T^2 \subset \mathbb{S}^3 \subset \mathbb{C}^2 \cong \mathbb{R}^4$ is defined as the length of this curve in $\mathbb{R}^4$). It is sufficient to verify that if $z = e^{i\varphi}$ and $w = e^{i\psi}$, then $|dz|^2 + |dw|^2 = (d\varphi)^2 + (d\psi)^2$. Clearly, $dz = ie^{i\varphi}d\varphi$ and $dw = ie^{i\psi}\,d\psi$. Therefore, $|dz|^2 + |dw|^2 = (d\varphi)^2 + (d\psi)^2$.

Three types of proper motions of the Lobachevsky plane. In Lobachevsky geometry, the following theorem is valid (we have already proved its analogs in Euclidean and spherical geometries).

THEOREM 8. *An arbitrary motion of the Lobachevsky plane can be represented as the composition of no more than three symmetries about lines.*

Proof. Recall the basic idea of the proof of this theorem in Euclidean geometry. Let g be an isometry. Suppose that $g(A) = B \neq A$. Consider the symmetry s about the line $l = \{ X \mid |AX| = |BX| \}$. We have $sg(A) = A$. In addition, if $g(Y) = Y$, then $Y \in l$, and hence $sg(Y) = Y$. Therefore, the isometry sg has more fixed points than g by at least one. Repeating this procedure no more than three times, we obtain an isometry with three noncollinear fixed points.

To apply this scheme of proof to Lobachevsky geometry, we must answer the following questions:

(i) Why the set of points equidistant from two given points is a hyperbolic line?
(ii) What is a symmetry about a line in hyperbolic geometry?
(iii) Why a motion of the hyperbolic plane with three noncollinear fixed points is the identity map?

Let us answer all these questions consecutively.

(i) We can assume that the given points are ia and ib ($a, b \in \mathbb{R}$). Then the set of points $(x, y) \in H$ equidistant from them has the equation

$$\frac{x^2 + (a-y)^2}{ay} = \frac{x^2 + (b-y)^2}{by},$$

i.e., $x^2 + y^2 = ab$.

(ii) We say that points A and A^* are *symmetric* (or *inverse*) with respect to a circle S if any circle through A and A^* is orthogonal to S. For a line S, this definition also makes sense; it is then equivalent to the usual definition of symmetry about the line.

The linear-fractional transformations preserve the relation of symmetry about a circle.

It can be verified that points A and A^* are symmetric with respect to a circle S if and only if the rays OA and OA^* coincide and $|OA| \cdot |OA^*| = R^2$, where R is the radius of the circle S.

We say that points A and A^* are *symmetric* with respect to a hyperbolic line in the Poincaré model if they are symmetric with respect to the Euclidean circle that

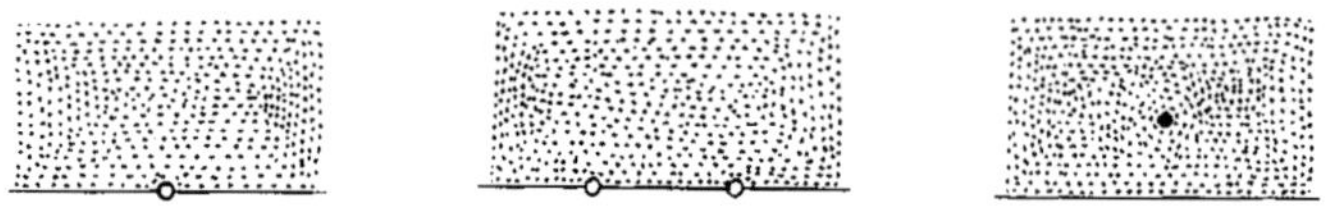

$D = 0$ $D > 0$ $D < 0$

FIGURE 5.27

contains this line. If the hyperbolic line is a part of a Euclidean line, then, as can easily be verified, any two points symmetric with respect to this line are equidistant from it. In addition, any symmetry about a hyperbolic line is an isometry.

(iii) We answer the third question in more detail than is strictly necessary. Namely, we shall describe all possible fixed point sets of motions of the Lobachevsky plane. We consider proper and improper motions separately.

(a) The equation $z = (a(-\bar{z}) + b)\,(c(-\bar{z}) + d)$ reduces to the form

$$-cz\bar{z} + dz + a\bar{z} - b = 0,$$

i.e., to the system

$$-c(x^2 + y^2) + dx + ax - b = 0, \quad dy - ay = 0.$$

If $a \neq d$, then the last equation has only one solution $y = 0$, and in this case $z \notin H$. If $a = d$, then the fixed point set is a hyperbolic line.

(b) The equation $z = (az + b)(cz + d)$ is quadratic. Its roots are

$$z_{1,2} = \frac{a - d \pm \sqrt{(d-a)^2 + 4bc}}{2c}.$$

We can assume that $ad - bc = 1$; then $D = (d-a)^2 + 4bc = (a+d)^2 - 4$, and

$$D = 0 \iff |a + d| = 2 \text{ (a \textit{parabolic} motion)},$$
$$D > 0 \iff |a + d| > 2 \text{ (a \textit{hyperbolic} motion)},$$
$$D < 0 \iff |a + d| < 2 \text{ (an \textit{elliptic} motion)}.$$

The fixed points of the transformation $z \mapsto (az+b)(cz+d)$ depend on the sign of D; they are shown in Figure 5.27. Note that only one fixed point (for $D < 0$) among all those shown in Figure 5.27 is a fixed point of a motion of the Lobachevsky plane; all other fixed points lie outside the upper half-plane H.

We have answered all three questions and have thereby completed the proof of the theorem. □

Let us consider proper motions of the Lobachevsky plane in more detail. Any proper motion can be represented as the composition of two symmetries about lines. In the Lobachevsky plane, there are three types of pairs of lines: two lines may be parallel, disjoint (and nonparallel), or intersecting. Let us show that compositions of symmetries about such pairs are parabolic, hyperbolic, or elliptic motions, respectively.

First, let us find the simplest arrangement of a pair of lines that can be achieved by means of motions.

1. For a pair of intersecting lines, it is convenient to use the Poincaré disk model. We can assume that the center of the disk is the intersection point of the lines. Then the lines under consideration are diameters of the disk. The composition of symmetries about these lines is a rotation about the center. This transformation has the form $z \mapsto e^{i\varphi} z$.

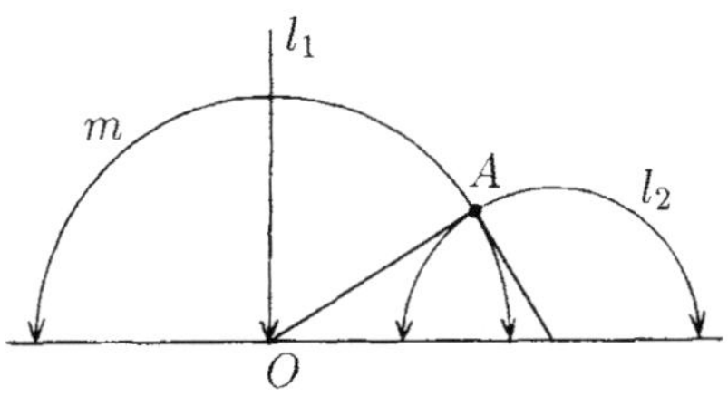

FIGURE 5.28

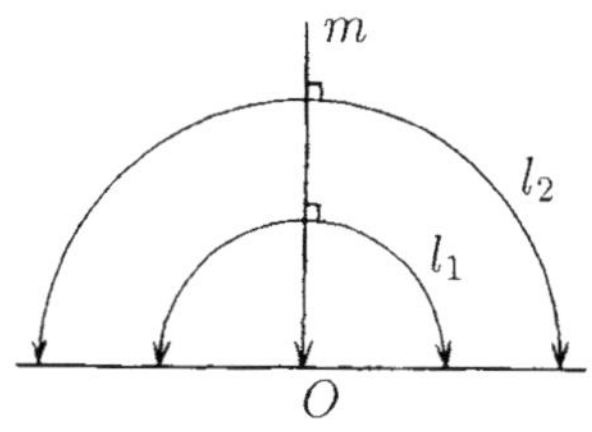

FIGURE 5.29

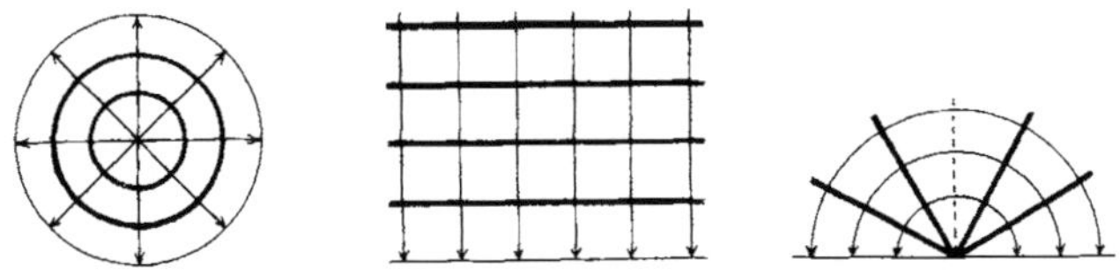

FIGURE 5.30

2. For a pair of parallel lines, the Poincaré upper half-plane model is more convenient. We can assume that the lines converge to the point ∞, i.e., they are rays parallel to the Oy axis. The composition of symmetries about these lines has the form $z \mapsto z + h$.

3. We call disjoint nonparallel lines *divergent* or *superparallel*. First, let us show that divergent lines l_1 and l_2 have a common perpendicular m. To this end, consider the Poincaré upper half-plane model. We can assume that the line l_1 is the ray Oy (Figure 5.28). Let us draw a tangent line OA from the point O to the Euclidean circle that contains the hyperbolic line l_2. It is easy to verify that the hyperbolic line m corresponding to the Euclidean circle of radius $|OA|$ centered at O is perpendicular to l_1 and l_2. The line m can be transformed into the ray Oy. Then the lines l_1 and l_2 are transformed into semicircles centered at O (Figure 5.29). The composition of the inversions with respect to the circles containing them has the form $z \mapsto kz$, where $k > 0$.

A proper motion can be represented as a composition of symmetries about lines l_1 and l_2 in many different ways. Given a proper motion, consider the set of all corresponding lines l_i. Such a set is called a *pencil of lines*. According to the type of the motion, the pencil is said to be *elliptic*, *parabolic* or *hyperbolic*. An elliptic pencil consists of lines passing through a fixed point; a parabolic pencil consists of pairwise parallel lines (i.e., of lines passing through a fixed point at infinity); and a hyperbolic pencil consists of lines orthogonal to a fixed line.

To each pencil of lines, we can assign the family of curves orthogonal to all lines in the pencil. This family is obtained as follows. Consider all the motions g which are compositions of symmetries about pairs of lines in the pencil. Take an arbitrary point A in the Lobachevsky plane and consider the set of all points of the form $g(A)$. Figure 5.30 shows the curves obtained for all three types of pencils of lines (the lines of the pencil are labeled by arrows). We use the Poincaré disk model for the elliptic pencil and the Poincaré upper half-plane model for the two other pencils.

The curves orthogonal to lines of an elliptic pencil are circles. The orthogonal curves for a parabolic pencil are called *horocycles* or *limit circles*; the latter term is due to the observation that each of these curves can be obtained by infinitely increasing the radius of the circle tangent to a given line at a given point. The orthogonal curves for a hyperbolic pencil are called *hypercycles* or *equidistant curves* because all points of an equidistant curve are at the same distance from the line to which the lines of the hyperbolic pencil are orthogonal. The term equidistant curve (hypercycle) is largely used to denote a set of points equidistant from a given line; this is a curve with two branches symmetric with respect to the given line.

Now we describe the proper motions in the model of Lobachevsky geometry on the upper sheet of the two-sheeted hyperboloid

$$[x,x] = -c^2, \qquad \text{where} \quad [x,x] = x_1^2 + x_2^2 - x_3^2.$$

First note that any linear transformation that preserves the pseudoscalar products $[x,y]$ and does not interchange the upper and lower sheets of the hyperboloid induces a motion of the Lobachevsky plane.

An arbitrary plane can be specified by the equation $[x,a] = p$, where a is a fixed vector and p is a fixed number. We define the *symmetry* about the plane $[x,a] = 0$ by setting $x \mapsto x' = x + \lambda a$, where λ is a number such that the vector $x + x'$ lies in the plane of symmetry, i.e., $[x + x', a] = 0$; thus this symmetry is given by $x \mapsto x - 2([x,a]/[a,a])a$. This definition makes sense if $[a,a] \neq 0$.

The planes tangent to the cone $[x,x] = 0$ are determined by equations of the form $[x,a] = 0$, where $[a,a] = 0$. These are the planes with respect to which symmetry is not defined. All other planes passing through the origin either intersect the two-sheeted hyperboloid $[x,x] = -c^2$ or they do not intersect the hyperboloid and are not tangent to the cone $[x,x] = 0$. The lines of the Lobachevsky plane only correspond to planes of the first kind. The symmetries about these planes do not interchange the sheets of the hyperboloid because they leave the intersection points of the plane with the hyperboloid fixed.

Simple calculations show that if $[a,a] \neq 0$, then the symmetry about the plane $[x,a] = 0$ preserves the pseudoscalar product; i.e., if $x \mapsto x'$ and $y \mapsto y'$, then we have $[x,y] = [x',y']$. Thus a symmetry about a plane intersecting the hyperboloid corresponds to a symmetry about a line in the Lobachevsky plane. The proper motion of the Lobachevsky plane corresponding to a composition of symmetries about a pair of planes is elliptic, parabolic, or hyperbolic depending on the position of the intersection line of the planes (namely, on whether the line intersects the hyperboloid, belongs to the cone, or lies outside the cone).

Suppose that the planes $[x,a_1] = 0$ and $[x,a_2] = 0$ intersect along a line containing some vector b, i.e., $[b,a_1] = [b,a_2] = 0$. Then the plane $[x,b] = p$ is mapped onto itself by the symmetries about these planes. Indeed, these symmetries take a vector x to $x' = x + \lambda a_i$; hence $[x',b] = [x,b]$. Therefore, the composition of the symmetries about the planes $[x,a_1] = 0$ and $[x,a_2] = 0$ maps the plane $[x,b] = p$ onto itself. This means that the circles, horocycles, and hypercycles are the sections of the hyperboloid by the planes $[x,b] = p$, where b ranges over the vectors inside the cone, on the cone, and outside the cone, respectively.

Problems

5.1. Prove that the symmetry about the hyperplane $(x,a) = 0$ is given by

$$x \mapsto x - 2\frac{(x,a)}{(a,a)}a.$$

5.2. Given a linear transformation A mapping each pair of orthogonal lines to a pair of orthogonal lines, prove that A is the composition of a homothety $x \mapsto kx$ and an isometry.

5.3. Prove that any proper motion of $\mathbb{R}^3$ with a fixed point can be represented as a composition of two symmetries about lines.

5.4. Let f be an isometry of $\mathbb{R}^n$ leaving the origin fixed, so that the fixed point set of f is a linear subspace of some dimension s. Prove that f can be represented as a composition of $n-s$ symmetries about hyperplanes, but it cannot be represented as a composition of a smaller number of symmetries.

5.5. Prove that any multivalued map $f\colon \mathbb{R}^2 \to \mathbb{R}^2$ (to each point, it assigns one or several points) preserving the distance 1 (i.e., such that if $|XY| = 1$, $X_1 \in f(X)$, and $Y_1 \in f(Y)$, then $|X_1Y_1| = 1$) is a single-valued map.

5.6. Give an example of an isometry $f\colon \mathbb{R}^2 \to \mathbb{R}^2$ and a bounded set $A \subset \mathbb{R}^2$ for which $f(A) \subset A$ but $f(A) \neq A$.

5.7. Give an example of a nonempty set A that can be represented as the union of two disjoint subsets isometric to A.

Spherical geometry. The sphere is assumed to have radius $R = 1$; a, b, c are the side lengths and α, β, γ are the angles of a spherical triangle.

5.8. Prove that $a + b + c < 2\pi$.

5.9. Find the angles of a *self-polar* triangle (this is a triangle that coincides with its polar triangle).

5.10. Does there exist a map of a domain on the sphere onto a domain on the Euclidean plane that takes the segments of spherical lines into segments of Euclidean lines?

5.11. Prove that any spherical triangle has inscribed and circumscribed circles.

5.12. Prove that (a) the medians and (b) the altitudes of a spherical triangle meet at one point.

5.13. Prove that the area of a spherical disk of radius r equals $4\pi \sin^2(r/2)$.

5.14. Prove that

$$\sin\frac{\alpha}{2} = \sqrt{\frac{\sin(p-b)\sin(p-c)}{\sin b \sin c}} \quad \text{and} \quad \cos\frac{\alpha}{2} = \sqrt{\frac{\sin p \sin(p-a)}{\sin b \sin c}},$$

where $p = (a+b+c)/2$.

5.15. Prove that

$$\cos\frac{\alpha+\beta}{2} = \frac{\cos\frac{a+b}{2}}{\cos\frac{c}{2}}\sin\frac{\gamma}{2} \quad \text{and} \quad \sin\frac{\alpha+\beta}{2} = \frac{\cos\frac{a-b}{2}}{\cos\frac{c}{2}}\cos\frac{\gamma}{2}.$$

5.16. Given $S = \alpha + \beta + \gamma - \pi$, prove that

(a) $$2\sin\frac{S}{2} = \frac{\sqrt{\sin p \sin(p-a)\sin(p-b)\sin(p-c)}}{\cos\frac{a}{2}\cos\frac{b}{2}\cos\frac{c}{2}};$$

(b) $$\tan^2\frac{S}{4} = \tan\frac{p}{2}\tan\frac{p-a}{2}\tan\frac{p-b}{2}\tan\frac{p-c}{2}.$$

5.17. Prove that in a spherical triangle with right angle γ, the following relations are valid: $\tan a = \tan\alpha \sin b$ and $\tan a = \tan c \cos\beta$.

5.18. Given the radius r of the circle inscribed in a spherical triangle, prove that

$$\tan r = \sqrt{\frac{\sin(p-a)\sin(p-b)\sin(p-c)}{\sin p}}.$$

5.19. Given the radius $\bar{R}$ of the circle circumscribed about a spherical triangle, prove that

$$\cot \bar{R} = \sqrt{\frac{\sin(\alpha - s)\sin(\beta - s)\sin(\gamma - s)}{\sin s}},$$

where $s = (\alpha + \beta + \gamma - \pi)/2$.

5.20. (a) Given a spherical line segment of length α, prove that the polars of all spherical lines intersecting this segment sweep out a set of area 4α.

(b) Given several spherical line segments with the sum of lengths less that π, prove that there exists a spherical line disjoint from each of these segments.

Models of Lobachevsky geometry.

5.21. Prove that a point z in the Poincaré unit disk model corresponds to the point $2z/(|z|^2 + 1)$ in the Klein model.

5.22. Prove the triangle inequality in the Poincaré upper half-plane model.

5.23. Find the minimal hyperbolic distance from a point (x, y) in the upper half-plane to the ray Oy.

5.24. Prove that all points on the Euclidean line $y = kx$ that lie in the upper half-plane are equidistant from the hyperbolic line Oy.

5.25. Prove that all proper motions of the Poincaré model in the unit disk have the form $z \mapsto (az + b)(\bar{b}z + \bar{a})$, where a and b are complex numbers such that $|a|^2 - |b|^2 = 1$.

5.26. Prove that a hyperbolic polygon whose angles are all less than π is convex.

Elementary hyperbolic geometry.

5.27. Prove that the relation $e^{-a/c} = \tan(\alpha/2)$ is equivalent to the relations

$$\cosh(a/c)\sin\alpha = 1, \qquad \sinh(a/c)\tan\alpha = 1, \qquad \tanh(a/c) = \cos\alpha.$$

5.28. Prove the relation $e^{-a/c} = \tan(\alpha/2)$ using one of the Poincaré models.

5.29. Prove that the sides and angles of a triangle with right angle γ are related by the formulas

$$\sinh a = \sinh c \sin\alpha, \qquad \tanh b = \tanh c \cos\alpha,$$
$$\cot\alpha \cot\beta = \cosh c, \qquad \cos\alpha = \cosh a \sin\beta.$$

5.30. Prove that the sides and angles of a triangle are related by

(a) $\cosh a \sin\beta = \cosh b \sin\alpha \cos\gamma + \cos\alpha \sin\gamma$;

(b) $\cosh a = \dfrac{\cos\alpha + \cos\beta\cos\gamma}{\sin\beta\sin\gamma}$.

5.31. Given a hyperbolic angle with a side passing through the center of the disk Δ in the Klein model, prove that if the angle is acute, then it is smaller than the corresponding Euclidean angle.

5.32. Prove that the triangle and the disk are convex figures.

5.33. Prove that if two triangles have equal respective angles, then the triangles are equal.

5.34. Prove that the orthogonal projection of a side of an acute angle to the other side is a bounded set.

5.35. (a) Given $\alpha < (n-2)\pi/n$, prove that there exists a regular n-gon with all angles α and all sides equal.

(b) Prove that if a is the side length of a regular n-gon with angle α and r is the radius of the inscribed circle, then

$$\cosh r = \frac{\cos\frac{\alpha}{2}}{\sin\frac{\pi}{n}} \quad \text{and} \quad \cosh\frac{a}{2} = \frac{\cos\frac{\pi}{n}}{\sin\frac{\alpha}{2}}.$$

(c) For which n does there exist a regular n-gon with angle $\alpha = 2\pi/n$ (i.e., with the sum of angles 2π)?

5.36. Given a quadrilateral in which three angles are $\pi/2$ and the sides joining the right angles have lengths a and b, find the fourth angle γ.

5.37. Write the area of the triangle with vertices $1, -1, \infty$ in the Poincaré upper half-plane model as a double integral and compute it.

5.38. Prove that if $w = (z-i)(z+i)$, then

$$\frac{|dz|}{\operatorname{Im} z} = \frac{2|dw|}{1-|w|^2}.$$

5.39. Prove that the length element in the Klein model has the form

$$ds^2 = \frac{dx^2 + dy^2 - (xdy - ydx)^2}{(1-x^2-y^2)^2}.$$

5.40. (a) Prove that in the Poincaré models in the upper half-plane and in the unit disk, the areas of infinitesimal rectangles with sides parallel to the coordinate axes are

$$\frac{dx\,dy}{y^2} \quad \text{and} \quad \frac{4dx\,dy}{(1-(x^2+y^2))^2},$$

respectively.

(b) Prove that in the Klein model, the infinitesimal rectangle formed by the lines $\rho = \text{const}$ and $\varphi = \text{const}$, where ρ, φ are the polar coordinates, has side lengths

$$\frac{d\rho}{1-\rho^2} \quad \text{and} \quad \frac{\rho\,d\varphi}{\sqrt{1-\rho^2}}.$$

5.41. Prove that the area of a disk of radius r is $4\pi\sinh^2(\frac{r}{2})$.

5.42. Prove that the length of the line joining the midpoints of two sides in a hyperbolic triangle is less than half the length of the opposite side.

5.43. Given a convex quadrilateral $ABCD$ with right angles A, B, and C and a point P on the segment $[C, D]$ such that $|BP| = |AD|$, prove that the rays BP and AD are parallel.

5.44. (a) Prove that the line passing through the midpoint of and perpendicular to the side AB of a triangle ABC is perpendicular to the line joining the midpoints of AC and BC.

(b) Given an isometric map $X \mapsto X'$ of one line onto another, prove that the midpoints of all segments $[X, X']$ either coincide or are collinear.

5.45. Prove that (a) the bisectors of the angles in a triangle, (b) the medians of a triangle, and (c) the altitudes of an acute-angled triangle meet at one point.

5.46 (Hyperbolic Menelaus theorem). Given a line l intersecting the sides BC, CA, AB of a triangle ABC at points A_1, B_1, C_1, respectively, prove that

$$\frac{\sinh|AC_1|}{\sinh|C_1B|}\cdot\frac{\sinh|BA_1|}{\sinh|A_1C|}\cdot\frac{\sinh|CB_1|}{\sinh|B_1A|}=1.$$

5.47 (Hyperbolic Ceva theorem). Given points A_1, B_1, C_1 on the sides BC, CA, AB of a triangle ABC, prove that the segments $[A, A_1]$, $[B, B_1]$, $[C, C_1]$ meet at one point if and only if one of the following equivalent relations holds:

(a) $\dfrac{\sin ACC_1}{\sin C_1CB}\cdot\dfrac{\sin BAA_1}{\sin A_1AC}\cdot\dfrac{\sin CBB_1}{\sin B_1BA}=1;$

(b) $\dfrac{\sinh|AC_1|}{\sinh|C_1B|}\cdot\dfrac{\sinh|BA_1|}{\sinh|A_1C|}\cdot\dfrac{\sinh|CB_1|}{\sinh|B_1A|}=1.$

5.48. Given numbers $\varphi_1, \dots, \varphi_n \in [0, \pi)$ whose sum is less than $(n-2)\pi$, prove that there exists a hyperbolic n-gon with angles $\varphi_1, \dots, \varphi_n$.

In problems 5.49–5.51, $p = (a + b + c)/2$.

5.49. Prove that

$$\sin^2\frac{\alpha}{2}=\frac{\sinh(p-b)\sinh(p-c)}{\sinh b\sinh c}\quad\text{and}\quad\cos^2\frac{\alpha}{2}=\frac{\sinh p\sinh(p-a)}{\sinh b\sinh c}.$$

5.50. Prove that the area $S = \pi - (\alpha + \beta + \gamma)$ of a hyperbolic triangle can be evaluated by the following formulas:

(a) $2\sin\dfrac{S}{2}=\dfrac{\sqrt{\sinh p\sinh(p-a)\sinh(p-b)\sinh(p-c)}}{\cosh\frac{a}{2}\cosh\frac{b}{2}\cosh\frac{c}{2}};$

(b) $\tan^2\dfrac{S}{4}=\tanh\dfrac{p}{2}\tanh\dfrac{p-a}{2}\tanh\dfrac{p-b}{2}\tanh\dfrac{p-c}{2}.$

5.51. Given the radius r of the circle inscribed in a triangle, prove that

$$r=\sqrt{\frac{\sinh(p-a)\sinh(p-b)\sinh(p-c)}{\sinh p}}.$$

5.52. Given the radius R of a circle circumscribed about a triangle (provided that such a circle exists), prove that

$$\tanh R=\sqrt{\frac{\sin\frac{S}{2}}{\sin(\alpha+\frac{S}{2})\sin(\beta+\frac{S}{2})\sin(\gamma+\frac{S}{2})}},$$

where $S = \pi - (\alpha + \beta + \gamma)$.

5.53. Given a convex hyperbolic quadrilateral $ABCD$ with right angles A and B and parallel rays AB and DC, prove that $e^{|CD|} = \sinh|AD|/\sinh|BC|$.

5.54. Given a convex hyperbolic quadrilateral $ABCD$ in which the ray AB is parallel to DC and the ray AD is parallel to BC, prove that $|AB| + |BC| = |AD| + |DC|$.

5.55. Given a convex hyperbolic quadrilateral $ABCD$ such that $|BA| + |AD| = |BC| + |CD|$, prove that the bisectors of the exterior angles A and C and the bisectors of the interior angles B and D belong to one pencil of lines.

Three types of proper motions of the Lobachevsky plane.

5.56. Prove that any pair of parallel lines can be transformed into any other pair of parallel lines by a motion.

5.57. Prove that any pencil of lines in the Klein model corresponds to a family of Euclidean lines having a common point (the point may lie inside or outside the disk or on its boundary).

5.58. Prove that all horocycles are pairwise congruent.

5.59. Find the length of an arc AB of a horocycle given the distance d between the points A and B.

5.60. Prove that the perpendiculars to the sides of a triangle through their midpoints belong to one pencil of lines.

5.61. Prove that the altitudes of a triangle belong to one pencil.

5.62. Given a point C moving along an arc AB of a circle, horocycle, or hypercycle, prove that the value $\alpha + \beta - \gamma$ is constant (α, β, γ are the angles of the triangle ABC).

5.63. Given a convex quadrilateral $ABCD$ with pairwise equal opposite sides, prove that the opposite sides of this quadrilateral lie on divergent lines.

5.64. Given an isometry f of the Lobachevsky plane such that the distance between the points X and $f(X)$ is the same for all X, prove that f is the identity transformation.

5.65. Given a transformation f of the Lobachevsky plane that increases the distance between any two points by the same factor of k, i.e., such that $d(f(X), f(Y)) = kd(X, Y)$ for any X and Y, prove that $k = 1$, i.e., f is an isometry.

CHAPTER 6

The Infinite-Dimensional World

This chapter briefly describes infinite-dimensional geometry and its applications to the theories of linear equations, linear inequalities, and quadratic functions.

6.1. Basic definitions

The spaces considered in this chapter are endowed with two structures, a linear structure and a topology; they may have infinite dimension.

DEFINITION 1. A set X with a system τ of subsets is called a *topological space* and denoted by (X, τ) if τ satisfies the following axioms: (i) $\varnothing$ and X belong to τ; (ii) the union of an arbitrary family of sets from τ belongs to τ; (iii) the intersection of any finite family of sets from τ belongs to τ.

In this case the system τ is called a *topology* and the sets from τ are called *open* sets. If a set from τ contains some point x, then it is said to be a *neighborhood of the point* x. A topological space (X, τ) is called *Hausdorff* if any two different points $x, y \in X$ have neighborhoods V_1 and V_2 such that $V_1 \cap V_2 = \varnothing$.

The topological notions that are most important to us are closedness, continuity, and compactness. Let us recall the definitions.

Let (X, τ) be a topological space. A subset C of X is called *closed* if its complement is open. The closure $\operatorname{Cl}(A, X)$ (or $\operatorname{Cl} A$, if it is clear in what space the closure is taken) is the smallest closed set containing A. A set $X_1 \subset X$ is said to be *everywhere dense* if $\operatorname{Cl} X_1 = X$. Let (Y, θ) be another topological space. A map $f \colon X \to Y$ is *continuous* if the preimage of any open subset of Y is open in X. A topological space is said to be *compact* if an arbitrary open cover (i.e., a cover of the space by open sets) contains a finite subcover.

DEFINITION 2. (a) A linear (vector) space[1] X is called a *topological vector space* if it is endowed with a topology with respect to which addition and multiplication by a number are continuous operations. A topological vector space X is *locally convex* if it is Hausdorff and every neighborhood of a point contains a convex neighborhood of this point.

The set of continuous linear functionals on X is called the *dual space* and denoted by X^*. On the space X^*, various topologies can be defined; the coarsest one consists of the open sets $V(x, a) = \{\, x^* \mid \langle x^*, x\rangle < a \,\}$ (here $\langle x^*, x\rangle$ is the value of the functional x^* at x) and their finite intersections and arbitrary unions.

(b) A space X is said to be *normed* (and denoted by $(X, \|\cdot\|)$) if, on this space, a *norm* is defined, i.e., a function $\|\cdot\| \colon X \to \mathbb{R}$ satisfying the following axioms:

(i) $\|x\| \geqslant 0$ for all $x \in X$, and $\|x\| = 0 \iff x = 0$;

[1] All spaces are considered over the field of real numbers.

(ii) $\|ax\| = |a|\,\|x\|$ for all $x \in X$ and $a \in \mathbb{R}$;
(iii) $\|x+y\| \leqslant \|x\| + \|y\|$ for all $x, y \in X$.

(c) A space X is called *pre-Hilbert* (and denoted by $(X, (\cdot\,,\cdot))$) if a function $(\cdot,\cdot)\colon X \times X \to \mathbb{R}$ is defined that satisfies the following axioms:

(i) $(x,x) \geqslant 0$ for all $x \in X$, and $(x,x) = 0 \iff x = 0$;
(ii) $(x,y) = (y,x)$ for all $x, y \in X$;
(iii) $(ax,y) = a(x,y)$ for all $x, y \in X$ and $a \in \mathbb{R}$;
(iv) $(x+y,z) = (x,z) + (y,z)$ for all $x, y, z \in X$.

A normed space can be made metric by defining the distance d as $d(x,y) = \|x-y\|$. This distance makes it possible to define the *norm topology* $\tau_{\|\cdot\|}$. A set U is open in the norm topology if for any $x \in U$, there exists a number $\varepsilon > 0$ such that y belongs to U whenever $y \in X$ and $\|x-y\| < \varepsilon$. A sequence x_n is called a *Cauchy sequence* if for any $\varepsilon > 0$, there exists a number N such that $d(x_n, x_m) < \varepsilon$ for all $n, m > N$. A metric space (X,d) is said to be *complete* if any Cauchy sequence converges in this space. A normed space complete with respect to the corresponding metric d is called a *Banach* space. A normed space is *separable* if it contains a countable everywhere dense set.

A pre-Hilbert space $(X, (\cdot,\cdot))$ is normed by the norm $\|x\| = (x,x)^{1/2}$. A complete (with respect to this norm) pre-Hilbert space is called a *Hilbert* space.

It remains to give one more important definition.

DEFINITION 3. Vector spaces X and Y are said to be *in duality* if there exists a bilinear form $\langle\cdot,\cdot\rangle\colon X \times Y \to \mathbb{R}$ such that the condition $\langle x,y\rangle = 0$ for all $y \in Y$ immplies $x = 0$, and the condition $\langle x,y\rangle = 0$ for all $x \in X$ implies $y = 0$.

The duality between X and Y determines *topologies* $\sigma(X,Y)$ and $\sigma(Y,X)$ on the spaces X and Y, respectively. Namely, the topology $\sigma(X,Y)$ consists of the neighborhoods $U(y,\varepsilon) := \{\, x \mid \langle x,y\rangle < \varepsilon \,\}_{y\in Y,\ \varepsilon>0}$ and their finite intersections and arbitrary unions, and the topology $\sigma(Y,X)$ consists of the neighborhoods $V(x,\varepsilon) := \{\, y \mid \langle x,y\rangle < \varepsilon \,\}_{x\in X,\ \varepsilon>0}$ and their finite intersections and arbitrary unions.

The topologies $\sigma(X,Y)$ and $\sigma(Y,X)$ turn X and Y into *locally convex linear topological spaces* (dual to each other). For a normed space X and its dual X^*, the topology $\sigma(X,X^*)$ is called *weak* and $\sigma(X^*,X)$, *weak**.

Note that any Hilbert space $(X, (\cdot\,,\cdot))$ is dual to itself with respect to the scalar product.

It can be proved that any locally convex space, in particular, a normed space X, is in duality with its dual X^* with respect to the bilinear form $(x,x^*) \to \langle x^*,x\rangle$, where $x \in X$, $x^* \in X^*$, and $\langle x^*,x\rangle$ is the value of the functional x^* at x.

Now, it is time to consider examples. We begin with the interpretation of a finite-dimensional space from the point of view of infinite-dimensional geometry.

EXAMPLES. 1. The coordinate space $\mathbb{R}^n$. It seems natural to consider the space $\mathbb{R}^n$ in parallel with the isomorphic (but nevertheless different) space $\mathbb{R}^{n*}$ of linear functionals on $\mathbb{R}^n$. The spaces $\mathbb{R}^n$ and $\mathbb{R}^{n*}$ form a dual pair. It is convenient to treat the vectors $x \in \mathbb{R}^n$ as columns and $y \in \mathbb{R}^{n*}$ as rows. This allows us, in particular, to write the bilinear form as a matrix product, namely, $\langle x,y\rangle = y \cdot x$. The duality topology of this space is the ordinary topology of the n-space. Intersections of open half-spaces form open parallelepipeds, which generate the topology in the usual way (a set $U \in \mathbb{R}^n$ is open if for any point $\xi = (\xi_1,\dots,\xi_n) \in U$, there exists a number

$\delta > 0$ such that $x = (x_1, \ldots, x_n) \in U$ whenever $|x_i - \xi_i| < \delta$ for $1 \leqslant i \leqslant n$). It can be shown that the duality topology is the only locally convex topology on $\mathbb{R}^n$.

We cannot pay due attention to Banach geometry; we only mention several points. Banach geometry arises when the vector space $\mathbb{R}^n$ is endowed with a metric that is invariant only with respect to the group of translations. The unit ball in any norm on $\mathbb{R}^n$ is a centrally symmetric bounded closed convex body B, and, conversely, any such body generates the norm $\|x\|_B = \inf\{\, t > 0 \mid x/t \in B \,\}$. (The function of B so defined is called the *Minkowski function of the set B*. Thus the norms in $\mathbb{R}^n$ are the Minkowski functions of centrally symmetric compact sets for which zero is an interior point.) All normed spaces with underlying set $\mathbb{R}^n$ are Banach spaces. There is an interesting scale of norms on $\mathbb{R}^n$:

$$\|x\|_p := \left(\sum_{i=1}^{n} |x_i|^p\right)^{1/p}, \quad 1 \leqslant p < \infty, \quad \|x\|_\infty = \max_{1 \leqslant i \leqslant n} |x_i|.$$

Banach geometry is a special geometric world. There are many spectacular results, but they are rather far away from the main topics of this book.

The space $\mathbb{R}^n$ is Hilbert if and only if its scalar product is defined by a symmetric positive definite matrix, i.e.,

$$(x, y) = \sum_{i,j=1}^{n} a_{ij} x_i y_j, \quad \text{where } a_{ij} = a_{ji} \text{ and } \sum_{i,j=1}^{n} a_{ij} x_i x_j > 0 \text{ for all } x \neq 0.$$

The only norm that is invariant with respect to the entire isometry group of the Euclidean vector space $\mathbb{R}^n$ is the usual norm $\|x\| = \left(\sum_{k=1}^{n} |x_k|^2\right)^{1/2}$.

2. Two separable infinite-dimensional Banach spaces play special roles; these are the space

$$l_2 = \left\{ x = (x_1, x_2, \ldots, x_n, \ldots) \mid \|x\| := \left(\sum_{k \in \mathbb{N}} x_k^2\right)^{1/2} < \infty \right\}$$

and the space $C([0,1])$ of all continuous functions on the interval $[0,1]$ with the norm $\|x(\cdot)\| = \max_{t \in [0,1]} |x(t)|$.

The former is universal for all separable Hilbert spaces (any separable Hilbert space is isometric to some subspace of l_2). The latter is universal even for all separable Banach spaces: any such space can be isometrically embedded in the space $C([0,1])$.

Let us give one more important example of a linear topological space.

3. The set $C^\infty(\mathbb{T})$ of infinitely differentiable 2π-periodic functions $x(\cdot)$ with the topology of uniform convergence of all derivatives ($x_n(\cdot) \to x(\cdot)$ if $x_n^{(k)}(\cdot)$ uniformly converges to $x^{(k)}(\cdot)$ for any $k \geqslant 0$) is another example of a linear topological space (of "test functions"). The dual space is the space of distributions ("generalized functions"), whose role in analysis is so important.

Historical comments. The theory of infinite-dimensional linear topological, Banach, Hilbert spaces and of continuous linear operators in such spaces is called (linear) *functional analysis*. Among the founders of functional analysis are remarkable mathematicians from all major mathematical schools of the first half of the twentieth century, such as Volterra, Peano (Italy), Hilbert, Schmidt (Germany), Hadamard, Fréchet (France), F. Riesz, von Neumann (Hungary), Helly, Hahn (Austria), Kolmogorov, Gelfand (USSR),

Wiener (USA), Banach, Steinhaus (Poland). (This list is, of course, incomplete, but a detailed exposition of the prehistory of functional analysis would lead us too far.)

The main stimuli to the development of linear functional analysis, which emerged as a synthesis of classical analysis, linear algebra, and geometry, were quantum mechanics, the theory of linear integral equations, and the theory of selfadjoint differential operators.

The notions of topological space and compactness were shaped under the influence of the research of Cantor, Hausdorff, Alexandroff, Urysohn, and other mathematicians; the Hilbert space was introduced by Hilbert as the space l_2 and, then, by Schmidt, von Neumann, and others axiomatically; the Banach spaces were defined by Halley (in the special case of convergent sequences) and, then, by Banach and Wiener; linear topological spaces first appeared in Kolmogorov's work, and locally convex linear topological spaces were introduced by von Neumann. The duality of linear spaces, which is the starting point of duality theory in convex analysis, is thoroughly developed in Bourbaki's *Éléments de mathématique*.

Now we pass to convex analysis, which can in a certain sense be considered a chapter of affine geometry.

The term *convex analysis* appeared quite recently, in the mid-1960s. This is the name of the area of mathematics between analysis and geometry that studies convex objects—sets, functions, and extremal problems. (Convexity proper exists in mathematics from time immemorial. Already Archimedes used the notion of convexity.)

The theory of convex bodies was extensively developed in the nineteenth century, with remarkable results on convex polyhedra obtained by Cauchy. But only at the end of the nineteenth century did Minkowski give an outline of the new area—*convex geometry*. In the 1930s convex geometry became fashionable. In particular, in Moscow it was the favorite subject of study at "school mathematical circles"[2] (largely due to Shnirel'man and Lyusternik). Convex geometry was the topic of a great many booklets and books in the 1930s–1940s. Then, interest in this topic decreased somewhat, but it reappeared in the 1960s.

Just at that time (starting with the late 1940s), the crucial role of convexity in extremal problems was understood. Linear and convex programing emerged, and mathematical economics, which is largely based on the ideas of convexity, began to develop; this led to the birth of a new direction called convex analysis. The "place of action" for the objects of convex analysis is locally convex (affine) spaces constructed on the basis of duality.

One of the most important theses of convex analysis is that *the basic objects of convex analysis (convex sets, convex functions, and convex extremal problems) admit dual descriptions, geometrical and analytical: one in the initial space and the other in the dual space.*

6.2. Statements of theorems

After this short introduction, we state the infinite-dimensional analogs of the theorems discussed in the second chapter (and proved in the two-dimensional case—see Theorems 1–5).

The first theorem describes properties of solutions to linear equations. Let X be a (separable) Hilbert space. A linear continuous operator $A\colon X \to X$ is said to be *compact* if the closure of the image of the unit ball (under the action of this

[2]EDITORIAL NOTE: These are special free weekly evening seminars for high-school students organized (on a voluntary basis) by university professors to encourage young people's interest in mathematics.

operator) is compact in X. We denote the operator $I - A$ by Λ. For an operator A, we define its dual operator by $(A^*x, y) = (x, Ay)$.

THEOREM 1. (a) *If X is a Hilbert space and $A\colon X \to X$ is a compact operator, then either the equation $\Lambda x = \xi$ is solvable for any ξ or* $\operatorname{Ker}\Lambda \neq \{0\}$, *i.e., the homogeneous equation $\Lambda x = 0$ has a nonzero solution.* (This is the Fredholm alternative.)

(b) *The equation $\Lambda x = \xi$ is solvable if and only if ξ is orthogonal to the kernel of $I - A^*$, where A^* is the dual operator.*

The meaning of Theorem 1 in the finite-dimensional case is as follows. Assertion (a) coincides with the well-known statement: *Either a system of n linear equations in n variables is solvable for an arbitrary right-hand side or the homogeneous system has a nonzero solution.* Assertion (b) is nothing but the duality relation for the subspaces of $\mathbb{R}^n$: each subspace is described, on the one hand, as the image of a linear operator and, on the other hand, as the orthogonal complement to the kernel of the dual operator.

Consider the separation theorems. Let X and Y be spaces in duality. (If X is a locally convex space, then we consider it in duality with its dual.) For $y \in Y$ and $b \in \mathbb{R}$ the set $\Gamma_{y,b} \subset X$ defined by the formula $\Gamma_{y,b} := \{\, x \mid \langle x, y\rangle = b \,\}$ is called a *hyperplane*. The sets $\Pi^-_{y,b} = \{\, x \mid \langle x, y\rangle \leqslant b \,\}$ and $\Pi^+_{y,b} = \{\, x \mid \langle x, y\rangle \geqslant b \,\}$ are called *half-spaces* (bounded by the hyperplane $\Gamma_{y,b}$). A point $\bar{x}$ is said to be *strictly separated* from a nonempty convex set $A \subset X$ if there exists a hyperplane such that the set A lies in one half-space bounded by this hyperplane and the point $\bar{x}$ lies in the other half-space (and does not belong to the hyperplane). The definition is given in the language of geometry. Algebraically, this means that there exists an $y \in Y$ such that

$$\langle \bar{x}, y\rangle > \sup\{\, \langle x, y\rangle \mid x \in A \,\}.$$

If there exists a nonzero $y \in Y$ such that the inequality is nonstrict, then the point $\bar{x}$ is said to be *separated* from A.

THEOREM 2 (First separation theorem). *If X is a locally convex space and an open convex set $U \subset X$ does not contain zero, then there exists a hyperplane passing through zero and disjoint from U (i.e., zero can be separated from U).*

THEOREM 3 (Second separation theorem). *In a locally convex space, a point that does not belong to a nonempty closed convex set can be strictly separated from this set.*

Let X and Y be spaces in duality. Consider a family $\{\, (y_\alpha, b_\alpha) \mid \alpha \in \mathcal{A} \,\}$ of elements of $Y \times \mathbb{R}$. We shall refer to the system $\langle x, y_\alpha\rangle = 0$ $(\leqslant 0)$ as a system of homogeneous equalities (inequalities) and to the system $\langle x, y_\alpha\rangle = b_\alpha$ $(\leqslant b_\alpha)$ as a system of nonhomogeneous equalities (inequalities).

A set $K \subset X$ is called a *convex cone* if for each $x_1, x_2 \in K$, all conical combinations $\alpha_1 x_1 + \alpha_2 x_2$, where $\alpha_i \in \mathbb{R}_+$, belong to K.

THEOREM 4 (Description of solutions to systems of equations and inequalities). *The solutions to systems of homogeneous equalities are closed subspaces, and vice versa; the solutions to systems of homogeneous inequalities are convex closed cones, and vice versa; the solutions to systems of nonhomogeneous equalities are closed linear varieties, and vice versa; the solutions to systems of nonhomogeneous inequalities are closed convex sets, and vice versa.*

This theorem reveals the duality phenomena mentioned in the introductory part of this chapter (see Section 6.1). Say, the notion of a (closed) affine variety can be defined in two different ways: geometrically, as a set of points in an affine space such that, together with any two points, it contains the entire line through these two points (and is closed), or analytically, as a solution of a system of nonhomogeneous equations (or, equivalently, as an intersection of hyperplanes). Similarly, an arbitrary closed convex set is (geometrically) a closed set (of points) such that, together with any two points, it contains the entire line segment joining these points, or (analytically) a solution of a system of nonhomogeneous linear inequalities (or an intersection of a system of half-spaces).

So far, we considered all these objects (subspaces, affine varieties, convex sets), so to speak, by themselves; now, we shall consider them together with their "twins" in the dual space.

THEOREM 5 (Krein–Mil'man). *A convex compact set in a locally convex space is the convex closure of its extreme points.*

In the finite-dimensional case, the words "convex closure" can be replaced by "convex hull".

To conclude, we state one more theorem; its finite-dimensional version was considered in the fourth chapter.

A continuous linear operator $A\colon X \to X$ mapping a Hilbert space $(X, (\cdot,\cdot))$ to itself is said to be *selfadjoint* if $(Ax, y) = (x, Ay)$ for any x and y from X. An orthonormal system e_i (finite or infinite) is a *basis* in X if $x \in X$ for any $x = \sum_i (x, e_i)e_i$. A nonzero vector $z \in X$ is said to be an *eigenvector* of A if there exists a number λ such that $Az = \lambda z$.

THEOREM 6 (Hilbert). *If X is a separable Hilbert space and $A\colon X \to X$ is a selfadjoint compact operator, then there exists a basis in X consisting of eigenvectors of A.*

6.3. Proofs of the theorems

First, we prove Theorem 2; next, we derive a number of corollaries of this theorem (including Theorems 3 and 4); after that, we prove Theorem 1 and, finally, the Krein–Mil'man and Hilbert theorems.

Proof of Theorem 2. A maximal subspace of X (i.e., a subspace contained in no proper subspace) different from X is called a *hyperplane*. The key step in the proof is the same as in the two-dimensional case. Let us state it as a lemma.

LEMMA. *If X is a locally convex space, U is an open convex set, and L is a vector subspace disjoint from U, then either L is a hyperplane or there exists a point $\eta \in X$ such that the vector subspace spanned by L and η does not intersect U.*

Indeed, consider the linear subspace spanned by L and an arbitrary point ξ in U. Let us denote this subspace by L_0. There are two possibilities: either L_0 coincides with the entire space (then L is a hyperplane) or there exists η not belonging to L_0. In the latter case, consider the two-dimensional plane spanned by the vectors ξ and η, endow it with the Euclidean structure, and consider a circle S in the obtained Euclidean plane. Without loss of generality, we can assume that ξ and η lie on this circle and are orthogonal. The half-space $L_- = \{\, x = t\xi + z \mid t < 0,\ z \in L \,\}$

is, obviously, disjoint from U (otherwise, L would intersect U), while the half-space $L_+ = \{x = t\xi + z \mid t > 0,\ z \in L\}$ intersects U, because $\xi \in U$. Let us "rotate" L_0 about L, i.e., consider the linear subspaces spanned by L and the vectors $(\cos\varphi, \sin\varphi)$; we denote the corresponding subspaces by L_φ. Let $\hat{\varphi}$ be the least upper bound of those φ between 0 and π for which L_φ intersects U. It is easy to show that $L_{\hat{\varphi}}$ does not intersect U.

Next, we use an argument called "transfinite induction." (In the finite-dimensional case, this is ordinary induction.)

Let us denote the family of all subspaces of X disjoint from U by $\mathcal{L}$ and order its elements by inclusion. The Zorn lemma (this is where we use transfinite induction) implies that this family has a maximal element. According to the lemma proved above, this element is a hyperplane. Since the closure of a subspace is a subspace, the maximal element cannot be everywhere dense (it would intersect U if it were); therefore, it is closed. Thus the maximal element is a maximal closed subspace not coinciding with X. It remains to note that such a subspace is a hyperplane in the sense of the definition given in Section 6.2. (Indeed, for an element ζ not belonging to a maximal proper closed subspace, we can construct a linear functional vanishing on the subspace and taking the value 1 at ζ. The union of the $(+1)$- and (-1)-level sets of the functional is a closed set not containing zero; therefore, there exists a neighborhood of zero disjoint from this set. Hence the functional is bounded, i.e., continuous.) This completes the proof of Theorem 2. □

Theorem 2 readily implies that *two disjoint convex sets one of which is open can be separated by a hyperplane.*

Indeed, if U is convex and open and V is convex, and if these sets are disjoint, then $U - V := \{x = u - v \mid u \in U,\ v \in V\}$ is open and does not contain zero. Applying Theorem 3, we immediately obtain the desired assertion: the hyperplane separating U and V is obtained by translation of the hyperplane separating $U - V$ from zero.

This assertion implies the second separation theorem.

Proof of Theorem 3. By assumption, there is a point that does not belong to the closed convex set. By the definition of a locally convex space, there exists a convex neighborhood of this point disjoint from the set. The assertion proved above implies that this neighborhood can be separated from the set by a hyperplane; this hyperplane strictly separates the point from the set, which proves Theorem 3. □

In the statement of Theorem 2, the open set does not contain zero. Using virtually the same argument as in the proof of Theorem 2, we could construct a hyperplane (passing through zero) that contains an *arbitrary subspace* disjoint from this open set, which readily implies the Hahn–Banach theorem on the extension of a linear functional in a normed (and even locally convex) space and, thereby, the nontriviality of the space dual to a normed space. (The standard proof of the Hahn–Banach theorem suggested originally by Helly is analytical, but it includes the same geometrical element as that used in our argument, namely, the "rotation" of a subspace about a subspace "of dimension lower by 1.")

Let us mention one more corollary. The *annihilator* of a subspace $L \subset X$ is the set

$$L^\perp := \{y \in Y \mid \langle x, y\rangle = 0\ \forall x \in L\}.$$

In a Hilbert space, the elements of the annihilator are the vectors $y \in X$ such that $(x, y) = 0$ for any $x \in L$, i.e., the elements of the orthogonal complement of L.

The correspondence between closed subspaces L and their orthogonal complements $L^{\perp}$ is a form of duality in Euclidean and Hilbert spaces.

LEMMA ON A NONTRIVIAL ANNIHILATOR. *Let L be a proper closed subspace of a locally convex space X. Then the annihilator of L is nontrivial (that is, $L^{\perp} \neq \{0\}$).*

This lemma is proved by strictly separating from L an element that does not belong to L.

Proof of Theorem 4. The proof is basically similar to the proof in the two-dimensional case. The sets $\langle x, y_\alpha \rangle = b_\alpha$ ($\leqslant b_\alpha$) are closed subspaces, cones, affine varieties, and convex sets; therefore, their intersections are also closed subspaces, cones, etc. We prove the converse assertions for convex sets. If a closed convex set contains all points of X, then it satisfies the inequality $\langle x, 0 \rangle \leqslant 0$. Suppose that it does not coincide with X. By the first separation theorem, it is contained in some half-space (separating a point not belonging to the set from the set itself). Next, consider the family of *all* half-spaces containing the set under consideration. The intersection of the elements of this family, certainly, contains this set. Moreover, the intersection coincides with the set, because if the intersection contained a point that does not belong to the set, then we could strictly separate this point from the set and thereby obtain a contradiction. This completes the proof of Theorem 4. □

Proof of Theorem 1. As we mentioned, each separable Hilbert space is isometric to l_2 or to a finite-dimensional Euclidean space. Since l_2 has an orthonormal basis (say, $\{e_i\}_{i \in \mathbb{N}}$, where each e_i is the vector whose ith coordinate equals 1 and all other coordinates are zero), all closed (infinite-dimensional) subspaces of Hilbert spaces have similar bases. We use the same notation e_i for the elements of such a basis in the subspace under consideration. The vector $\sum_i (x, e_i) e_i$ is the orthogonal projection of x to this subspace. Note that the operator adjoint to a compact operator is compact.

The proof of Theorem 1 involves several steps.

Step 1. The image of X under the map Λ is closed in X.

Indeed, let y belong to the closure of $\operatorname{Im} \Lambda$. Then there exists a sequence $\{y_n\}$ such that $y_n = \Lambda x_n$ for $n \in \mathbb{N}$ and $y_n \to y$. We can assume that each vector x_n is orthogonal to $\operatorname{Ker} \Lambda$ (otherwise, we subtract from x_n its projection to $\operatorname{Ker} \Lambda$); in addition, all these vectors are bounded in norm (otherwise, we could divide each x_n by $\|x_n\|$ and obtain the bounded sequence $x_n/\|x_n\|$ that contains a subsequence converging to a vector η of norm 1, orthogonal to $\operatorname{Ker} \Lambda$ and belonging to $\operatorname{Ker} \Lambda$, but such a vector does not exist). Therefore, we can select a subsequence x_{n_k} such that Ax_{n_k} converges; hence x_{n_k} converges, and y belongs to the image of Λ.

Step 2. We have the obvious alternative: either the image ΛX is the entire X or it is a proper subspace of X. In the latter case, ΛX is closed in X by what was proved at step 1. By the lemma on a nontrivial annihilator, there exists a nonzero element $y \in X$ such that $(x - Ax, y) = 0$ for all $x \in X$, i.e., $y = A^* y$, as required.

Step 3. Suppose that $\Lambda X = X$. Let us prove that $\operatorname{Ker} \Lambda = 0$.

Indeed, if $\Lambda x_1 = 0$ for some $x_1 \neq 0$, then we can construct a sequence $\{x_k\}_{k \in \mathbb{N}}$ of elements such that $\Lambda x_{k+1} = x_k$ for $k \in \mathbb{N}$. Setting $L_k = \operatorname{Ker} \Lambda^k$, we obtain an infinite system of expanding subspaces ($L_k \neq L_{k+1}$, because x_{k+1} belongs to $L_{k+1} \setminus L_k$). Next, let us select a sequence of unit vectors $y_k \in L_k$ orthogonal to

L_{k-1}. For $l > k$, we have

$$\|Ay_l - Ay_k\| \overset{\text{id}}{=} \|y_l - (y_k + \Lambda y_l - \Lambda y_k)\| \geqslant 1,$$

because the vector $y_k + \Lambda y_l - \Lambda y_k$ belongs to L_{l-1} (by definition). Therefore, the sequence $\{y_k\}$ does not contain a convergent subsequence, which contradicts the compactness of A. Thus $\operatorname{Im}\Lambda = X$ implies $\operatorname{Ker}\Lambda = \{0\}$.

Step 4. Conclusion of the proof.

Let ΛX be a proper subspace of X. Then (Step 2) the kernel of the operator Λ^* in nontrivial. Therefore (Step 3), the image of Λ^* does not coincide with X, and (Step 2) the kernel of Λ is nontrivial. This concludes the proof of Theorem 1. □

Proof of Theorem 5. We say that S is a supporting set for $C \subset X$ if S is a closed affine set intersecting C and such that if some interior point of a segment $[x, y]$ contained in C lies in S, then the entire segment lies in S. Let us prove that among supporting sets there are one-point sets and supporting points are nothing but extreme points (this makes us recall the definition of an extreme point).

The existence of an extreme point is proved in two steps.

Step 1. There exists a supporting hyperplane.

Indeed, it is easy to see that, for an arbitrary vector $x^* \in X^*$, the hyperplane $H = H(x^*) = \{\, x \mid \langle x^*, x\rangle = \max_{x\in C}\langle x^*, x\rangle \,\}$ is supporting.

Step 2: transfinite induction. Consider the family of *all* supporting sets contained in H and denote it by $\mathcal{S}$. Let us order the elements of $\mathcal{S}$ by inclusion. By the Zorn lemma, there exists a maximal chain $\mathcal{M}$. The intersection of all supporting sets from $\mathcal{M}$ belongs to $\mathcal{M}$ (because $\mathcal{M}$ is maximal). It remains to show that this intersection (we denote it by $\widehat{S}$) consists of one point. If $\widehat{S}$ contains two elements ξ and η, then we separate them by a functional $\hat{x}^*$ (i.e., take $\hat{x}^*$ such that $\langle \hat{x}^*, \xi\rangle < \langle \hat{x}^*, \eta\rangle$) and put $\bar{S} := \widehat{S} \cup \Gamma$, where $\Gamma = \{\, x \mid \langle \hat{x}^*, x\rangle = \sup_{x\in\widehat{S}\cup C}\langle \hat{x}^*, x\rangle \,\}$. Obviously, $\bar{S}$ is a closed set meeting C; it is also easy to see that it is a supporting set, which contradicts the maximality of $\widehat{S}$.

What remains to be proved is easy. Let $\operatorname{extr} C$ denote the set of all extreme points of the set C. If the closure of $\operatorname{extr} C$ (we denote it by C') does not coincide with C (i.e., is a proper subset of C), then we strictly separate a point of $C \setminus C'$ from C' by a functional $\bar{x}^*$, consider the supporting hyperplane $H(\bar{x})$, and find an extreme point C not belonging to C' on this hyperplane. This contradiction proves Theorem 5. □

It remains to prove the Hilbert theorem. As mentioned, we can assume that X is l_2. Consider the extremal problem

$$|(Ax, x)| \to \max, \quad (x, x) \leqslant 1. \tag{1}$$

Take $\{x^n\}_{n\in\mathbb{N}}$, where $x^n = (x_k^n)_{k\in\mathbb{N}} \in l_2$. We have

$$\sum_k (x_k^n)^2 \leqslant 1 \quad \text{for any } n. \tag{2}$$

Let us apply the diagonal process to construct a subsequence n_l for which we have $x_k^{n_l} \to \bar{x}_k$. Inequalities (2) imply that $\bar{x} = (\bar{x}_k)$ lies in the unit ball. Since A is compact, $\bar{x}$ is a solution of (1). Certainly, $\|\bar{x}\| = 1$.

A little geometrical fragment now follows. Let us prove that $\bar{x}$ is an eigenvector. Suppose it is not; then the vectors $\bar{x}$ and $A\bar{x}$ determine a two-dimensional plane. The section of the unit sphere of the Hilbert space by this plane is a circle passing through the vector $\bar{x}$. Let us draw a perpendicular to $\bar{x}$ in the plane under consideration, take the unit vector $\bar{\xi}$ on this perpendicular, and slightly perturb the vector $\bar{x}$, namely, consider the vectors $e_\alpha = \bar{x}\cos\alpha + \bar{\xi}\sin\alpha$. All these vectors have unit length, while the function $f(\alpha) := (Ae_\alpha, e_\alpha)$ does not have local extremum at zero (because $\bar{x}$ is not an eigenvector and, therefore, $f'(0) = 2(A\bar{x}, \bar{\xi}) \neq 0$), which contradicts the definition of $\bar{x}$.

The final part of the proof is easy. We have constructed the first eigenvector; let us denote it by z^1. We have $Az^1 = \lambda_1 z^1$. Recall that $|\lambda_1|$ is a maximum for problem (1). Consider the orthogonal complement $X_1 = \{ x \mid (x, z^1) = 0 \}$. It is invariant under the operator A; indeed, if $x \in X_1$, then

$$(Ax, z^1) = (x, Az^1) = (x, \lambda_1 z^1) = 0, \qquad \text{i.e.,} \quad Ax \in X_1.$$

The space X_1 is also a Hilbert space. We can formulate problem (1) for X_1 and, if (Ax, x) is not identically zero on X_1, find an eigenvector in X_1. Repeating this construction, we shall either obtain X_k such that $(x, x) = 0$ for all $x \in X_k$ at some kth step or will be able to continue forever. Our "geometric" argument implies that, in the former case, $Ax = 0$ for all $x \in X_k$, and the required basis in X is formed by the basis of X_k and the eigenvectors constructed; in the latter case, we obtain an infinite system of eigenvectors $\{z^n\}_{n\in\mathbb{N}}$ with nonzero eigenvalues decreasing in absolute value. Consider the closure of the linear hull of all these eigenvectors and take its orthogonal complement Ξ. By construction, we have $(Ax, x) = 0$ for all $x \in \Xi$, i.e., A is the zero operator on Ξ. Supplementing the basis of Ξ by $\{z^k\}$, we obtain the required basis of X. This completes the proof of the theorem.

6.4. Concluding comments

Geometry as a science was born between the eighth and seventh centuries B.C. According to a legend, the father of geometry, who first tried to realize what a proof was, was the Ionian merchant Thales. He is believed to have proved the first theorems, including the theorem about the equality of the base angles in an isosceles triangle. Among the great ancient geometers to be mentioned are Pythagoras, Euclid, Archimedes, and Apollonius. Pythagoras organized the first scientific school (whose members, according to the legend, proved the theorem named after Pythagoras). Elementary geometry on the plane and in space was first developed in Euclid's *Elements*; the theory of measuring areas and volumes was created by Eudoxus (soon after that, Archimedes developed methods for measuring the area of the surface of a sphere and the volume of a ball) and the theory of conic sections, by Apollonius and his predecessors.

With the beginning of the Renaissance, science in Europe rushed towards new achievements. In the seventeenth century, Fermat and Descartes constructed algebraic models of geometry on the plane, and Desargues and Pascal started developing the principles of projective geometry. Galileo, Kepler, and Newton began to create the science of nature. (Since then, geometry has become the language of the natural sciences. It was, in particular, discovered that the trajectories of planets and projectiles (bodies thrown at an angle to the horizon) can be described by conic sections.)

The convergence of algebra and geometry, in particular, led to the birth (in the work of Euler) of affine geometry and to the geometrization of the theory of linear equations. In the early twentieth century, this was crowned by Fredholm's fundamental theory of linear equations in infinite-dimensional spaces.

During the more than two thousand years that passed from Euclid's times, the concept of Euclidean space has evolved, on the one hand, to the notion of infinite-dimensional Hilbert space and, on the other hand, to the notion of smooth (Riemannian) manifold.

The Hilbert theorem on compact quadratic forms in a Hilbert space is one of the extreme points of the theory that was initiated by Apollonius' *Conics*.

Many centuries of thinking about Euclid's fifth postulate and the development of projective and differential geometry and of complex analysis led Gauss, Lobachevsky, Bolyai, Beltrami, Cayley, Klein, Poincaré, and other mathematicians to realizing and proving the consistency of Lobachevsky geometry. Riemann's general idea of constructing the geometric objects that are now known as Riemannian manifolds was complemented by Klein's Erlangen Program. As a result, Euclidean, Lobachevsky, and Riemannian geometries were incorporated into the class of manifolds of constant curvature. Lobachevsky geometry proved to be one of the most important components of the modern view of the world we live in: relativity theory is based on this geometry.

The creation of convex geometry in the work of Minkowski led to the emergence of, on the one hand, linear functional analysis, and on the other hand, convex analysis, which is a basis for the theory of extremal problems.

We tried to explain all this in our book.

We did not consider many important aspects of geometry. In particular, we said nothing about the development of geometry during the past fifty years, while the recent advances in geometric science are fundamental. Surveying the evolution of science in the past decades, V. I. Arnold wrote: "At the highest level, model mathematics and theoretical physics merge again into one science. The geometry of vector bundles and connections, symplectic topology and the classification of 4-manifolds, the theory of knots and braids, supersymmetry and the theory of Lie algebras, quantum groups and more general Hopf algebras, singularities of hypersurfaces and hyper-Kähler structures, asymptotics of hypergeometric integrals and the computation of the numbers of rational curves on algebraic surfaces, nonlinear partial differential equations and the spectral theory of differential operators—all these areas of the purest mathematics [almost all belong to geometry! (*Authors' Note*)] turn out to be different aspects of quantum field theory."

Many of these advances in modern geometry are to become the subject of study in subsequent geometry courses.

Addendum

1. Geometry and physics

Projectiles move along parabolas (Galileo and Newton).

> The Great Book of Nature is written in the language of geometry.
>
> Galileo

It is rightfully believed that the father of modern natural sciences is Galileo Galilei. He was the first to ask Nature serious questions: for the first time in the history of science, he performed physical experiments. Letting bodies slide along polished slanted surfaces and measuring time against his pulse, he discovered the basic laws of mechanics.

Galileo combined his scientific research with teaching—he taught mathematics to students at the universities of Pisa and Padua. He admired Euclid and Archimedes and knew about Appolonius' theory of conic sections. Using both experiment and mathematics, Galileo managed to describe the trajectories of projectiles.

A complete theory of the motion of material bodies was constructed by Newton who translated the laws of classical mechanics (in particular, those discovered by Galileo) into the language of calculus.

According to Newton's laws of mechanics, a projectile of mass m is acted upon by the force $F = -mg$ directed downward, opposite to the Oy axis perpendicular to the surface of the earth (which is assumed flat); see Figure A.1. Therefore, according to Newton's laws, the trajectory $(x(t), y(t), z(t))$ of a body thrown from the origin at angles (the angle of departure) to the Ox axis in the plane Oxy with an initial velocity v satisfies the equations

$$mx'' = 0, \quad my'' = -g, \quad mz'' = 0$$

and the initial conditions $x(0) = y(0) = z(0) = 0$, $x'(0) = v\cos\varphi$, $y'(0) = v\sin\varphi$, and $z'(0) = 0$.

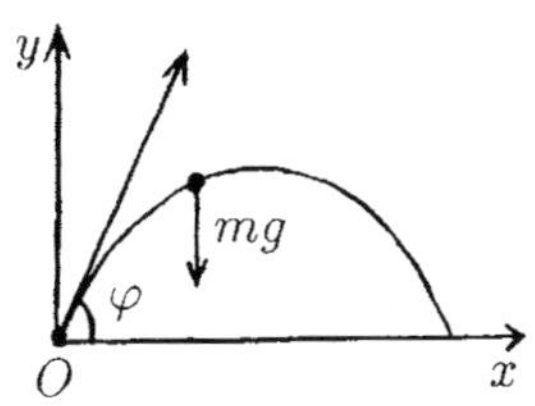

FIGURE A.1

Integrating these equations, we obtain

$$x(t) = vt\cos\varphi, \quad y(t) = vt\sin\varphi - \frac{gt^2}{2}, \quad z(t) = 0,$$

which gives the equation

$$y = x\tan\varphi - \frac{x^2}{2v^2\cos^2\varphi}$$

for the trajectory of the projectile. This is the equation of a parabola, one of the most important geometric figures that compose the language in which the laws of nature are written.

The planets move along ellipses, and the asteroids, along second-order curves (Kepler and Newton).

> I write my book. Whether it will be read by my contemporaries or descendants, it will find its reader in any case. Was not the Lord awaiting a contemplator of his creation for six thousand years? [1]
>
> Kepler

After years of hard work on Tycho Brahe's observations, Kepler first drew the somewhat unexpected conclusion that the planets of the solar system move along circles (and that was remarkable, because the circle was a basic geometric line of ancient mathematics, and it was believed that everything reasonable must perform a circular motion, or at least a combination of circular motions), but—oh, shame!—the Sun itself was not at the center!

That seemed very strange to Kepler, and he began to recalculate everything again, even more carefully. And then he found that actually, the planets move along ellipses, but of very small eccentricities; that was why he initially took them for circles. As to the Sun itself, it is positioned at a focus. At present, this assertion is known to all of us as *Kepler's first law.*

The first law (together with the second one, known as the *area law*) was published by Kepler in 1609.

A complete theory of the motion of planets was constructed by Newton. He deduced Kepler's laws from his own laws of mechanics and the law of universal gravitation. Let us describe this deduction.

We begin with the formulas that express Newton's second law and the law of universal gravitation for the planar motion of a planet or comet of mass m moving under the attraction to the Sun, ignoring the action of other cosmic bodies. We obtain the following picture.

By the law of universal gravitation, the force F with which the Sun acts on the planet or comet is equal to $-k(x(t), y(t))/r^3(t)$, where $r(t) = (x^2(t)+y^2(t))^{1/2}$. Applying Newton's second law, we obtain

$$mx'' = -\frac{kx}{(x^2+y^2)^{3/2}}, \quad my'' = -\frac{ky}{(x^2+y^2)^{3/2}}. \tag{1}$$

In the polar coordinates $x = r\cos\varphi$ and $y = r\sin\varphi$, we have

$$x' = r'\cos\varphi - r\sin\varphi\varphi', \quad y' = r'\sin\varphi + r\cos\varphi\varphi', \tag{2}$$

[1] The epigraph is taken from Kepler's book *Harmony of the World*, where he reviewed his studies of the motion of planets.

whence

$$(3) \qquad \text{(a) } x' \sin\varphi - y' \cos\varphi = -r\varphi', \quad \text{(b) } x'^2 + y'^2 = r'^2 + r^2\varphi'^2;$$

from this and (1) we obtain

$$(4) \qquad x'' \sin\varphi - y'' \cos\varphi = 0.$$

Differentiation of (1)–(4) yields the two famous conservation laws

$$(5) \qquad m((x'^2 + y'^2)/2) + k/r = H,$$

$$(6) \qquad r^2\varphi' = L/m.$$

Equality (5) expresses the *law of conservation of energy*, and (6), Kepler's *area law* mentioned above (the constant L has a physical meaning, but we shall not discuss it).

Relations (5), (6), and (3) (b) give the equality

$$\frac{m}{2}\left(r'^2 + \left(\frac{L}{mr}\right)^2\right) + \frac{k}{r} = H,$$

which is equivalent to the differential equation

$$(7) \qquad \frac{dr}{dt} = \sqrt{\frac{2}{m}\left(H - \frac{k}{r}\right) - \frac{L^2}{m^2r^2}}\,.$$

From the area law (6), we have

$$(8) \qquad \frac{d\varphi}{dt} = \frac{L}{mr^2}\,;$$

formulas (7) and (8) lead to the integrable differential equation

$$\frac{d\varphi}{dr} = \frac{L/(mr^2)}{\sqrt{2/m(H - k/r) - L^2/m^2r^2}}\,.$$

Integrating this equation, we finally obtain

$$r = \frac{p}{1 + e\cos\varphi}\,, \quad p = \frac{L^2}{m|k|}\,, \quad e = \sqrt{\frac{2HL^2}{mk^2} + 1}\,, \quad L = \sqrt{m|k|p}\,.$$

Therefore, the orbit is elliptic if $H < 0$, parabolic if $H = 0$, and hyperbolic if $H > 0$ (recall the chapter about conics and quadrics). Is this not the apotheosis of the theory of conic sections!

Geometry and special relativity (Einstein and Minkowski).

> The world is simple. Very simple. But no more than that!
>
> Albert Einstein

Let us try to understand the basic formulas of special relativity in the case of a (relatively) simple example.

Imagine the following situation. A train passes a rectilinear section of a railroad at velocity v. On the platform of the third car, a smoker S is standing. At the moment when the platform of the third car shoots past an assistant station-master M, it is passed by a passenger P walking inside the train at constant velocity v'. The station-master, the smoker, and the passenger are assumed to have chronometers.

It is required to determine the velocity V of the passenger P with respect to the assistant station-master M.

But wait, any sane man would say, why S, why chronometers? Obviously, $V = v + v'$!

And Newton would answer the same. As to Einstein, he obtained a completely different formula, namely $V = (v+v')/(1+vv')$. Apparently, that was the occasion on which the following satire was written:

Nature and Nature's laws lag hid in night.
God said: "Let Newton be!" And all was light.
But Satan brought Einstein to the fore.
Now all is dark, just as before.

(The first two lines of this verse are by the English poet Alexandre Pope (1688–1744).)

Being armed with a thorough knowledge of geometry, let us try to understand whether the situation with relativity theory is as dark as it seems to the author of the satire.

First, we should mention that Newton was not aware of one important fact that was experimentally established only in the beginning of the twentieth century. It was discovered that *light had a constant velocity in an arbitrary inertial coordinate system* (i.e., in a coordinate system moving at a constant velocity), and nothing can move faster than light.

Thus, if at that initial moment, all three characters—one at rest (M), the second moving at the velocity of the train (S), and the third having his own additional velocity (P)—would catch three sunbeams in a mirror and reflect them along the road, then these sunbeams (however surprising it may seem) would neither fall behind nor outrun each other.

This unavoidably implies that the chronometers of M, S, and P behave differently! and, moreover, that coordinates and time cannot be considered independently.

Suppose that we have three equivalent coordinate systems: one (fixed) is associated with M, the second (moving) with S, and the third (moving "even faster") with P. At every moment of time, we have three readings t, t', t'' of the chronometers of M, S, P, respectively, and every chronometer has its own coordinates (in each of the coordinate systems).

Suppose that, at some moment of time, say t according to the chronometer of M and t' according to the chronometer of S, the passenger P has coordinates x in the system associated with M and x' in the system associated with S. We assume that Galileo's law of inertia is valid, according to which all inertial coordinate systems are equivalent (i.e., if there are curtains on the windows of the platform, then S cannot determine whether the train is moving or not); it is then natural to suppose that the *relation between the pairs* (x, t) *and* (x', t') *is linear.* This is the most important postulate of relativity theory. (If coordinate transformations were nonlinear, then things would be fairly obscure indeed.) Finally, we assume that the time scale is such that the velocity of light equals one.

Now, all assumptions have been made, and we can start with the derivation of the formula for the sum of velocities.

Suppose that S reflects a sunbeam at the initial moment of time and the position of this sunbeam after time t (measured by M) is determined by coordinates (x, t)

in the fixed coordinate system and by coordinates (x', t') in the system of the train. The conjecture asserting that the velocity of light is constant then implies that we have $x/t = x'/t' = 1$, i.e., the linear transformation relating coordinates must preserve the form $x^2 - t^2$. Such transformations are familiar to us—they are the hyperbolic rotations (which are in no way more complicated than ordinary rotations)

$$t' = t\cosh a - x\sinh a, \tag{9a}$$

$$x' = -t\sinh a + x\cosh a. \tag{9b}$$

Now, consider the moment of time when the chronometer of M reads t; let the position of the smoker S at this moment in the coordinate system associated with the train be $(0, t')$ (the smoker continues smoking in the train, so his coordinate x' is still zero, and his chronometer reads t'). Then (9b) implies that

$$v = x/t = \tanh a. \tag{10}$$

Only two steps remain to the formula for the addition of velocities. At time t (according to the chronometer of M), the passenger has coordinates (Vt, t) in the fixed coordinate system and $(v't', t')$ in the coordinate system associated with the train. This, (9a), and (9b) give

$$t = t'\cosh a + v't'\sinh a,$$
$$Vt = t'\sinh a + v't'\cosh a.$$

Therefore,

$$V = \frac{\sinh a + v'\cosh a}{\cosh a + v'\sinh a} \overset{\text{Id}}{=} \frac{v + v'}{1 + vv'}.$$

That is it, Einstein's formula is proved. And so things are not that obscure. All is fairly simple. Even very simple. But no more than that!

Let us ask the question: Is it possible to express the essence of special relativity by a single sentence (certainly, intendend for Masters in Geometry)?

This is quite possible. Consider the map $x \mapsto (x+v)/(1+xv)$; we have already encountered it several times. This map takes the interval $[-1, 1]$ to itself, the point 0 to the point v, and the point v' to $(v+v')/(1+vv')$, i.e., to the sum of (the velocities) v and v'.

Thus, on the one hand, we have obtained a translation of the Lobachevsky line in the model on the interval $(-1, 1)$; on the other hand, the same map determines a formula for the sum of velocities in special relativity. Consider the "space of velocities" corresponding to one-dimensional motions. To be more precise, we take some coordinate system for a fixed system and assign a point on the interval $(-1, 1)$ to each of the other systems moving along one line; the summation of velocities corresponds to a translation of the Lobachevsky line. The same is true of the two-dimensional case (plane motions) and of the general, three-dimensional, case.

Thus one of the interpretations of special relativity is given by the sentence: *The space of velocities in special relativity is nothing but the Lobachevsky line in the one-dimensional case, the Lobachevsky plane in the two-dimensional case, and the Lobachevsky space in the three-dimensional case.*

This view on the Lobachevsky plane is adopted in the book [**DSS**], where the reader can find many interesting things about geometric interpretations of diverse physical phenomena in the language of Lobachevsky geometry.

2. Polyhedra and polygons

Convex polyhedra. The theory of convex polyhedra is the cradle of convex geometry. We pay much attention to it in this section of the *Addendum.*

The convex hull of N points in the affine n-space among which there are $n+1$ affinely independent points is called an *n-dimensional convex polyhedron.*

A hyperplane Π^{n-1} is said to be a *face hyperplane* of an n-dimensional convex polyhedron $M^n \subset \mathbb{R}^n$ if M^n lies "on one side" of Π^{n-1} and the intersection of M^n with Π^{n-1} is an $(n-1)$-dimensional polyhedron. The intersection of a face hyperplane with the n-dimensional polyhedron M^n is an $(n-1)$-dimensional face (or merely $(n-1)$-face) of the polyhedron M^n. An $(n-2)$-face of the polyhedron can be defined by induction as an $(n-2)$-face of its $(n-1)$-face, etc. Zero-dimensional faces are called *vertices* and one-dimensional, *edges.*

Polyhedra in multidimensional spaces often have unexpected properties. One of the most striking examples is a convex polyhedron having an arbitrary number of vertices but no diagonals!

THEOREM 1. *For any $N \geqslant 5$, there exists a convex polyhedron in $\mathbb{R}^4$ such that it has N vertices and any two of its vertices are joined by an edge.*

Proof. Consider the curve $\gamma(t) = (t, t^2, t^3, t^4)$ in $\mathbb{R}^4$ and choose N different points $A_i = (t_i, t_i^2, t_i^3, t_i^4)$ on this curve. Let us show that the convex hull of these points has the required properties. Take points A_i and A_j and consider the polynomial

$$(t-t_i)^2(t-t_j)^2 = a_0 + a_1 t + a_2 t^2 + a_3 t^3 + a_4 t^4.$$

Let us show that the hyperplane

$$a_0 + a_1 x_1 + a_2 x_2 + a_3 x_3 + a_4 x_4 = 0$$

intersects γ at exactly two points A_i and A_j, and all other points of γ lie on one side of this hyperplane. Indeed,

$$a_0 + a_1 t + a_2 t^2 + a_3 t^3 + a_4 t^4 = (t-t_i)^2(t-t_j)^2 \geqslant 0.$$

Thus all the points $A_1, \dots, A_N$ lie on one side of the hyperplane under consideration, and only the points A_i and A_j lie on the hyperplane itself. Therefore, A_iA_j is an edge of the polyhedron which is the convex hull of the points $A_1, \dots, A_N$. □

The Euler–Poincaré formula for the alternating sum of the numbers of faces (of various dimensions) of a convex polyhedron. In 1758 Euler proved that for a convex polyhedron in $\mathbb{R}^3$, the number v of vertices, the number e of edges, and the number f of faces are related by $v - e + f = 2$. Descartes (1620) knew this formula, but had not published it. Later, Poincaré generalized Euler's formula to n-dimensional CW-complexes, in particular, to convex n-dimensional polyhedra as follows.

THEOREM 2. *Let N_k be the number of k-faces of a convex n-dimensional polyhedron M^n. Then*

$$\sum_{k=0}^{n-1} (-1)^k N_k = (-1)^{n-1} + 1. \tag{1}$$

Proof. We argue by induction on n. For $n = 2$, the assertion is obvious: the number of vertices of a convex polygon equals the number of its sides. Suppose that (1) is valid for any convex $(n-1)$-dimensional polyhedron, where $n \geqslant 3$. For the given convex polyhedron M^n, we can choose a line l such that the projections of the vertices of M^n on this line are pairwise different. Indeed, there are only finitely many lines joining pairs of vertices of the polyhedron; for l, we can take any line that is not orthogonal to any of these lines. It can be constructed as a line passing through the origin and lying in none of the hyperplanes through the origin that are orthogonal to the lines under consideration.

Let $a_1 < a_2 < \cdots < a_v$ be the projections of the vertices of the polyhedron to the line l. For each i, take a point b_i on the line l between a_i and a_{i+1}. Let us draw hyperplanes $H_1, \ldots, H_{2v-1}$ orthogonal to the line l and passing through the points $a_1, b_1, a_2, b_2, \ldots, a_{v-1}, b_{v-1}, a_v$, respectively. For $i = 2, 3, \ldots, 2v-2$, the set $M_i = M^n \cap H_i$ is a convex $(n-1)$-dimensional polyhedron. (The point a_i belongs to H_{2i-1}, and the point b_i belongs to H_{2i}.)

Let F^k be a k-face of the polyhedron M^n. We put $\psi(F^k, M_i) = 1$ if the polyhedron M_i contains at least one interior point of F^k and $\psi(F^k, M_i) = 0$ otherwise. The projection of the face F^k to the line l is a segment $[a_s, a_t]$. We have $\psi(F^k, M_i) = 1$ for $i = 2s, 2s+1, \ldots, 2t-2$; therefore,

$$\sum_{i=2}^{2v-2} (-1)^i \psi(F^k, M_i) = 1.$$

Summing these equalities over all k-faces F^k, we obtain

$$\sum_{k\text{-faces}} \sum_{i=2}^{2v-2} (-1)^i \psi(F^k, M_i) = N_k.$$

Next, consider the alternating sum of the obtained equalities over k:

$$\sum_{k=1}^{n-1} (-1)^k \sum_{k\text{-faces}} \sum_{i=2}^{2v-2} (-1)^i \psi(F^k, M_i) = \sum_{k=1}^{n-1} (-1)^k N_k. \tag{2}$$

The idea of the proof of the Euler–Poincaré formula is to change the order of summation, apply the induction hypothesis to the polyhedra M_i, and use the equality $v = N_0$.

Let $N_k(M_i)$ be the number of k-faces of the polyhedron M_i. Then

$$\sum_{1\text{-faces}} \psi(F^1, M_i) = N_0(M_i) - \varepsilon_1(i), \tag{3}$$

where $\varepsilon_1(i) = 1$ for odd i and $\varepsilon_1(i) = 0$ for even i. Indeed, if i is even, then all vertices of M_i are interior points of edges of M^n, and if $i = 2s-1$ is odd, then so are all the vertices except one, a_s, which is also a vertex of M^n. Now suppose that $k > 1$; then

$$\sum_{k\text{-faces}} \psi(F^k, M_i) = N_{k-1}(M_i). \tag{4}$$

Equalities (3) and (4) imply

$$\sum_{k=1}^{n-1} (-1)^k \sum_{k\text{-faces}} \psi(F^k, M_i) = \sum_{k=1}^{n-1} (-1)^k N_{k-1}(M_i) + \varepsilon_1(i).$$

According to the induction hypothesis,

$$\sum_{k=1}^{n-1}(-1)^k N_{k-1}(M_i) = -((-1)^{n-2}+1) = -1+(-1)^{n-1}.$$

Therefore,

$$\sum_{k=1}^{n-1}(-1)^k \sum_{k\text{-faces}} \psi(F^k, M_i) = (-1)^{n-1} + \varepsilon_2(i),$$

where $\varepsilon_2(i) = \varepsilon_1(i) - 1$. Hence

$$\sum_{i=2}^{2v-2}(-1)^i \sum_{k=1}^{n-1}(-1)^k \sum_{k\text{-faces}} \psi(F^k, M_i) = (-1)^{n-1}\sum_{i=2}^{2v-2}(-1)^i + \sum_{i=2}^{2v-2}(-1)^i\varepsilon_2(i)$$
$$= (-1)^{n-1} + \sum_{i=2}^{2v-2}(-1)^i\varepsilon_2(i);$$

here $\varepsilon_2(i) = 0$ if i is odd and $\varepsilon_2(i) = -1$ if i is even. Thus

$$\sum_{i=2}^{2v-2}(-1)^i \sum_{k=1}^{n-1}(-1)^k \sum_{k\text{-faces}} \psi(F^k, M_i) = (-1)^{n-1} - v + 1.$$

Comparing this expression with (2), we obtain

$$\sum_{k=1}^{n-1}(-1)^k N_k = (-1)^{n-1} - N_0 + 1, \quad \text{i.e.,} \quad \sum_{k=0}^{n-1}(-1)^k N_k = (-1)^{n-1} + 1. \quad \square$$

Dual polyhedra. Let $A_1, \dots, A_N$ be the vertices of a convex polyhedron M in $\mathbb{R}^n$ for which the origin O is an interior point (M then contains a ball of radius ε centered at O). Put $a_i = \overrightarrow{OA_i}$ and consider the set M^* determined by the system of inequalities $(x, a_i) \leqslant 1$ with $i = 1, \dots, N$. This finite system of inequalities is equivalent to the infinite system $(x, a) \leqslant 1$, where a is an arbitrary convex linear combination of the vectors a_i, i.e., $a = \sum \lambda_i a_i$, where $\sum \lambda_i = 1$ and $\lambda_i \geqslant 0$. Otherwise, a can be defined as a vector $\overrightarrow{OA}$, where $A \in M$.

The polyhedron M contains a sphere of radius ε centered at O. Therefore, if $x \in M^*$, then $(x, a) \leqslant 1$ for all vectors a of length ε. Thus $|x| \leqslant \varepsilon^{-1}$, i.e., the set M^* is bounded. On the other hand, the polyhedron M lies in a sphere of some radius R centered at O. Hence, if $|x| \leqslant R^{-1}$, then $(x, a_i) \leqslant 1$. Thus the set M^* contains a ball of radius R^{-1} centered at O. Taking into account the fact that the set M^* is determined by a finite system of linear inequalities, we conclude that M^* is a convex polyhedron.

If $a \in M$ and $b \in M^*$, then $(b, a) \leqslant 1$. Moreover,

$$(5) \qquad M^* = \{b \mid (b,a) \leqslant 1 \,\forall a \in M\} \quad \text{and} \quad M = \{a \mid (b,a) \leqslant 1 \,\forall b \in M^*\}.$$

Only the inclusion $\{a \mid (b,a) \leqslant 1 \,\forall b \in M^*\} \subset M$ needs to be proved. Let $a' \notin M$. Then there exists a hyperplane $(x, h) = 1$ that separates a' from M; in other words, h satisfies $(a', h) > 1$ and $(a, h) \leqslant 1$ for all $A \in M$, i.e., $h \in M^*$ and $(a', h) > 1$.

Since equalities (5) are symmetric, we have $M = (M^*)^*$. For this reason, the polyhedra M and M^* are said to be *dual* (with respect to the point O).

Two faces of a polyhedron are called *incident* if one of them is contained in the other.

THEOREM 3. *There is a one-to-one correspondence between the k-faces of an n-dimensional convex polyhedron M and the $(n-k-1)$-faces of its dual polyhedron M^* preserving incidence (if $F_1 \subset F_2$, then $F_1^* \supset F_2^*$).*

Proof. For a face F of a polyhedron M, consider all supporting hyperplanes of M whose intersections with M coincide with F, i.e., all hyperplanes of the form $(x, b) = 1$ such that $(a, b) \leqslant 1$ for all $a \in M$, but the equality $(a, b) = 1$ is only attained at points $a \in M$ belonging to the face F. The endpoints of the vectors b corresponding to these supporting hyperplanes form a dual face F^*.

Let V and V^* be the affine varieties generated by the faces F and F^*. It is easy to verify that if $\dim V = k$, then $\dim V^* = n - k - 1$. Indeed,

$$V^* = \{x \mid (x, a) = 1 \,\forall a \in V\}.$$

Choosing a point $a_0 \in V$, we obtain the following system of homogeneous linear equations:

$$(x, a - a_0) = 0 \qquad \text{for} \quad a \in V;$$

together with nonhomogeneous equation $(x, a_0) = 1$, it determines V^*. The solution of the system of homogeneous equations is a space of dimension $n - k$, and the nonhomogeneous equation cuts out a hyperplane of dimension $n - k - 1$ in this space.

The preservation of incidence of faces follows from the observation that if $U \subset V$, then $U^* \supset V^*$. □

The Gram–Sommerville formula for the alternating sum of solid angles at the faces of a convex polyhedron. Let M^n be a convex n-dimensional polyhedron, and let F_i^k be one of its k-faces $(0 \leqslant k \leqslant n-1)$. The solid angle at the face F_i^k can be defined as follows. Consider an interior point A of F_i^k (if $k = 0$, then $A = F_i^k$) and the sphere $S(A, \varepsilon)$ of radius ε centered at A. For a sufficiently small ε, the ratio of the $(n-1)$-dimensional volumes of the figures $M^n \cap S(A, \varepsilon)$ and $S(A, \varepsilon)$ does not depend on ε. We call this ratio the *angle in absolute measure* at the face F_i^k, or a *solid angle in absolute measure.*[2] The angle in absolute measure at a face of dimension $n - 1$ is equal to $1/2$.

Let σ_k be the sum of solid angles in absolute measure at all k-faces of a convex polyhedron M^n. If $n = 2$, i.e., M^n is a convex m-gon, then the solid angle at its vertex equals half the angle (in radians) at this vertex, and the solid angle at an edge (a side) of the polygon is $1/2$; thus $\sigma_0 = (m-2)/2$ and $\sigma_1 = m/2$, and therefore $\sigma_1 - \sigma_0 = 1$. In 1874 Gram proved that for convex polyhedra in $\mathbb{R}^3$, the relation $\sigma_2 - \sigma_1 + \sigma_0 = 1$ holds. A similar formula for n-dimensional convex polyhedra was proved by Sommerville (1927), but his proof contained a gap, which was removed by Grünbaum (1967).

THEOREM 4 (Gram's formula). *Let σ_k be the sum of solid angles in absolute measure at all k-faces of an n-dimensional convex polyhedron M^n. Then*

$$\sigma_{n-1} - \sigma_{n-2} + \cdots + (-1)^{n-1}\sigma_0 = 1, \quad \textit{i.e.,} \quad \sum_{k=0}^{n-1}(-1)^k \sigma_k = (-1)^{n-1}.$$

[2]Ordinary solid angles differ from the solid angles in absolute measure in normalization: the complete solid angle equals the $(n-1)$-dimensional volume of the unit $(n-1)$-sphere, whereas in the absolute measure it is 1.

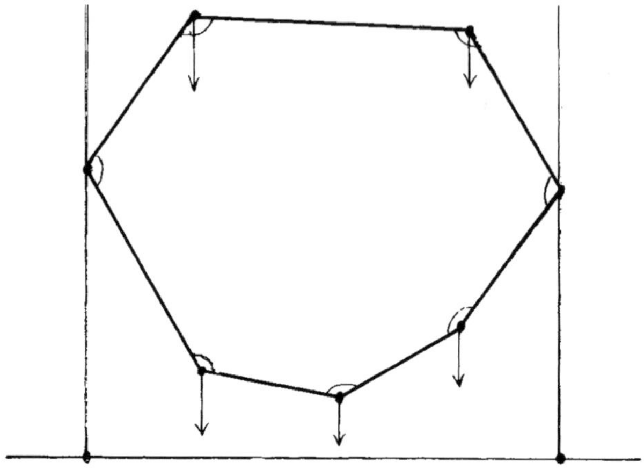

FIGURE A.2

Proof. (Shepard) On each face F_i^k, take an interior point A_i^k and consider the cone formed by the rays from the point A_i^k through all points of the polyhedron M^n. Let O be the origin. Shifting each cone by the vector $\overrightarrow{A_i^k O}$, we obtain cones with common vertex O. It is required to prove that if we assign the weight $(-1)^k$ to every cone for which the dimension of the corresponding face is k, then each point in the space (except the boundary points of the cones) will be covered by the cones $(-1)^{n-1}$-fold, and the sum of weights of all cones containing the point under consideration will be $(-1)^{n-1}$.

Let x be a vector not belonging to the boundaries of the cones under consideration; the vector x is then parallel to none of the k-faces of the polyhedron M^n, where $1 \leqslant k \leqslant n-1$. The cone corresponding to a face F_i^k contains x if and only if the points $A_i^k + \varepsilon x$ with arbitrary sufficiently small $\varepsilon > 0$ belong to M^n. Otherwise, this property can be described as follows. Let $\pi_x(M^n)$ be the orthogonal projection of the polyhedron M^n on the hyperplane orthogonal to the vector x. The polyhedron $\pi_x(M^n)$ is covered by two layers of projections of the faces of M^n. The first layer is formed by the projections of those faces for which the corresponding cones contain the vector x, and the second by the projections of the faces for which the corresponding cones contain the vector $-x$. The cones that correspond to the faces projected on the boundary of the polyhedron $\pi_x(M^n)$ contain neither x nor $-x$ (Figure A.2).

Let N_k' be the number of k-faces of the $(n-1)$-dimensional polygon $\pi_x(M^n)$, and let N_k'' be the number of k-faces in the first layer of the cover of $\pi_x(M^n)$, i.e., the number of k-faces of M^n corresponding to cones containing the vector x. Let us put $N_k = N_k' + N_k''$. An argument similar to the one used in the proof of the Euler–Poincaré formula (see Theorem 2 on p. 140) shows that $\sum_{k=0}^{n-1}(-1)^k N_k = 1$, i.e.,

$$\sum_{k=0}^{n-1}(-1)^k N_k'' = 1 - \sum_{k=0}^{n-1}(-1)^k N_k'.$$

We have $N_{n-1}' = 0$ and, according to the Euler–Poincaré formula,

$$\sum_{k=0}^{n-2}(-1)^k N_k' = (-1)^{n-2} + 1.$$

Therefore, $\sum_{k=0}^{n-1}(-1)^k N_k'' = (-1)^{n-1}$, as required. □

The Gauss–Bonnet theorem. Let A be a vertex of a convex three-dimensional polyhedron. We define its *curvature* $k(A)$ to be the difference between 2π and the sum of the angles at the vertex A of all faces incident to the vertex A. This definition makes sense not only for vertices, but also for all points on the surface of the polyhedron. But both the interior points of faces and the interior points of edges have zero curvature. This is so because a small neighborhood of an edge point A can be "unfolded" to become flat (by a small neighborhood, we mean the set of points on the surface of the polyhedron that are at distances ε or less from A). A small neighborhood of a vertex also can be "unfolded" after it is cut. But then an "angular crack" appears, which is precisely the curvature of the vertex.

THEOREM 5. *The total curvature of a polyhedron (i.e., the sum of curvatures of all its vertices) equals*

$$2\pi(N_0 - N_1 + N_2),$$

where N_0 is the number of vertices, N_1 is the number of edges, and N_2 is the number of faces.

Proof. The sum of curvatures of all vertices is equal to $2\pi N_0 - \sigma$, where σ is the sum of all angles of the faces of the polyhedron. The sum of angles of the ith face equals $(n_i - 2)\pi$, where n_i is the number of sides of the ith face. Therefore,

$$\sigma = \sum_{i=1}^{N_2}(n_i - 2)\pi = \sum n_i - 2\pi N_2.$$

Clearly, $\sum n_i = 2N_1$, because each edge belongs to exactly two faces and, hence, each edge is included twice in $\sum n_i$. □

We have not used convexity in the proof of Theorem 5; the theorem is valid for nonconvex polyhedra too. But in the case of convex polyhedra, we can apply the Euler formula $N_0 - N_1 + N_2 = 2$.

Historical comments. Descartes knew that the total curvature of a convex polyhedron is 4π, but his proof was lost. The other expression for the curvature ($2\pi(N_0 - N_1 + N_2)$) was also known to Descartes. Thus Descartes was aware of the formula $N_0 - N_1 + N_2 = 2$ long before Euler, but Descartes' notes on this topic were not published until 1860. For this reason, the formula $N_0 - N_1 + N_2 = 2$ is usually called after Euler rather than Descartes.

The theorem that the total curvature of the surface of a polyhedron is $2\pi(N_0 - N_1 + N_2)$ is sometimes called the Gauss–Bonnet theorem because they proved a similar theorem for smooth surfaces. (The theorem for polyhedra can be obtained from the theorem for smooth surfaces by passing to the limit.)

Theorem 5 can be generalized to n-dimensional polyhedra. Its proof uses Gram's formula (see Theorem 4 on p. 143); for this reason, we consider angles in absolute measure, i.e., assume the complete angle to be 1.

First, consider a four-dimensional polyhedron M. For a vertex V_i and an edge (1-face) E_j of this polyhedron, we put $k_0(V_i) = 1 - s_3(V_i)$, where $s_3(V_i)$ is the sum of the solid angles with vertex V_i of all 3-faces incident to the vertex V_i, and $k_1(E_j) = 1 - s_2(E_j)$, where $s_2(E_j)$ is the sum of the dihedral angles with edge E_j of all 3-faces incident to the edge E_j.

THEOREM 6 (Grünbaum–Shepard).

$$\sum_i k_0(V_i) - \sum_j k_1(E_j) = N_0 - N_1 + N_2 - N_3 = 0.$$

Proof. Let us write Gram's formula for each 3-face F_k of the polyhedron M:

$$\sigma_0(F_k) - \sigma_1(F_k) + \sigma_2(F_k) = 1,$$

where $\sigma_0(F_k)$ is the sum of solid angles at the vertices of the polyhedron F_k, $\sigma_1(F_k)$ is the sum of dihedral angles at the edges of F_k, and $\sigma_2(F_k)$ is half the number of 2-faces in F_k. Summing these equalities for all 3-faces, we obtain

$$\sum \sigma_0(F_k) - \sum \sigma_1(F_k) + \sum \sigma_2(F_k) = N_3,$$

i.e., $s_3 - s_2 + N_2 = N_3$, where s_3 is the sum of all solid angles (over all 3-faces of the polyhedron M) and s_2 is the sum of all dihedral angles. By definition, we have $\sum k_0(V_i) = N_0 - s_3$ and $\sum k_1(E_j) = N_1 - s_2$. Therefore,

$$\sum k_0(V_i) = N_0 - s_3 - \sum k_1(E_j) = N_0 - N_1 + s_2 - s_3 = N_0 - N_1 + N_2 - N_3. \quad \square$$

The following assertion about an n-dimensional polyhedron M is proved similarly.

THEOREM 7. *Let $k_i(F^{n-i}) = 1 - s_{n-1-i}(F^{n-i})$, where $s_{n-1-i}(F^{n-i})$ is the sum of all solid angles at the $(n-i)$-faces F^{n-i} over all $(n-1)$-faces of the polyhedron M that contain F^{n-i}, and let $k_i(M)$ be the sum of $k_i(F^{n-i})$ over all $(n-i)$-faces of M. Then*

$$\sum_{i=0}^{n-3} k_i(M) = \sum_{i=0}^{n-1} (-1)^i N_i.$$

The Minkowski theorem. Let M be a convex m-dimensional polyhedron. Suppose that $n_1, \dots, n_k$ are the outward normal unit vectors to its $(m-1)$-faces and $F_1, \dots, F_k$ are the $(m-1)$-dimensional volumes of these faces. It is easy to verify that $\sum F_i n_i = 0$. Indeed, if Π is an arbitrary hyperplane and n is the unit normal to this hyperplane, then (n, n_i) is the cosine of the angle between Π and the hyperplane of the face F_i; therefore, $(n, n_i)F_i$ is the $(m-1)$-dimensional volume of the projection of the face F_i on Π taken with the corresponding sign. The projections of faces of the polyhedron M cover the projection of M twice; we have $(n, n_i) > 0$ for one of the two faces covering a point and $(n, n_i) < 0$ for the other. Therefore, $\sum (n, n_i)F_i = 0$, i.e., $(n, \sum F_i n_i) = 0$ for any unit vector n. Hence $\sum F_i n_i = 0$.

For any convex m-dimensional polyhedron, the vectors $n_1, \dots, n_k$ must generate the entire m-dimensional space. Indeed, if the vectors $n_1, \dots, n_k$ were contained in some hyperplane Π, then the polyhedron would be unbounded in the normal direction to Π.

Minkowski proved that these two conditions on the vectors n_i and the numbers F_i are not only necessary but also sufficient.

THEOREM 8 (Minkowski). *Let $n_1, \dots, n_k$ be unit vectors in $\mathbb{R}^m$ generating the entire space $\mathbb{R}^m$. If $F_1, \dots, F_k$ are positive numbers such that $\sum F_i n_i = 0$, then there exists a convex m-dimensional polyhedron for which the volumes of the $(m-1)$-faces equal $F_1, \dots, F_k$ and the vectors $n_1, \dots, n_k$ are the outward unit normals to these faces.*

Proof. First, note that the vectors $n_1, \ldots, n_k$ are not directed into one half-space, i.e., there exists no vector n such that $(n, n_i) > 0$ for all i. Indeed, if $(n, n_i) > 0$ for all i, then $(n, \sum F_i n_i) = \sum F_i(n, n_i) > 0$ because $F_i > 0$. On the other hand, $(n, \sum F_i n_i) = 0$, because $\sum n_i F_i = 0$.

Choose positive numbers $h_1, \ldots, h_k$. Let us show that the set

$$M(h_1, \ldots, h_k) = \{x \in \mathbb{R}^m | (x, n_i) \leqslant h_i, \ i = 1, \ldots, k\}$$

is an m-dimensional convex polyhedron. To this end, we must verify that it is m-dimensional and bounded. The dimension of the set under consideration equals m, because it contains the ball $|x| \leqslant \min h_i$. It is bounded because, for any vector $e \neq 0$, there exists a vector n_i such that $(e, n_i) > 0$, and hence points of the ray te belong to $M(h_1, \ldots, h_k)$ only for $t \leqslant h_i/(e, n_i)$.

We denote the volume of $M(h_1, \ldots, h_k)$ by $V(h_1, \ldots, h_k)$ and the $(m-1)$-dimensional volume of the intersection of $M(h_1, \ldots, h_k)$ with the hyperplane $(x, n_i) = h_i$ by $F_i(h_1, \ldots, h_k)$ (the latter volume is only nonzero if $M(h_1, \ldots, h_k)$ has an $(m-1)$-face with outward normal n_i). It is easy to verify that

$$\frac{\partial V(h_1, \ldots, h_k)}{\partial h_i} = F_i(h_1, \ldots, h_k).$$

Indeed, if $F_i(h_1, \ldots, h_k) > 0$, then the replacement of h_i by $h_i + \varepsilon$ adds a figure whose volume approximately equals $\varepsilon F_i(h_1, \ldots, h_k)$ (at small $\varepsilon > 0$) to the polyhedron $M(h_1, \ldots, h_k)$. If $F_i(h_1, \ldots, h_k) = 0$ and the hyperplane $(x, n_i) = h_i$ is disjoint from the polyhedron, then the volume does not change; finally, if $F_i(h_1, \ldots, h_k) = 0$ and the polyhedron intersects $(x, n_i) = h_i$ in a face of dimension less than $m - 1$, then a set whose volume is a higher-order infinitesimal is added.

Consider all polyhedra $M(h_1, \ldots, h_k)$ for which the positive numbers $h_1, \ldots, h_k$ satisfy $\sum F_i h_i = 1$. Let us show that the family of all these polyhedra contains a polyhedron of maximal volume. Since the numbers $F_1, \ldots, F_k$ are positive, the intersection of the hyperplane $\sum F_i h_i = 1$ with the positive orthant $x_1 \geqslant 0, \ldots, x_k \geqslant 0$ is a $(k-1)$-simplex. On this simplex, the continuous function $V(h_1, \ldots, h_k)$ has a maximum; but we must prove that the maximum is attained at an interior point of the simplex. Let $h_1^0, \ldots, h_k^0$ be the point of maximum on the simplex. The maximum is positive; therefore, the set $M(h_1^0, \ldots, h_k^0)$ is a convex $(m-1)$-dimensional polyhedron (this should be mentioned, because the preceding remark that this set contains the ball $|x| \leqslant \min h_i$ is false for zero h_i). Take an arbitrary vector $a \in \mathbb{R}^m$. The inequalities $(x, n_i) \leqslant h_i^0 \, (i = 1, \ldots, k)$ and $(x + a, n_i) \leqslant h_i^0 \, (i = 1, \ldots, k)$ determine polyhedra that are obtained from each other by translations by $\pm a$. The inequality $(x + a, n_i) \leqslant h_i^0$ is equivalent to the inequality $(x, n_i) \leqslant h_i'$, where $h_i' = h_i^0 - (a, n_i)$. Let us choose a such that the origin is strictly inside the corresponding polyhedron, i.e., $h_i' > 0 \, (i = 1, \ldots, k)$; this polyhedron has the same maximal volume. In addition,

$$\sum F_i h_i' = \sum F_i h_i^0 - \sum F_i(a, n_i) = \sum F_i h_i^0 - (a, \sum F_i n_i) = \sum F_i h_i^0,$$

because $\sum F_i n_i = 0$.

Thus, the continuously differentiable function $V(h_1, \ldots, h_k)$ subject to the constraint $\sum F_i h_i = 1$ has a point of maximum $h_1^0, \ldots, h_k^0$ with positive coordinates h_i^0. This means that the hypersurface $V(h_1, \ldots, h_k) = V(h_1^0, \ldots, h_k^0)$ is tangent to the hyperplane $\sum F_i h_i = 1$ at the point $h_1^0, \ldots, h_k^0$, i.e., the normal vectors to the hyperplane and to the hypersurface at the point $h_1^0, \dots, h_k^0$ are proportional:

$$\frac{\partial V(h_1^0, \dots, h_k^0)}{\partial h_i} = \lambda F_i, \quad i = 1, \dots, k.$$

As shown above,

$$\frac{\partial V(h_1^0, \dots, h_k^0)}{\partial h_i} = F_i(h_1^0, \dots, h_k^0),$$

where $F_i(h_1^0, \dots, h_k^0)$ is the $(m-1)$-dimensional volume of the face with outer normal vector n_i. Thus the volumes of the faces of the polyhedron $M(h_1^0, \dots, h_k^0)$ are proportional to the required volumes $F_1, \dots, F_k$; therefore, the required polyhedron can be obtained from $M(h_1^0, \dots, h_k^0)$ by applying a homothety with factor $\lambda^{-1/(m-1)}$. □

The Cauchy theorem on rigid convex polyhedra. Two polyhedra are called *equivalent* if their faces are in a one-to-one correspondence preserving the dimensions of the faces and the incidence relation, i.e., such that the faces of dimension k correspond to faces of the same dimension k and, if a face F_1 is contained in a face F_2, then the corresponding face F_1' is contained in the face F_2'.

THEOREM 9 (Cauchy). *If two convex polyhedra M and M' in $\mathbb{R}^3$ are equivalent, each face of M is isometric to the corresponding face of M', and the isometries between faces map all vertices to the corresponding vertices, then the polyhedra themselves are isometric.*

REMARK. A similar assertion for nonconvex polyhedra is false. As an example, we can take a cube, construct two equal pyramids, exterior and interior, whose bases coincide with one of the faces of the cube, and consider the two polyhedra, one obtained by adding the exterior pyramid to the cube and another obtained by subtracting the interior pyramid.

Proof. Let O and O' be corresponding vertices of the polyhedra M and M'. Consider two spheres of small radius ε centered at O and O'. The intersections of these spheres with the polyhedra M and M' are convex spherical polygons $A_1 \dots A_n$ and $A_1' \dots A_n'$ with equal respective sides. To prove the Cauchy theorem, we need two lemmas about properties of spherical polygons.

LEMMA 1. *Let $A_1 \dots A_n$ and $A_1' \dots A_n'$ be convex spherical polygons such that their respective sides, except A_1A_n and $A_1'A_n'$, have equal lengths and satisfy the inequalities $\angle A_2 \leqslant \angle A_2', \dots, \angle A_{n-1} \leqslant \angle A_{n-1}'$. Then $|A_1A_n| \leqslant |A_1'A_n'|$.*

Proof. We argue by induction on n. For $n = 3$, the required assertion follows from the first spherical law of cosines (see p. 87)

$$\cos a = \cos b \cos c + \sin b \sin c \cos \alpha,$$

which implies that the side a increases as the angle α increases.

Now suppose that the required assertion is proved for convex $(n-1)$-gons. We start by proving the required assertion for pairs of convex n-gons that have equal angles at the vertices A_i and A_i', where $1 < i < n$. The triangles $A_{i-1}A_iA_{i+1}$ and $A_{i-1}'A_i'A_{i+1}'$ are equal, because their two respective sides and the angles between these sides are equal. Hence their sides $A_{i-1}A_{i+1}$ and $A_{i-1}'A_{i+1}'$ and the angles at these sides are equal; therefore, the induction hypothesis applies to the polygons $A_1 \dots A_{i-1}A_{i+1} \dots A_n$ and $A_1' \dots A_{i-1}'A_{i+1}' \dots A_n'$.

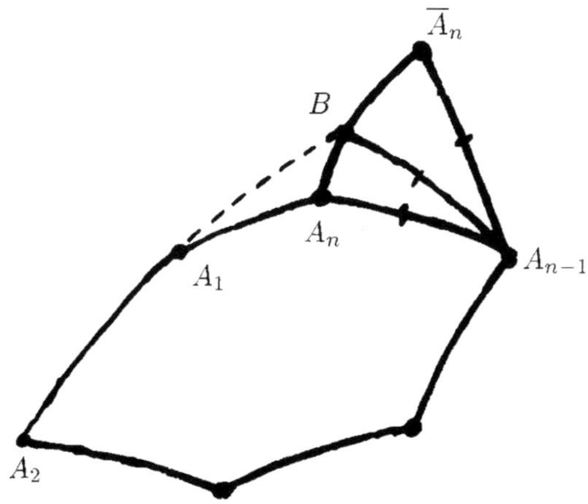

FIGURE A.3

To prove the induction step, we would like to make equal the angles at two respective vertices, say at A_{n-1} and A'_{n-1}. For this purpose, we rotate the side $A_{n-1}A_n$ so that the angle at the vertex A_{n-1} increases and becomes equal to the angle at the vertex A'_{n-1}. If the resulting polygon $A_1 \dots A_{n-1}\bar{A}_n$ is convex, then the proof is complete. It remains to consider the case where the resulting polygon is nonconvex (Figure A.3). In this case, the extension of the side A_1A_2 intersects the segment $[A_n, \bar{A}_n]$ at some point B. Clearly,

$$|A_1B| = |A_2B| - |A_1A_2|, \qquad |A'_1A'_n| \geqslant |A'_2A'_n| - |A'_1A'_2| = |A'_2A'_n| - |A_1A_2|.$$

To the convex $(n-1)$-gons $A_2A_3 \dots A_{n-1}B$ and $A'_2A'_3 \dots A'_{n-1}A'_n$, the induction hypothesis applies; therefore, $|A_2B| \leqslant |A'_2A'_n|$, and hence $|A_1B| \leqslant |A'_1A'_n|$. In addition, the triangles $A_1A_{n-1}A_n$ and $A_1A_{n-1}B$ have equal respective sides incident to the vertex A_{n-1}, and $\angle A_1A_{n-1}A_n < \angle A_1A_{n-1}B$. Therefore, $|A_1A_n| < |A_1B| \leqslant |A'_1A'_n|$. □

LEMMA 2. *Let $A_1 \dots A_n$ and $A'_1 \dots A'_n$ be nonequal convex spherical polygons with equal respective sides. Suppose that for each i, the angle A_i is labeled by $+$ if $\angle A_i > \angle A'_i$ and by $-$ if $\angle A_i < \angle A'_i$. Then the sign of the label changes strictly at least four times when we go around the polygon $A_1 \dots A_n$.*

Proof. Suppose that only one angle, A_2 say, is labeled. Then, by Lemma 1, either $|A_1A_n| > |A'_1A'_n|$ or $|A_1A_n| < |A'_1A'_n|$, which contradicts the assumptions of the lemma. The same arguments show that not all labels are the same; among them there must be both pluses and minuses. Thus it only remains to eliminate the situation where some segment $[P, Q]$ separates these pluses and minuses (Figure A.4).

Suppose that such a segment exists. Let P' and Q' be the corresponding points on the sides of the polygon $A'_1 \dots A'_n$. Applying Lemma 1 to the polygons M_1 and M'_1, we obtain $|PQ| > |P'Q'|$, and applying the same lemma to the polygons M_2 and M'_2, we obtain $|PQ| < |P'Q'|$. □

Now we can turn to the proof of the Cauchy theorem. Suppose that two corresponding dihedral angles in the polyhedra M and M' under consideration are not equal. Let us label the edges of M by $+$ and $-$ depending on whether the dihedral angles at these edges are larger or smaller than the corresponding dihedral

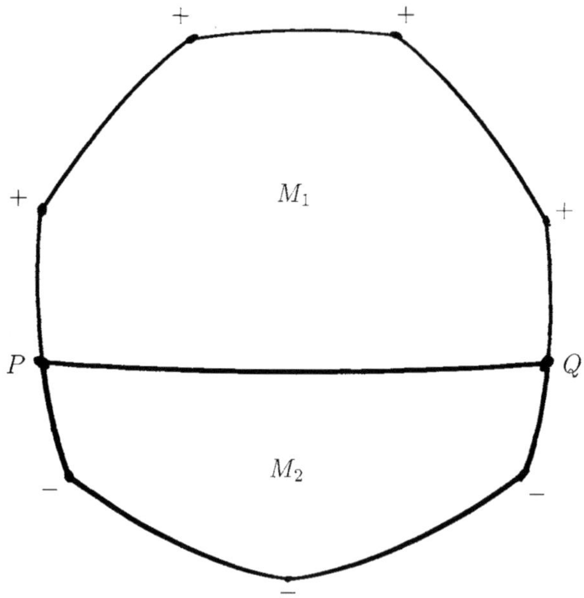

FIGURE A.4

angles of M'. If some vertex is incident to a labeled edge, then, according to Lemma 2, this vertex is incident to at least four labeled edges, and the sign changes at least four times in passing around this vertex.

First, consider the case where all edges of the polyhedron are labeled. The total number of changes of sign in passing around the vertices is equal to the total number of sign changes in going around the face boundaries; we denote this number by N. Let a_k be the number of k-side faces. The number of changes of sign in traversing the boundary of a k-sided face is even and does not exceed k; therefore,

$$N \leqslant 2a_3 + 4a_4 + 4a_5 + 6a_6 + \cdots .$$

Clearly, $a_3 + a_4 + a_5 + \cdots = N_2$ and $3a_3 + 4a_4 + 5a_5 + \cdots = 2N_1$, where N_1 is the number of edges and N_2 is the number of faces. Hence

$$4N_1 - 4N_2 = 2a_3 + 4a_4 + 6a_5 + 8a_6 + \cdots \geqslant N.$$

According to the Euler formula, $4N_1 - 4N_2 = 4N_0 - 8$, whence $N \leqslant 4N_0 - 8 < 4N_0$. On the other hand, in going around each vertex, the sign changes no less than four times; therefore, $N \geqslant 4N_0$. We have arrived at a contradiction.

If not all edges are labeled, we can use the same argument. Suppose that N_1' is the number of labeled edges, N_0' is the number of vertices incident to labeled edges, N_2' is the number of regions into which the labeled edges partition the surface of the polyhedron, and a_k' is the number of these regions bounded by k labeled edges. The network of labeled edges may be disconnected, i.e., it may consist of several components not connected to each other by labeled edges. But the two most important properties are retained, namely, $a_2' = 0$ and at least two labeled edges pass through any vertex of the network, i.e., there are no "free" labeled edges that share no endpoints with other labeled edges (the first property is obvious and the second follows from Lemma 2). Therefore, the same argument leads us to a contradiction if we are able to prove that $4N_1' - 4N_2' \leqslant 4N_0' - 8$, i.e., $N_0' - N_1' + N_2' \geqslant 2$.

Let us successively add nonlabeled edges to labeled ones; at each step, we add one edge with at least one vertex belonging to a labeled or already added edge. At every step, the number of edges increases by 1, and the number of vertices plus the number of regions either increases by 1 or does not change. Indeed, if we add a new vertex, then the number of regions does not change, and if the added edge joins two old vertices, then either this edge splits one region into two or this edge joins the vertices of two different components and breaks no region. Thus, at each step, the value $\tilde{N}_0 - \tilde{N}_1 + \tilde{N}_2$ either does not change or decreases by 1, and in the very end, when all edges are added, it equals 2. Therefore, $N_0' - N_1' + N_2' \geqslant 2$, as required. □

Regular polyhedra. A *regular n-gon* is a convex n-gon all of whose sides and angles are equal. The angle of a regular n-gon is $(n-2)\pi/n$. This implies that a regular n-gon is unique up to similarity. The existence of a regular n-gon is fairly evident. A regular n-gon can be defined differently as a convex n-gon which is transformed into itself under a rotation about some point through the angle $2\pi/n$.

The definition of a regular polyhedron in $\mathbb{R}^3$ requires care: the equality of all faces and all polyhedral angles is certainly not the right condition. Indeed, take an arbitrary right parallelepiped and consider one of the two tetrahedra whose edges are diagonals of the parallelepiped. It is easy to verify that all faces and all trihedral angles of this tetrahedron are equal.

Regular polyhedra in $\mathbb{R}^3$ can be defined in different ways. One of the definitions is as follows: a convex polyhedron in $\mathbb{R}^3$ is called *regular* if its dihedral angles are all equal and the faces are equal regular polygons. Instead of the equality of dihedral angles, we can require that for all vertices the endpoints of the incident edges form equal regular polygons. Indeed, for a convex polyhedron whose faces are equal regular polygons, both conditions are equivalent to the requirement that all polyhedral angles be equal and regular. (A convex n-hedral angle is called *regular* if it is transformed into itself under a rotation about some axis through the angle $2\pi/n$.)

Let us show that any regular polyhedron contains an interior point equidistant from all its vertices. Let us draw perpendiculars to two adjacent faces through the centers of these faces. They meet at a point whose distances to the centers of the faces are $a\tan\varphi$, where a is the distance from the center of a face to its side and φ is half the dihedral angle between the faces of the polyhedron. Now we draw perpendiculars to all faces through their centers and, on each perpendicular, mark a point (inside the polyhedron) at distance $a\tan\varphi$ from the center of the face. For each pair of adjacent faces, the marked points coincide; therefore, all these points coincide. As a result, we obtain a point equidistant from all vertices of the polyhedron. This point is called the *center* of the regular polyhedron.

Let us classify the regular polyhedra in $\mathbb{R}^3$. The *Schläfli symbol* of a regular polyhedron is the ordered pair of numbers $\{r_1, r_2\}$, where r_1 is the number of sides in a face of the polyhedron and r_2 is the number of sides in the polygon formed by the endpoints of the edges incident to one vertex.

THEOREM 10. (a) *Regular polyhedra in* $\mathbb{R}^3$ *can only have Schläfli symbols* $\{3,3\}$ (*tetrahedron*), $\{3,4\}$ (*octahedron*), $\{4,3\}$ (*cube*), $\{5,3\}$ (*dodecahedron*), *and* $\{3,5\}$ (*icosahedron*).

(b) *For each Schläfli symbol mentioned above, a regular polyhedron with this symbol exists and is unique up to similarity.*

Proof. (a) Suppose that a_3 is the edge length of a regular polyhedron with Schläfli symbol $\{r_1, r_2\}$, R_3 is the radius of the sphere circumscribed about this polyhedron, a_2 is the side length of the regular polygon formed by the endpoints of the edges incident to one vertex, and R_2 is the radius of the circle circumscribed about this polygon. Let us show that the values $\rho_3 = (a_3/2R_3)^2$ and $\rho_2 = (a_2/2R_2)^2$ are related by

$$\rho_3 = 1 - \frac{\cos^2(\pi/r_1)}{\rho_2}. \tag{6}$$

Let O be the center of the regular polyhedron. Consider a vertex A of this polyhedron and the regular r_2-gon $B_1 \dots B_{r_2}$ formed by the endpoints of the edges incident to A. If $\angle B_1OA = 2\varphi$, then $a_3 = 2R_3 \sin\varphi$ and $R_2 = a_3 \cos\varphi$. The face of the polyhedron under consideration is a regular r_1-gon with side a_3 and the length of its shortest diagonal is a_2; hence $a_2 = 2a_3 \cos(\pi/r_1)$. Therefore,

$$\rho_3 = \left(\frac{a_3}{2R_3}\right)^2 = \sin^2\varphi, \qquad \rho_2 = \left(\frac{a_2}{2R_2}\right)^2 = \frac{\cos^2(\pi/r_1)}{\cos^2\varphi}.$$

These formulas obviously imply (6).

It is easy to verify that $\rho_2 = \sin^2(\pi/r_2)$, i.e., ρ_2 only depends on r_2, and (6) shows that ρ_3 only depends on $\{r_1, r_2\}$. For this reason, it is convenient to introduce the notations $\rho\{r_2\} = \rho_2$ and $\rho\{r_1, r_2\} = \rho_3$, which only depend on the Schläfli symbol of the polyhedron rather than on the polyhedron itself.

Formula (6) and the obvious inequality $\rho_3 > 0$ imply that

$$\rho_2 > \cos^2\frac{\pi}{r_1} \geqslant \cos^2\frac{\pi}{3} = \frac{1}{4}, \qquad \text{i.e.,} \quad \sin\frac{\pi}{r_2} > \frac{1}{2};$$

hence $r_2 \leqslant 5$. Simple calculations show that

$\rho\{3\} = 3/4$, and $\rho\{3\} > \cos^2(\pi/r_1)$ only if $r_1 \leqslant 5$;
$\rho\{4\} = 1/2$, and $\rho\{4\} > \cos^2(\pi/r_1)$ only if $r_1 \leqslant 3$;
$\rho\{5\} = (5 - \sqrt{5})/8$, and $\rho\{5\} > \cos^2(\pi/r_1)$ only if $r_1 \leqslant 3$.

(b) First, we prove the uniqueness (up to similarity) of a regular polyhedron with a given Schläfli symbol. Consider two regular polyhedra with the same Schläfli symbol. Applying a similarity transformation, we can position them so that a face of one polyhedron coincides with a face of the other and the polyhedra lie on the same side of the plane containing this common face. The number $(a_3/2R_3)^2 = \rho\{r_1, r_2\}$ is completely determined by the Schläfli symbol; therefore, the spheres circumscribed about the polyhedra under consideration have equal radii, and the centers of the polyhedra coincide. Hence the polyhedra themselves coincide.

Let us prove the existence of the required regular polyhedra. The existence of a regular tetrahedron and the cube is obvious. The convex hull of the centers of the faces of a cube is an octahedron.

The convex hull of the centers of the faces of an icosahedron is a dodecahedron (and vice versa). Thus, it suffices to prove the existence of an icosahedron. Let us orient the edges of an octahedron so that for each face, the orientation determine some direction of going around its boundary (Figure A.5). Next, we divide the edges in the ratio $\lambda : (1 - \lambda)$ in accordance with their orientations. Let us show that λ can be chosen so that the division points form the vertices of an icosahedron. For any λ, the polyhedron obtained has edges of two types, depending on whether or not they belong to faces of the octahedron. If the edge length of the octahedron

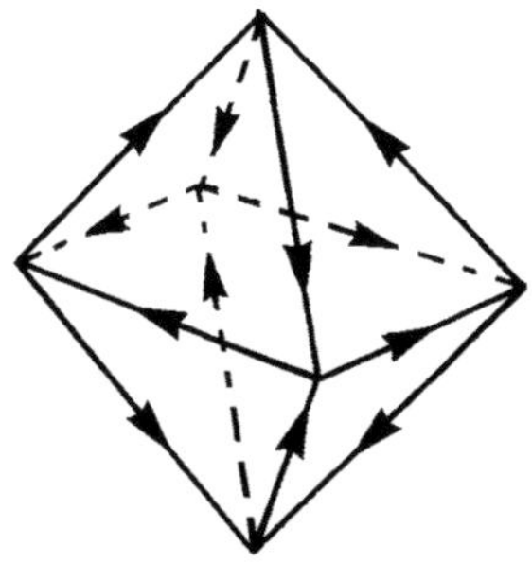

FIGURE A.5

is 1, then the squared length of an edge of the first type is

$$\lambda^2 + (1-\lambda)^2 - 2\lambda(1-\lambda)\cos 60^\circ = 3\lambda^2 - 3\lambda + 1,$$

and the squared length of an edge of the second type is

$$2(1-\lambda)^2 = 2 - 4\lambda + 2\lambda^2$$

(such an edge is the hypotenuse of a right triangle with legs of length $1-\lambda$). Thus, if $\lambda = (\sqrt{5}-1)/2$, then all the faces are regular triangles. The dihedral angles of the polyhedron obtained are equal because (for any λ) all division points are equidistant from the center of the octahedron. □

A regular polyhedron in $\mathbb{R}^n$ can be defined by induction. Namely, a convex polyhedron in $\mathbb{R}^n$ is said to be *regular* if all $(n-1)$-faces are equal regular polyhedra, and the endpoints of the edges incident to vertices form equal regular $(n-1)$-dimensional polyhedra. The Schläfli symbol is also defined by induction: if a 2-face of a regular polyhedron is an r_1-gon and the Schläfli symbol of the polyhedron formed by the endpoints of the edges incident to a vertex equals $\{r_2, \ldots, r_n\}$, then the *Schläfli symbol* of the polyhedron equals $\{r_1, r_2, \ldots, r_n\}$.

The existence of a center in a regular polyhedron is proved in the same way as in the three-dimensional case.

Put $\rho\{r_1, r_2, \ldots, r_n\} = \rho_n = (a_n/2R_n)^2$, where a_n is the edge length and R_n is the radius of the circumscribed sphere. Formula (6) then takes the form

$$\rho\{r_1, r_2, \ldots, r_n\} = 1 - \frac{\cos^2(\pi/r_1)}{\rho\{r_2, \ldots, r_n\}}. \tag{7}$$

This formula is proved as in the three-dimensional case with the only difference that instead of faces, 2-faces should be considered.

THEOREM 11. (a) *Regular polyhedra in $\mathbb{R}^4$ can only have Schläfli symbols* $\{3,3,3\}$, $\{3,3,4\}$, $\{4,3,3\}$, $\{3,4,3\}$, $\{5,3,3\}$, *and* $\{3,3,5\}$.

(b) *Regular polyhedra in $\mathbb{R}^n$ with $n > 4$ can only have Schläfli symbols* $\{3,\ldots,3\}$, $\{3,\ldots,3,4\}$, *and* $\{4,3,\ldots,3\}$.

(c) *For each Schläfli symbol mentioned above, a regular polyhedron with this symbol exists and is unique up to similarity.*

Proof. (a) Formula (7) implies that $\rho\{r_2, r_3\} > \cos^2(\pi/r_1) \geqslant 1/4$. Applying (6), we can calculate $\rho\{r_2, r_3\}$ for all regular polyhedra. The calculations show that

$\rho\{3,3\} = 2/3$, and hence $r_1 \leqslant 5$;
$\rho\{4,3\} = 1/3$, and hence $r_1 \leqslant 3$;

$\rho\{5,3\} = (3-\sqrt{5})/6 < 1/4$;
$\rho\{3,4\} = 1/2$, and hence $r_1 \leqslant 3$;
$\rho\{3,5\} = (5-\sqrt{5})/10$, and hence $r_1 \leqslant 3$.

(b) First, consider regular polyhedra in $\mathbb{R}^5$. It is easy to verify that the inequality $\rho\{r_1, r_2, r_3\} > 1/4$ holds only for the Schläfli symbols $\{3,3,3\}$ and $\{3,3,4\}$. We have $\rho\{3,3,3\} = 5/8$ (so the inequality $5/8 > \cos^2(\pi/r_1)$ only holds if $r_1 \leqslant 4$) and $\rho\{3,3,4\} = 1/2$ (so the inequality $1/2 > \cos^2(\pi/r_1)$ only holds if $r_1 \leqslant 3$).

For polyhedra in $\mathbb{R}^n$ with $n \geqslant 5$, we obtain
$\rho\{3,\dots,3\} = \frac{n+1}{2n}$, and $\frac{n+1}{2n} > \cos^2(\pi/r_1)$ only if $r_1 \leqslant 4$;
$\rho\{4,3,\dots,3\} = 1/n < 1/4$;
$\rho\{3,\dots,3,4\} = 1/2$, and $1/2 > \cos^2(\pi/r_1)$ only if $r_1 \leqslant 3$.

(c) The uniqueness of a regular polyhedron with a given Schläfli symbol in $\mathbb{R}^n$ is proved in the same way as in $\mathbb{R}^3$.

A regular simplex in $\mathbb{R}^n$ is obtained as the intersection of the hyperplane $x_1 + \dots + x_{n+1} = 1$ with the positive orthant $x_1 \geqslant 0, \dots, x_{n+1} \geqslant 0$; we can also easily construct a cube and use it to obtain the dual polyhedron with vertices at the centers of the $(n-1)$-faces of the cube. Thus it only remains to prove the existence of polyhedra with Schläfli symbols $\{3,4,3\}$, $\{5,3,3\}$, and $\{3,3,5\}$ in $\mathbb{R}^4$.

It is easy to verify that the polyhedron with vertices $(\pm 2,0,0,0)$, $(0,\pm 2,0,0)$, $(0,0,\pm 2,0)$, $(0,0,0,\pm 2)$, and $(\pm 1,\pm 1,\pm 1,\pm 1)$ is regular and has Schläfli symbol $\{3,4,3\}$. The polyhedron with the same vertices plus the vertices $(\pm\tau, \pm 1, \pm\tau^{-1}, 0)$, where $\tau = (\sqrt{5}+1)/2$, and all points obtained from them by even permutations of coordinates is regular and has Schläfli symbol $\{5,3,3\}$. The dual polyhedron has Schläfli symbol $\{3,3,5\}$. □

The Cauchy formula. Let $\mathbb{S}^{n-1}$ be the unit sphere in $\mathbb{R}^n$. Take a continuous function $f : \mathbb{S}^{n-1} \to \mathbb{R}$ on the sphere, choose a polyhedron inscribed in $\mathbb{S}^{n-1}$, and consider the integral sum $\sum f(\xi_k)V_k$, where V_k is the $(n-1)$-dimensional volume of the kth face and ξ_k is an arbitrary point on this face. As the polyhedron tends to the sphere, i.e., the maximal distance between the points on the faces tends to zero, the integral sums approach a limit. This limit is called the *integral* of the function f over the sphere $\mathbb{S}^{n-1}$ and denoted by

$$\int_{\mathbb{S}^{n-1}} f(\xi)\,d\mu.$$

A very important property of an integral over the sphere is its invariance with respect to motions:

$$\int_{\mathbb{S}^{n-1}} f(g\xi)\,d\mu = \int_{\mathbb{S}^{n-1}} f(\xi)\,d\mu$$

for an arbitrary orthogonal transformation $g : \mathbb{R}^n \to \mathbb{R}^n$. The invariance obviously follows from the definition of integral sums.

THEOREM 12 (Cauchy formula). *Suppose that $M \subset \mathbb{R}^n$ is a convex polyhedron, $\xi \in \mathbb{S}^{n-1}$, and $f(\xi)$ is the $(n-1)$-dimensional volume of the projection of M to the hyperplane orthogonal to the unit vector ξ. Then*

$$\int_{\mathbb{S}^{n-1}} f(\xi)\,d\mu = \beta_{n-1}P,$$

where β_{n-1} is the volume of the unit $(n-1)$-ball and P is the "surface area" of the polyhedron M, i.e., the sum of the $(n-1)$-dimensional volumes of all its faces.

Proof. Consider one of the $(n-1)$-dimensional faces of M separately. Let $f_i(\xi)$ be the volume of the projection of this face on the hyperplane orthogonal to the vector ξ. Clearly, $f_i(\xi) = |(\xi, n_i)|V_i$, where V_i is the volume of the face and n_i is the unit normal vector to the face. The invariance of integrals over the sphere implies that

$$\int_{\mathbb{S}^{n-1}} |(\xi, n_i)|\, d\mu = c_n$$

is a constant only depending on the dimension n. Therefore,

$$\int_{\mathbb{S}^{n-1}} f_i(\xi)\, d\mu = c_n V_i.$$

Each point in the projection of the polyhedron M (except for points from a set of measure zero) is covered by exactly two projections of faces. This means that

$$\int_{\mathbb{S}^{n-1}} f(\xi)\, d\mu = \int_{\mathbb{S}^{n-1}} \frac{1}{2} \sum f_i(\xi)\, d\mu = \frac{c_n}{2} \sum V_i = \frac{c_n}{2} P.$$

It is more convenient to evaluate the constant for the unit ball rather than for a polyhedron. Thus we pass to the limit and assume that M is the unit ball. We then have $f(\xi) = \beta_{n-1}$, and hence

$$\int_{\mathbb{S}^{n-1}} f(\xi)\, d\mu = \beta_{n-1} P,$$

where P is the $(n-1)$-dimensional volume of the sphere $\mathbb{S}^{n-1}$; in the case under consideration, it coincides with the surface area of the unit ball M. □

COROLLARY. *If a convex polyhedron lies inside another convex polyhedron, then the surface area of the inner polyhedron is less than the surface area of the outer polyhedron.*

The Steiner–Minkowski formula. Let $M \subset \mathbb{R}^n$ be a convex polyhedron. For a nonnegative number r, we write M_r to denote the set of points at a distance at most r from M, i.e., the union of all balls of radius r centered at points of M. It is easy to verify that the set M_r is convex. Indeed, if $a, b \in M$, $|u|, |v| \leqslant r$, and $0 \leqslant \lambda \leqslant 1$, then the distance between the points $\lambda(a+u) + (1-\lambda)(b+v)$ and $\lambda a + (1-\lambda)b \in M$ equals $|\lambda u + (1-\lambda)v| \leqslant r$.

THEOREM 13 (Steiner–Minkowski formula). *The volume V_r of M_r depends on r polynomially; namely,*

$$V_r = V_0 + Pr + \cdots + \beta_n r^n,$$

where V_0 is the volume of the polyhedron M, P is the surface area of M, and β_n is the volume of the unit n-ball.

Proof. We argue by induction on n; our proof uses the Cauchy formula (we have written and proved this formula for convex polyhedra, but it can be carried over to all convex figures by passing to the limit). For $n = 1$, the Steiner–Minkowski formula is obvious. Suppose that it is proved for $(n-1)$-dimensional polyhedra and consider a convex polyhedron M in $\mathbb{R}^n$. Let $p_\xi(M)$ be the projection of M on the hyperplane orthogonal to a unit vector ξ. Clearly, $p_\xi(M)_r = p_\xi(M_r)$. Let $V_r(\xi)$ be the $(n-1)$-dimensional volume of the set $p_\xi(M_r)$. By the induction hypothesis,

$$V_r(\xi) = V_0(\xi) + \cdots + \beta_{n-1} r^{n-1}.$$

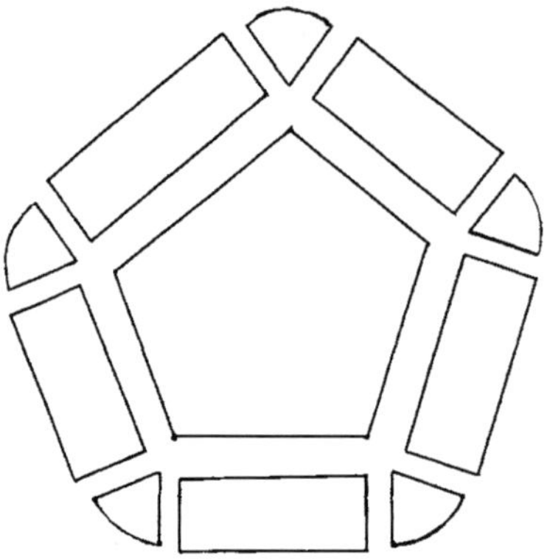

FIGURE A.6

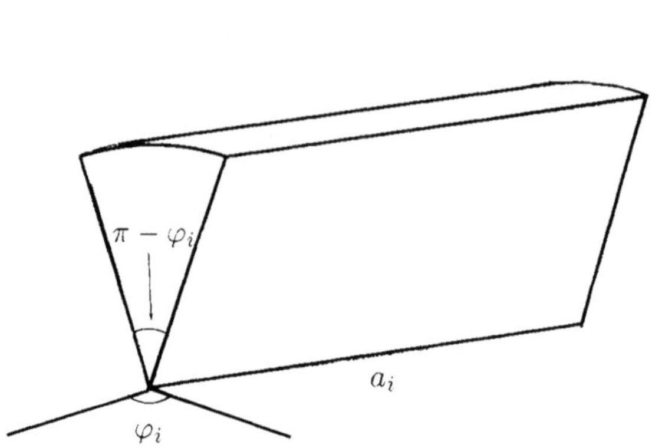

FIGURE A.7

Let P_r be the boundary surface area of the body M_r. According to the Cauchy formula, we have

$$\begin{aligned}\beta_{n-1}^{-1}P_r &= \int_{\mathbb{S}^{n-1}} V_r(\xi)\,d\mu = \int_{\mathbb{S}^{n-1}} V_0(\xi)\,d\mu + \cdots + \int_{\mathbb{S}^{n-1}} \beta_{n-1}r^{n-1}\,d\mu \\ &= \beta_{n-1}^{-1}P_0 + a_1 r + \cdots + a_{n-1}r^{n-1},\end{aligned}$$

where $a_1, \ldots, a_{n-1}$ are constants.

It is clear from geometric considerations that $dV_r/dr = P_r$; therefore,

$$\begin{aligned}V_r &= V_0 + \int_0^r P_s\,ds = V_0 + \int_0^r (P_0 + a_1' s + \cdots + a_{n-1}' s^{n-1})\,ds \\ &= V_0 + P_0 r + b_1 r^2 + \cdots + b_{n-1}r^n.\end{aligned}$$

Here $P_0 = P$ is the surface area of the polyhedron M, the coefficients $b_1, \ldots, b_{n-2}$ only depend on the polyhedron, and the last coefficient b_{n-1} does not depend even on the polyhedron. For a degenerate polyhedron consisting of one point, we have $V_r = \beta_n r^n$; therefore, $b_{n-1} = \beta_n$. □

Steiner proved the formula for V_r in the cases of $n = 2$ and 3; in the general case, it was proved by Minkowski. For $n = 2$, the proof is clear from Figure A.6. For $n = 3$, we similarly obtain, first, the parts constituting the ball and, secondly, the prisms whose bases are the faces of the polyhedron. In addition, we obtain cylindrical sectors, one of which is shown in Figure A.7. An edge of length a_i with dihedral angle φ_i corresponds to a cylindrical sector of volume $a_i(\pi - \varphi_i)r^2/2$.

Polygons in $\mathbb{R}^m$. In this subsection, we discuss some properties of polygons in $\mathbb{R}^m$; but first, let us mention the following important property of sets of points in $\mathbb{R}^m$.

THEOREM 14. *Let $\{A_1, \ldots, A_n\}$ and $\{B_1, \ldots, B_n\}$ be two sets of points in Euclidean m-space such that $|A_iA_j| = |B_iB_j|$ for all i and j. Then there exists an isometry that maps each point A_i to the corresponding point B_i.*

Proof. The point A_1 can be mapped to B_1 by using either the identity map (if $A_1 = B_1$) or a symmetry about a hyperplane (if $A_1 \neq B_1$). Suppose that we have constructed an isometry f mapping the points $A_1, \ldots, A_k$ to $B_1, \ldots, B_k$, respectively. If $f(A_{k+1}) \neq B_{k+1}$, we apply the symmetry g about the hyperplane consisting of all points equidistant from $f(A_{k+1})$ and B_{k+1}. For $i = 1, \ldots, k$, the

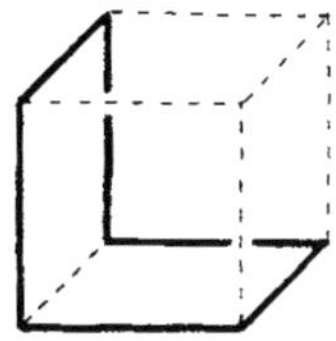

FIGURE A.8

point $f(A_i) = B_i$ is equidistant from $f(A_{k+1})$ and B_{k+1}. Therefore, the isometry $g \circ f$ maps the points $A_1, \dots, A_{k+1}$ to $B_1, \dots, B_{k+1}$, respectively. □

Let $A_1, \dots, A_n$ be pairwise different points in the Euclidean m-space. The closed polygonal line with sides $A_1A_2, \dots, A_{n-1}A_n, A_nA_1$ is called the *spatial polygon* $A_1 \dots A_n$. A spatial polygon is said to be *k-equilateral* if the length of the segment $[A_i, A_{i+j}]$ does not depend on i for $j = 1, \dots, k$.

If a convex plane polygon is 2-equilateral, then it is k-equilateral for any k. However, the hexagon shown in Figure A.8 is 2-equilateral but not 3-equilateral.

As is known, an arbitrary 2-equilateral pentagon in $\mathbb{R}^3$ is plane. This assertion admits the following generalization.

THEOREM 15. *The dimension of a minimal affine space containing a k-equilateral $(2k + 1)$-gon is even.*

Proof. Let $A_1, \dots, A_{2k+1}$ be a k-equilateral polygon. It is d-equilateral for all $d \geqslant 1$, because any two of its vertices are joined by a polygonal line with no more than k sides. Put $B_i = A_{i+1}$. Then $|B_iB_j| = |A_iA_j|$ for all i and j. By Theorem 14, there exists an isometry f that maps the points A_i to B_i for all $i = 1, \dots, 2k + 1$. Obviously, $(2k+1)^{-1}(A_1 + \dots + A_{2k+1})$ is a fixed point of this map. Let us denote this point by O and take it for the origin. Then the isometry is a linear map. Let us denote the linear hull of the points $A_1, \dots, A_{2k+1}$ by Π. Clearly, f maps Π onto itself; consider the restriction of the isometry f to Π.

Suppose that the dimension of Π is odd. A linear isometry in an odd-dimensional space leaves some straight line fixed (the characteristic polynomial of an orthogonal matrix of odd order has at least one root equal to ± 1), and the orthogonal complement of this line is also invariant. Thus, if the image of at least one point A_i under the projection to the invariant line is not zero, then the images of all points are the same, and hence the arithmetic mean of all points does not coincide with the point O. On the other hand, if they all are projected to zero, then the linear hull of the points under consideration lies in the orthogonal complement of the line, which is an even-dimensional space. This contradicts the assumption that their linear hull is Π and completes the proof of the theorem. □

This result readily implies that a 2-equilateral pentagon in three-dimensional Euclidean space is plane because the dimension of the affine hull of all its points is even and no higher than three. There are nonplane 2-equilateral pentagons; for instance, we can take the vertices of a regular four-dimensional simplex and join them by a closed polygonal line in arbitrary order.

Interesting examples of k-equilateral polygons in Euclidean $2k$-spaces can be obtained by using the curve

$$x(t) = (\cos t, \sin t, \cos 2t, \sin 2t, \dots, \cos kt, \sin kt).$$

Namely, the n-gon with vertices at the points

$$A_j = x(\tfrac{2\pi j}{n}), \quad 1 \leqslant j \leqslant n,$$

is k-equilateral.

3. Additional questions of projective geometry

The complex projective space $\mathbb{C}P^n$. The complex projective space is defined analogously to the real one. The points of the *complex projective space* $\mathbb{C}P^n$ are $(n+1)$-tuples $(z_1, \dots, z_{n+1})$, $z_i \in \mathbb{C}$, and the numbers $z_1, \dots, z_{n+1}$ are not all zero; the $(n+1)$-tuples $(z_1, \dots, z_{n+1})$ and $(\lambda z_1, \dots, \lambda z_{n+1})$, where $\lambda \in \mathbb{C}$, are considered equivalent.

In the space $\mathbb{C}P^n$, the points of the form $(x_1, \dots, x_{n+1})$ with $x_i \in \mathbb{R}$ form a subset which is naturally identified with $\mathbb{R}P^n$. Thus $\mathbb{C}P^n$ can be regarded as $\mathbb{R}P^n$ completed by complex points.

Without considering the complex projective plane, it is impossible, for example, to elucidate the somewhat mysterious situation with the intersection of circles. Two ellipses can intersect at four points, but two circles cannot intersect at more than two points. How can this be explained? The point is that any two circles intersect at two fixed points of a complex line at infinity that do not belong to the real projective plane. More precisely, the following assertion is valid.

Any circle passes through the points $(1, i, 0)$ *and* $(1, -i, 0)$.

Indeed, the equation of a circle in homogeneous coordinates has the form

$$(x - az)^2 + (y - bz)^2 = R^2 z^2. \tag{1}$$

To find the intersection points of the circle with the line at infinity $z = 0$, we substitute the value $z = 0$ in (1). As a result we obtain $x^2 + y^2 = 0$, i.e., $y = \pm ix$. Thus any circle passes through the points $(1, \pm i, 0)$ lying on the line at infinity. This completes the proof of the assertion. □

The k-dimensional planes in $\mathbb{C}P^n$ can be defined similarly to k-dimensional planes in $\mathbb{R}P^n$. The geometry of the space $\mathbb{C}P^n$ is to a certain degree analogous to the geometry of the space $\mathbb{R}P^n$. For instance, any two straight lines in $\mathbb{C}P^2$ meet at one point. This is so because a system of equations

$$a_1 x + b_1 y + c_1 z = 0, \qquad a_2 x + b_2 y + c_2 z = 0$$

has a unique (up to proportionality) solution over both fields $\mathbb{R}$ and $\mathbb{C}$. But the properties of configurations of points and lines on the complex projective plane and on the real projective plane can and do differ substantially. One of the most interesting phenomena of this kind is the existence in $\mathbb{C}P^2$ of a counterexample to the following theorem of Sylvester for $\mathbb{R}P^2$.

SYLVESTER'S THEOREM. *Let points $X_1, \dots, X_n$ in the real projective plane be such that any line X_iX_j contains at least one point X_k different from X_i and X_j. Then all points $X_1, \dots, X_n$ lie on one line.*

Proof. Sylvester's theorem is projective, but its simplest proof uses the metric. We can pass from the projective plane to the Euclidean plane as follows. Consider a line l passing through none of the points X_i. We can assume that this is the line at infinity. Let us remove l from the projective plane and introduce Cartesian coordinates on the remaining set. Suppose that not all points X_i lie on one line. Let us draw all the lines X_iX_j and find the smallest nonzero distance d from the given

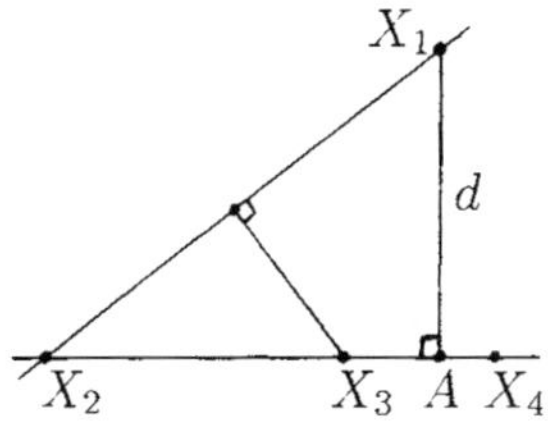

FIGURE A.9

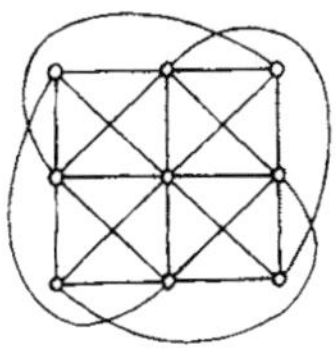

FIGURE A.10

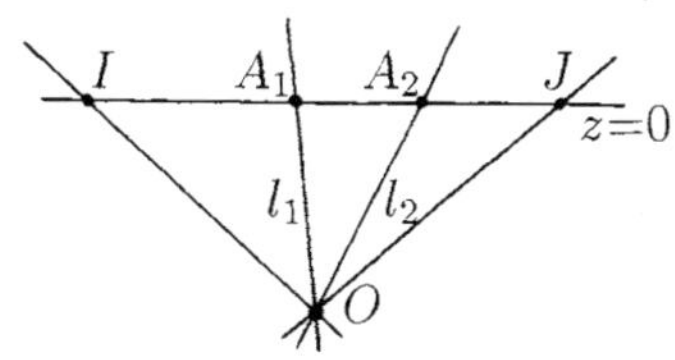

FIGURE A.11

points to these lines. To be definite, suppose that the distance from the point X_1 to the line X_2X_3 is d. By assumption, the line X_2X_3 contains one more point X_4. From X_1, let us draw a perpendicular $[X_1, A]$ to the line X_2X_3. Two of the points X_2, X_3, X_4 lie on this line on one side of the point A. To be definite, suppose that the points X_2 and X_3 lie on one side of A and $|AX_3| < |AX_2|$ (Figure A.9). Then the distance from X_3 to X_1X_2 is less than d. We arrived at a contradiction. □

On the complex projective plane, there exists a configuration of nine points for which Sylvester's theorem is not valid. The following assertion holds.

Suppose that $\varepsilon^3 = 1$ *but* $\varepsilon \neq 1$. *Consider the following nine points on the complex projective plane:*

$$\begin{array}{lll} (0,1,-1), & (0,\varepsilon^2,-\varepsilon), & (0,\varepsilon,-\varepsilon), \\ (-1,0,1), & (-\varepsilon^2,0,1), & (-\varepsilon,0,1), \\ (1,-1,0), & (-\varepsilon,1,0), & (-\varepsilon^2,1,0). \end{array}$$

These points are not collinear, but any line containing two of these points contains a third one.

Let us prove this. It is easy to verify that each of the twelve lines

$$\begin{array}{rrr} x=0, & y=0, & z=0, \\ x+y+z=0, & x+y+\varepsilon z=0, & x+y+\varepsilon^2 z=0, \\ x+\varepsilon y+z=0, & x+\varepsilon^2 y+z=0, & \varepsilon x+y+z=0, \\ \varepsilon^2 x+y+z=0, & x+\varepsilon y+\varepsilon^2 z=0, & x+\varepsilon^2 y+\varepsilon z=0 \end{array}$$

contains precisely three of the given points. The configuration of these lines is schematically shown in Figure A.10. Through each pair of given points, there passes exactly one of the twelve lines specified above, and this line contains yet another given point. This concludes the proof of the assertion. □

Using the points $I, J = (1, \pm i, 0) \in \mathbb{C}P^2$, we can express metric parameters in terms of cross ratios, i.e., in a projectively invariant way.

For example, consider the lines l_1 and l_2 defined by the equations

$$x = y \tan\varphi_1 + r_1 \quad \text{and} \quad x = y\tan\varphi_2 + r_2,$$

respectively, in the plane $\mathbb{R}^2$. The angle between these lines is $\pm(\varphi_1 - \varphi_2)$. The lines l_1 and l_2 correspond to the lines $x = y\tan\varphi_1 + r_1 z$ and $x = y\tan\varphi_2 + r_2 z$ in

$\mathbb{C}P^2$. They intersect the line at infinity $z = 0$ at the points

$$A_1 = (1, \tan\varphi_1, 0) \quad \text{and} \quad A_2 = (1, \tan\varphi_2, 0),$$

respectively. Let O be the intersection point of l_1 and l_2 (Figure A.11). Then

$$[l_1, l_2, OI, OJ] = [A_1, A_2, I, J] = \frac{\tan\varphi_1 - i}{\tan\varphi_1 + i} : \frac{\tan\varphi_2 - i}{\tan\varphi_2 + i}.$$

Clearly,

$$\frac{\tan\varphi - i}{\tan\varphi + i} = \frac{\sin\varphi - i\cos\varphi}{\sin\varphi + i\cos\varphi} = -e^{-2i\varphi}.$$

Therefore,

$$[l_1, l_2, OI, OJ] = e^{\pm 2i\alpha}, \tag{2}$$

where α is the angle between l_1 and l_2.

Formula (2) is called *Laguerre's formula.* Given points I and J, we can apply it to express the angle (which is a metric parameter) in terms of the cross ratio (a projective parameter).

The polar line of a point with respect to a curve in $\mathbb{C}P^2$. From the point of view of projective geometry, the harmonic mean differs from the arithmetic and geometric means in that it is preserved by projective transformations of the line. Recall that a number x_0 is the *harmonic mean* of real numbers $x_1, \dots, x_n$ if

$$\frac{n}{x_0} = \frac{1}{x_1} + \dots + \frac{1}{x_n}.$$

This definition does not exhibit the projective invariance of the harmonic mean. To obtain an invariant definition, we must add the origin a to the set of points $x_0, x_1, \dots, x_n$, because the harmonic mean changes with the origin. We call a point x_0 the *harmonic mean* of points $x_1, \dots, x_n$ on a line with origin a if

$$\frac{n}{x_0 - a} = \frac{1}{x_1 - a} + \dots + \frac{1}{x_n - a}. \tag{3}$$

Consider a point t on the given line different from all the points $a, x_0, x_1, \dots, x_n$. Let us multiply both sides of (3) by $t - a$ and subtract n from both sides of the result. Since $(t-a)(w-a) = (t-w)(w-a) + 1$, we obtain

$$\frac{n(t - x_0)}{x_0 - a} = \frac{t - x_1}{x_1 - a} + \dots + \frac{t - x_n}{x_n - a}, \qquad \text{i.e.,} \quad \sum_{i=1}^{n} \frac{t - x_i}{x_i - a} : \frac{t - x_0}{x_0 - a} = n.$$

Thus a point x_0 is the harmonic mean of points $x_1, \dots, x_n$ on a line with origin a if, for an arbitrary point t on this line, we have

$$[x_1, x_0, t, a] + \dots + [x_n, x_0, t, a] = n.$$

This definition is invariant with respect to projective transformations of the line.

A remarkable property of the harmonic mean was discovered by Newton's student *Cotes* (it was Cotes to whom Newton entrusted the preparation of the second edition of *Philosophiae Naturalis Principia Mathematica*).

COTES' THEOREM. *Let a line rotating about a point A intersect an nth-order curve at n points $X_1, \dots, X_n$. Then the harmonic mean of the points $X_1, \dots, X_n$ moves along some (straight) line.* (The point A is assumed to be the origin; it must lie outside the curve under consideration.)

Proof. We can assume that the point A has coordinates $(0,0)$. A curve of order n is given by an equation of the form $\sum a_{ij}x^iy^j = 0$, where $i,j \geq 0$ and $i+j \leq n$. This curve $\sum a_{ij}x^iy^j = 0$ does not pass through A; hence $a_{00} \neq 0$. A line through A is given by an equation of the form $y = kx$ (the only exception is the line $x = 0$). Substituting this relation in the equation of the curve, we obtain

$$0 = \sum a_{ij}k^jx^{i+j} = a_{00} + (a_{10} + ka_{01})x + \cdots ,$$

where the dots denote the terms $\alpha_2x^2, \ldots, \alpha_nx^n$, which are not essential in what follows. If $x_1, \ldots, x_n$ are the roots of the obtained polynomial, then $(x_1, kx_1), \ldots, (x_n, kx_n)$ are the intersection points of the curve under consideration with the line $y = kx$. The harmonic mean of these points has coordinates (x_0, kx_0), where

$$\frac{n}{x_0} = \frac{1}{x_1} + \cdots + \frac{1}{x_n} = \frac{x_2x_3\cdots x_n + \cdots + x_1x_2\cdots x_{n-1}}{x_1\cdots x_n} = -\frac{a_{10}+ka_{01}}{a_{00}};$$

to write the last equality, we have applied Viète's theorem. All the points

$$\left(-\frac{na_{00}}{a_{10}+ka_{01}}, -\frac{kna_{00}}{a_{10}+ka_{01}}\right)$$

lie on the line $a_{10}x + a_{01}y + na_{00} = 0$; this equation indeed determines a line, because $a_{00} \neq 0$; if $a_{10} = a_{01} = 0$, then this is the line at infinity. (The line at infinity is obtained if the sum $1/x_1 + \cdots + 1/x_n$ is identically zero. For a conic, this means that A is the center of the conic.) □

To an nth-order curve defined by $\sum a_{ij}x^iy^j = 0$ in $\mathbb{R}^2$ we can assign the set of points defined by the same equation in $\mathbb{C}^2$. In addition, to this curve we can assign the sets of points in $\mathbb{R}P^2$ and in $\mathbb{C}P^2$ defined by the equation $\sum a_{ij}x^iy^jz^{n-i-j} = 0$. All these sets of points are called *algebraic curves.*

We can define the harmonic mean for complex numbers by the same formula as for real numbers. After that, we can define the harmonic mean of n points on the complex line $\mathbb{C}^1$ with a given origin and the harmonic mean of the intersection points of an nth-order curve in $\mathbb{C}^2$ with a line through a given point A.

On the real projective plane, a line does not always intersect an nth order curve at precisely n points; the intersection points may be imaginary. But on the complex projective plane, a line always intersect an nth-order curve at precisely n points (with multiplicity taken into account). Therefore, on the complex projective plane, for any line through a point A, the harmonic mean of the intersection points of this line with an nth-order curve is defined. All harmonic means lie on the line $a_{10}x + a_{01}y + na_{00}z = 0$. This line is called the *polar line* of the point A with respect to the given curve.

For a curve on the real plane (on the real projective plane), the polar line of a point A with respect to a given curve is also defined as the line determined by the equation $a_{10}x + a_{01}y + na_{00} = 0$ (respectively, by the equation $a_{10}x + a_{01}y + na_{00}z = 0$).

The polar line of a point A with respect to a conic (on the real plane) has a very simple geometric meaning if two tangent lines AP and AQ to the given conic can be drawn from the point A (Figure A.12). The tangent line AP intersects the conic at two coinciding points. The harmonic mean of two equal numbers x and x is x. Therefore, P is the harmonic mean of the intersection points of the line AP with the conic. Thus PQ is the polar line of the point A with respect to the given conic.

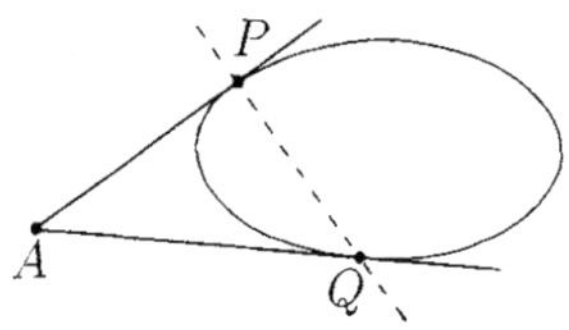

FIGURE A.12

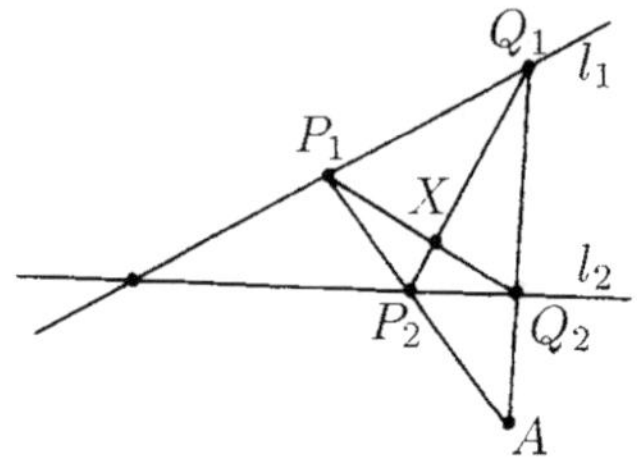

FIGURE A.13

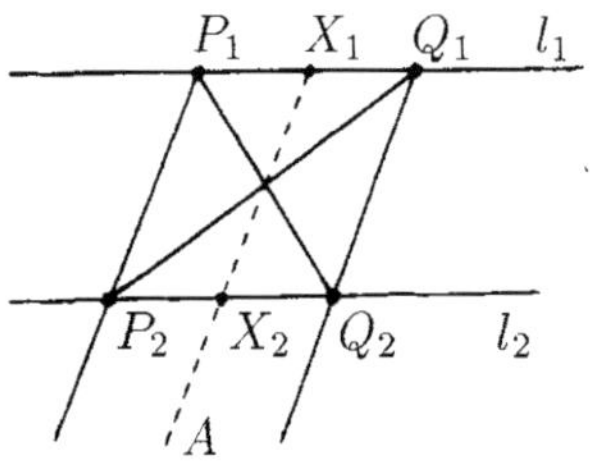

FIGURE A.14

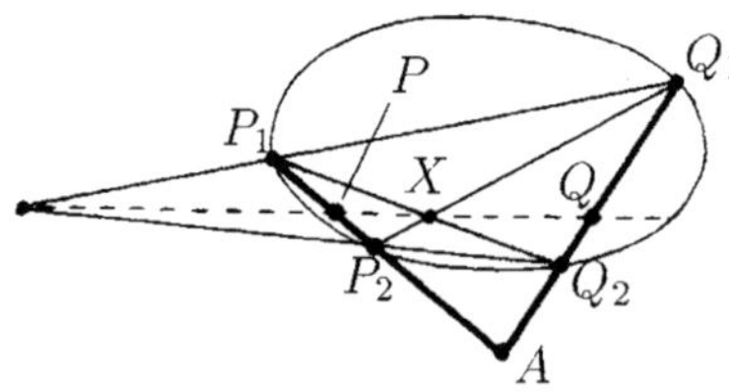

FIGURE A.15

There is another, more general, geometric method for constructing the polar line of a point with respect to a conic. First, consider the case of a degenerate second order curve consisting of two intersecting lines.

THEOREM 1. *If two lines through a point A intersect lines l_1 and l_2 at points P_1, Q_1, and P_2, Q_2* (Figure A.13) *and X is the intersection point of the lines P_1Q_2 and P_2Q_1, then X lies on the polar line of A with respect to the curve $l_1 \cdot l_2 = 0$.*

Proof. We can assume that the point A and the intersection point of the lines l_1 and l_2 are at infinity, i.e., $P_1P_2 \parallel Q_1Q_2$ and $l_1 \parallel l_2$ (Figure A.14). Let the line AX intersect l_1 and l_2 at points X_1 and X_2. Then X is the midpoint of the segment $[X_1, X_2]$. It remains to prove that if the origin on the line AX is at the point at infinity A, then the harmonic mean of X_1 and X_2 is the midpoint of the segment $[X_1, X_2]$. The condition

$$\frac{t - x_1}{x_1 - a} : \frac{t - x_0}{x_0 - a} + \frac{t - x_2}{x_2 - a} : \frac{t - x_0}{x_0 - a} = 2$$

with $a = \infty$ can be written as

$$\frac{t - x_1}{t - x_0} + \frac{t - x_2}{t - x_0} = 2,$$

i.e., $x_0 = (x_1 + x_2)/2$. □

REMARK. The intersection point of the lines l_1 and l_2 also lies on the polar line of the point A with respect to the curve $l_1 \cdot l_2 = 0$.

Now, consider a nondegenerate second-order curve.

THEOREM 2. *If two lines through a point A intersect a conic at points P_1, P_2 and Q_1, Q_2* (Figure A.15) *and X is the intersection point of the lines P_1Q_2 and P_2Q_1, then X lies on the polar line of the point A with respect to this conic.*

Proof. The polar lines of the point A with respect to all conics through the points P_1, P_2, Q_1, and Q_2 are the same. Indeed, if P and Q are the harmonic means of the points P_1, P_2 and Q_1, Q_2 (see Figure A.15), then PQ is the polar for

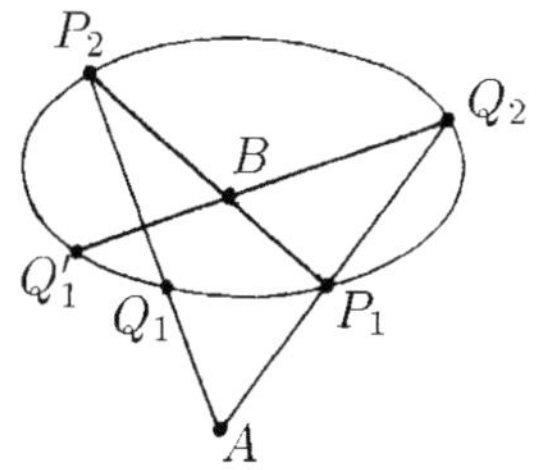

FIGURE A.16

an arbitrary conic through the points P_1, P_2, Q_1, and Q_2. According to Theorem 1, the point X belongs to the polar line of A with respect to the curve $P_1Q_1 \cdot P_2Q_2 = 0$; therefore, it belongs to the polar line of A with respect to the initial conic. □

THEOREM 3. *If the polar line of a point A with respect to a given conic passes through a point B, then the polar line of B passes through A.*

Proof. Let us draw a line passing through B and intersecting the conic at points P_1 and P_2. The lines AP_1 and AP_2 intersect the conic at points Q_1 and Q_2, and the line BQ_2 intersects the conic at a point Q_1' (Figure A.16). According to Theorem 2, the line Q_1Q_2 intersects P_1P_2 at a point lying on the polar line of A. Therefore, if the line P_1P_2 is not polar for the point A, then Q_1Q_2 intersects it at B, and hence $Q_1' = Q_1$. We can assume that the line P_1P_2 differs from the polar line of the point A. Then, according to Theorem 2, the lines P_1Q_2 and P_2Q_1 intersect at a point on the polar line of B; on the other hand, their intersection point is precisely A. □

Projective duality. In this subsection, we consider the "points–lines" duality in $\mathbb{R}P^2$ and in $\mathbb{C}P^2$.

In the preceding subsection, we showed that if a point B lies on the polar line of a point A with respect to some conic, then A lies on the polar line of B. This property enables us to construct a point A whose polar line coincides with an arbitrary given line a by using only a straightedge. To do this, we choose points B_1 and B_2 on a; if b_1 and b_2 are the polar lines of these points and A is the intersection point of b_1 and b_2, then the polar line of A passes through B_1 and B_2 and, therefore, coincides with a.

We have defined polar lines only for points not lying on the conic. From continuity considerations, the polar line of a point on the conic should be tangent to the conic at this point. Thus, given a conic on the projective plane, to each point A we can assign a line $A^\perp$—the polar line of this point with respect to the conic, and to each line a we can assign the point $a^\perp$ whose polar line is a. If lines a_1 and a_2 meet at a point B, then the points $a_1^\perp$ and $a_2^\perp$ lie on the line $B^\perp$. For a configuration of points A and lines a, the configuration of lines $A^\perp$ and points $a^\perp$ is said to be *dual*.

The relation between the coordinates of a point and the equation of its polar line has the simplest form for the conic $x^2 + y^2 + z^2 = 0$. Note that although this conic has no real points, the polar line with respect to it is defined not only on the complex projective plane, but also on the real plane, because the equation of the conic is real.

THEOREM 4. *The polar line of a point $(x_0 : y_0 : z_0)$ with respect to the conic $x^2 + y^2 + z^2 = 0$ is given by the equation $x_0x + y_0y + z_0z = 0$.*

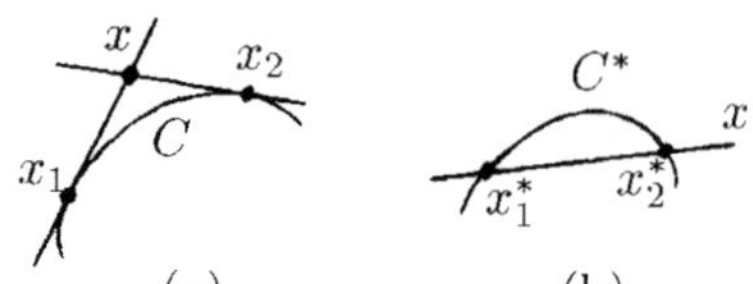

FIGURE A.17

Proof. The tangent lines to the conic $x^2+y^2+z^2=0$ from the point $(x_0:y_0:z_0)$ touch the conic at points $(x_1:y_1:z_1)$ and $(x_2:y_2:z_2)$. The tangent line at the point $(x_i:y_i:z_i)$ is given by the equation $x_ix+y_iy+z_iz=0$. The tangents

$$x_1x+y_1y+z_1z=0, \qquad x_2x+y_2y+z_2z=0$$

pass through the point $(x_0:y_0:z_0)$, whence $x_ix_0+y_iy_0+z_iz_0=0$. Therefore, the line $x_0x+y_0y+z_0z=0$ passes through the points $(x_1:y_1:z_1)$ and $(x_2:y_2:z_2)$, and hence this line is polar for the point $(x_0:y_0:z_0)$. □

The geometric meaning of Theorem 4 is very simple: if to a point $A\in\mathbb{R}P^2$ we assign a line a in $\mathbb{R}^3$, then to the polar line of A the plane orthogonal to a is assigned.

For an algebraic curve $C\subset\mathbb{C}P^2$, a curve $C^*\subset\mathbb{C}P^2$ dual to C with respect to a conic Γ can be defined. Namely, to each $x\in C$ we assign the tangent line $l(x)$ to C at x, and to the line $l(x)$, we assign the point x^* whose polar line with respect to Γ coincides with $l(x)$. The curve C^* consisting of the points x^* is called the curve *dual* to C with respect to the conic Γ. The curve dual to a curve in $\mathbb{R}^2$ is defined similarly.

THEOREM 5. *If C is a smooth curve, then $(C^*)^*=C$.*

Proof. Suppose that tangent lines $l(x_1)=(x_1^*)^\perp$ and $l(x_2)=(x_2^*)^\perp$ to the curve C intersect at a point x (Figure A.17 (a)). This means that the points x_1^* and x_2^* of the curve C^* lie on the line $x^\perp$ (Figure A.17 (b)). Let the point x_1 tend to x_2. In the limit, the points x_1 and x_2 coincide with x, and the line $x^\perp$ coincides with the tangent line to the curve C^* at the point x^*. Thus the point $(x^*)^*$ of the curve $(C^*)^*$ coincides with x, and hence $(C^*)^*=C$. □

As an example, consider the curve dual to $y=x^3$ with respect to the conic $x^2+y^2+z^2=0$ (we assume that the plane $\mathbb{R}^2$ is embedded in $\mathbb{C}P^2$ using the map $(x,y)\mapsto(x,y,1)$). The tangent line to $y=x^3$ at a point (x_0,y_0) has the equation $y-y_0=3x_0^2(x-x_0)$, i.e., $-3x_0^2x+y+2x_0^3=1$. Thus the point $(x_0,x_0^3,1)\in\mathbb{C}P^2$ is assigned the point $(-3x_0,1,2x_0^3)\in\mathbb{C}P^2$, which is equivalent to the point $(-3x_0^{-1}/2,x_0^{-3}/2,1)=(x_1,y_1,1)$. The dual curve is almost the same as the initial one: its equation is $(-\frac{2}{3}x_1)^3=2y_1$. But there is a very essential difference between them. The point $(0,0,1)$ is assigned the point $(0,1,0)$. In a neighborhood of $(0,1,0)$, it is convenient to use the coordinates $(x_1,1,z_1)$. We then have $x_1=-3x_0$ and $z_1=2x_0^3$; i.e., the dual curve has the equation $27z_1^2+x_1^3=0$. Thus the dual to the inflection point (Figure A.18 (a)) is a cusp point (Figure A.18 (b)). It is also clear that the dual to the self-intersection point is a double tangent (Figure A.19); more precisely, to the self-intersection point x, two points x_1^* and x_2^* are assigned, and the tangent lines at these points are the same. In passing to the dual curve, a line l intersecting C at points $x_1,\dots,x_n$ corresponds to the point $l^\perp$ through which the tangent lines to C^* at the points $x_1^*,\dots,x_n^*$ (Figure A.20) pass. According to

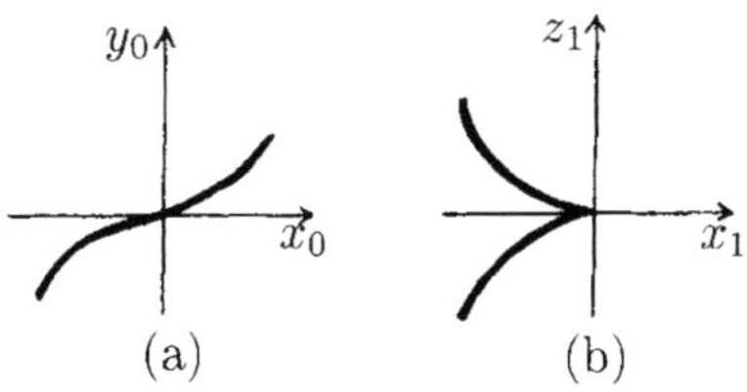

FIGURE A.18

FIGURE A.19

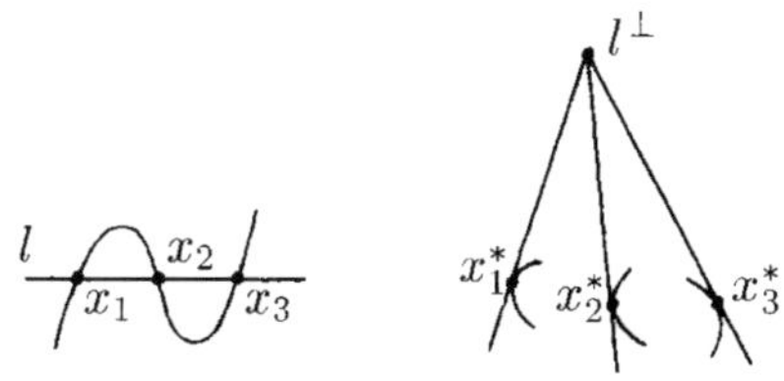

FIGURE A.20

Theorem 5, this means that the order of the curve C^* equals the number of tangent lines to the curve C that can be drawn from a point in general position.

From each point not lying on a conic, exactly two tangent lines to this conic can be drawn (the tangent lines may be imaginary). So it is natural to expect that an arbitrary conic is dual to a conic. Before proving that is indeed so, we introduce notation convenient for writing the equations of conics. Let $x = (x_1, x_2, x_3)$. Then any conic in $\mathbb{C}P^2$ can be given by an equation of the form $xAx^T = 0$, where A is a 3×3 symmetric matrix. In this notation, the conic $x_1^2 + x_2^2 + x_3^2 = 0$ is $xx^T = 0$.

THEOREM 6. *The conic $xBx^T = 0$, where $B = A^{-1}$, is dual to the conic $xAx^T = 0$ with respect to the conic $xx^T = 0$.*

Proof. Suppose that a point p belongs to the conic $xAx^T = 0$, i.e., $pAp^T = 0$. The tangent line at p to this conic has the equation $pAx^T = 0$, i.e., $qx^T = 0$, where $q = pA$. Thus the point p corresponds to the point $p^* = pA$ on the dual curve. Therefore, if $B = A^{-1}$, then $p^*B(p^*)^T = pABAp^T = pAp^T = 0$. (Recall that the matrix A is symmetric, i.e., $A^T = A$.) $\square$

For the ellipse $x^2/a^2 + y^2/b^2 = 1$, the matrix A has the form

$$\begin{pmatrix} a^{-2} & 0 & 0 \\ 0 & b^{-2} & 0 \\ 0 & 0 & -1 \end{pmatrix}; \quad \text{therefore,} \quad B = \begin{pmatrix} a^{2} & 0 & 0 \\ 0 & b^{2} & 0 \\ 0 & 0 & -1 \end{pmatrix},$$

i.e., this ellipse is dual to the ellipse $a^2x^2 + b^2y^2 = 1$. In particular, if $ab = 1$, we obtain the pair of dual ellipses $f_1 = x^2/a^2 + a^2y^2 - 1 = 0$ and $f_2 = a^2x^2 + y^2/a^2 - 1 = 0$ (Figure A.21). Consider the curve $f_1f_2 = \varepsilon$, where ε is a sufficiently small number. Depending on the sign of ε, this curve has one of the two forms shown in Figure A.22. The corresponding dual curves are shown in Figure A.23.

Pappus' and Desargues' theorems state that certain points belong to certain lines. For assertions of such kind, the passage to dual configurations yields dual assertions. If an assertion states that some points belong to certain conics and lines, then the dual assertion can be obtained by considering the dual conics. Thus it is always sufficient to prove one of two dual assertions. The duality of the following two assertions is obvious.

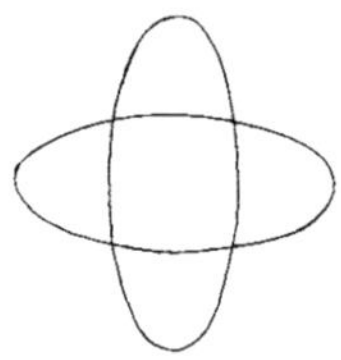

FIGURE A.21

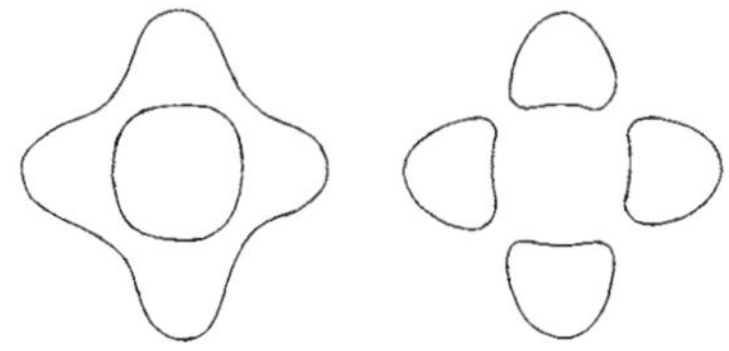

FIGURE A.22

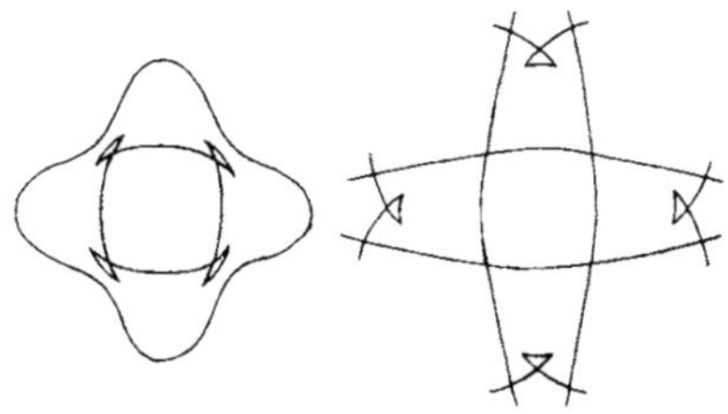

FIGURE A.23

PASCAL'S THEOREM. *If points A, B, C, D, E, F lie on a conic, then the intersection points P, Q, R of the line AB with the line DE, of BC with EF, of CD with FA, respectively, are collinear.*

BRIANCHON'S THEOREM. *If lines a, b, c, d, e, f are tangent to a conic, line p joins the intersection point of a and b with the intersection point of d and e, line q joins the intersection point of b and c with the intersection point of e and f, line r joins the intersection point of c and d with the intersection point of f with a, then the lines p, q, r meet at one point.*

(In other words, the diagonals of the hexagon circumscribed about a conic meet at one point.)

Fixed points of projective transformations of the line and Steiner's construction. Every projective transformation of a line is completely determined by the images of three points. In particular, given the images of three points under some projective transformation, we can find the fixed points of this transformation. *Jacob Steiner* (1796–1863) suggested the following method for constructing fixed points of a projective transformation of a line by using only a straightedge provided a conic (say, a circle) is drawn on the plane.

Choose a point M on the conic. To a point X on the given line l we can assign the point X' on the conic at which the conic meets the line MX (Figure A.24). Suppose that the image of the point X under the projective transformation is X_1. Assigning the line $O_1'X'$ to $O'X_1'$, we obtain a projective correspondence (the point O does not vary, and the point X is variable). Indeed, the correspondence $MX \mapsto MX'$ is projective, and $\angle X'O'Y' = \angle XMY$ and $\angle X_1'O_1'Y_1' = \angle X_1MY_1$. The line $O_1'O'$ is assigned to $O'O_1'$; therefore, all intersection points of the variable lines $O'X_1'$ and $O_1'X'$ lie on one line Δ (see Theorem 2 on p. 50). The intersection points of Δ with the conic correspond to the fixed points of the projective transformation. Thus the line Δ only depends on the transformation itself and does not depend on the choice of the point O (if the line Δ does not intersect the conic, the complexification must be considered).

The construction of the line Δ in the case where the images O_1, X_1, and Y_1 of points O, X, and Y are given is shown in Figure A.24. The points A' and B' correspond to the fixed points A and B of the projective transformation under

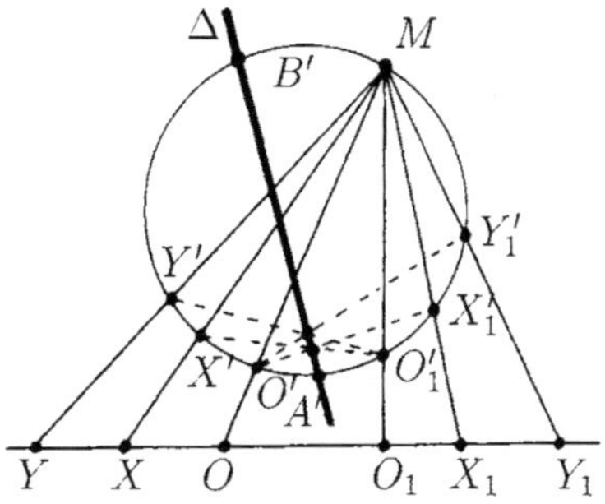

FIGURE A.24

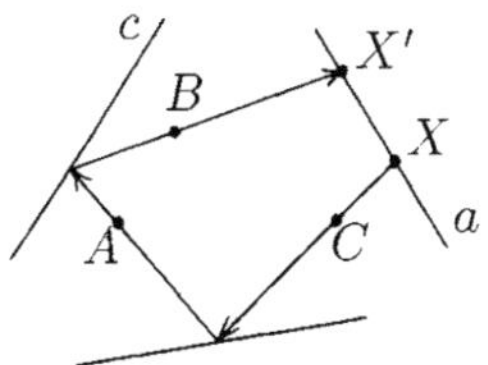

FIGURE A.25

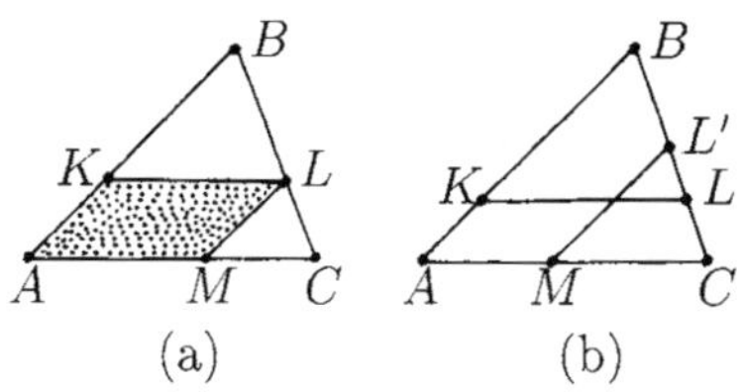

FIGURE A.26

consideration. By applying Steiner's construction, many construction problems can be solved. Let us give two examples of such problems.

1. *Given three points and three lines on the plane, construct a triangle whose vertices lie on the given lines and sides pass through the given points.*

Consider the projective transformation of one of the lines which is the composition of the projections

$$a \xrightarrow{C} b \xrightarrow{A} c \xrightarrow{B} a$$

(Figure A.25). For a vertex of the required triangle, we can take an arbitrary fixed point of this transformation.

2. *Inscribe a parallelogram $AKLM$ of a given*[3] *area in a given triangle ABC* (Figure A.26 (a)).

It is required to find a point L for which $|AK| \cdot |AM|$ takes a certain value, i.e., $|BL| \cdot |LC|$ takes a certain value k.

Consider the transformation $L \mapsto L'$ of the line BC such that $|BL'| \cdot |LC| = k$, i.e., $b - l' = k/(l - c)$ (Figure A.26 (b)). The problem reduces to constructing a fixed point of this projective transformation.

There is a convenient criterion for a projective transformation of a line to have fixed points.

THEOREM 7. *Suppose that f is a projective transformation of a line, $I = f(\infty)$, $f(J) = \infty$, and O is the midpoint of the segment $[I, J]$. The transformation f has no fixed points if and only if the points $f(O)$ and J lie on one side of the point O.*

Proof. Let us introduce a coordinate system on the line with origin at O. Then $f(\infty) = p$ and $f(-p) = \infty$. We must prove that the equation $f(x) = x$ has no solutions if and only if $pq < 0$, where $q = f(0)$. The conditions $f(\infty) = p$,

[3]The area can be specified by, for example, a square. Not to complicate the problem, we assume that the area is specified by a parallelogram with sides parallel to the lines AB and AC.

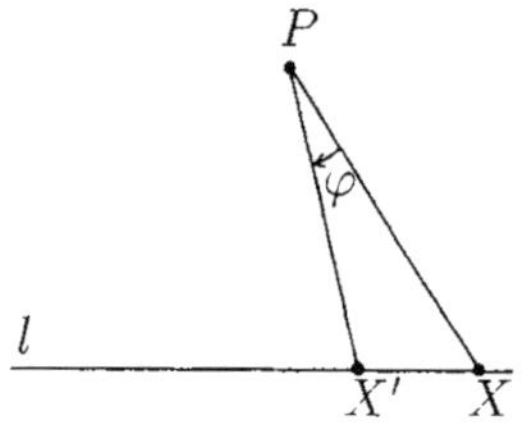

FIGURE A.27

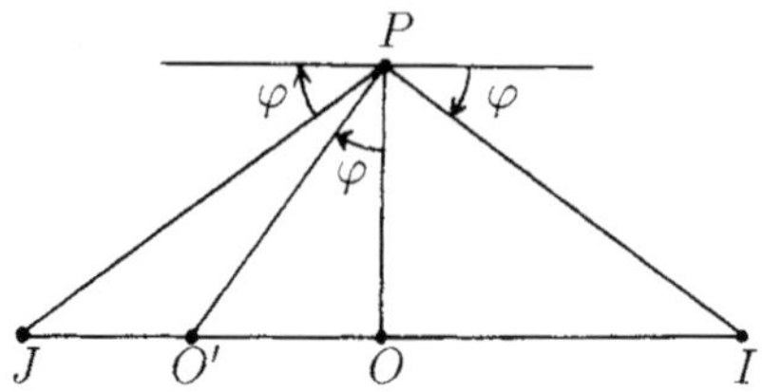

FIGURE A.28

$f(-p) = \infty$, and $f(0) = q$ imply $f(x) = (px + pq)/(x + p)$. Therefore, the equation $f(x) = x$ is equivalent to the equation $x^2 = pq$, which has no solutions if and only if $pq < 0$. □

The proof of Theorem 7 gives yet another method for constructing fixed points of a projective transformation of a line: first, we construct the points $I = f(\infty)$ and $J = f^{-1}(\infty)$; next, we find the midpoint O of the segment $[I, J]$ and the point $O' = f(O)$; a fixed point X satisfies the relation $|OX|^2 = |OI| \cdot |OO'|$.

Let P be a point outside a line l. The transformations of the projective line without fixed points have a simple geometric description. We say that a transformation $X \mapsto X'$ of the line l is *induced by rotation* through an angle φ about the point P if it acts as shown in Figure A.27.

THEOREM 8. *A nonidentity projective transformation of a line has no fixed points if and only if it is induced by rotation.*

Proof. Obviously, a nonidentity transformation of a line l induced by a rotation is projective and has no fixed points. Consider a projective transformation f of l without fixed points. We set $I = f(\infty)$ and $J = f^{-1}(\infty)$; O is the midpoint of the segment $[I, J]$, and $O' = f(O)$. According to Theorem 7, the points O' and J lie on one side of O. Let us draw a perpendicular to l from O and consider its segment $[O, P]$ of length $\sqrt{|OI| \cdot |OO'|}$ (Figure A.28). The triangles $O'PO$ and PIO are similar; therefore, $\angle O'PO = \angle PIO = \angle PJO = \varphi$. The transformation induced by the rotation through the angle φ about P maps the points ∞, O, and J to the points I, O', and ∞, respectively. Hence this transformation coincides with f. □

Projective involutions and harmonic quadruples of points and lines. Let G be an arbitrary set. A transformation $f\colon G \to G$ is called an *involution* if $f(f(x)) = x$ for all $x \in G$; it is assumed that f is not the identity.

The involutive transformations of the projective line admit various geometric descriptions; we shall discuss two of them.

We can identify the projective line with a conic C by means of the stereographic projection from a point $M \in C$ (Figure A.29). After that, we can assume that an arbitrary transformation $X \mapsto X_1$ of the projective line induces a transformation $X' \mapsto X_1'$ of the conic.

THEOREM 9. *A projective transformation $X \mapsto X_1$ of the projective line is an involution if and only if all lines $X'X_1'$ pass through one point A.*

Proof. We shall use the line Δ introduced in the description of Steiner's construction; as a byproduct, we shall prove that Δ is the polar line of A.

Take two points O' and X' on the conic C and consider their images O_1' and X_1' under a projective involution. By the definition of involution, the image of $Y' = X_1'$ is $Y_1' = X'$. Therefore, the line Δ passes through the intersection points

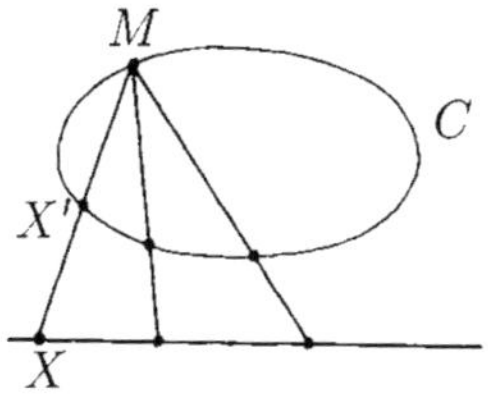

FIGURE A.29

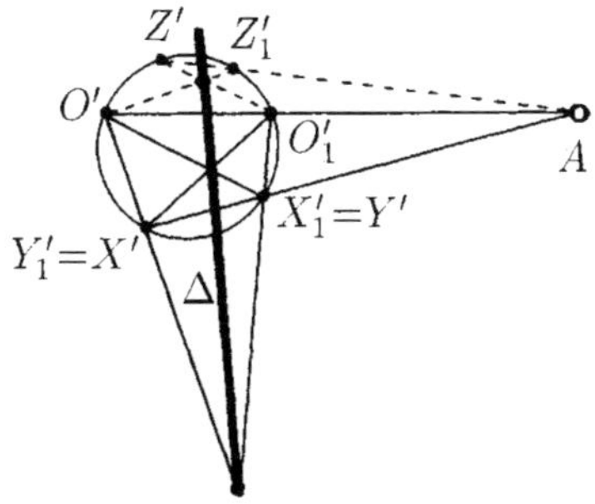

FIGURE A.30

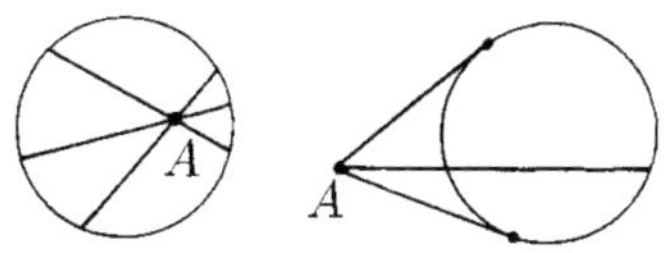

(a) (b)

FIGURE A.31

of the pairs of lines $O'X'_1$, O'_1X and $O'X'$, $O'_1X'_1$ (Figure A.30). This means that Δ is the polar line of the point A at which the lines $O'O'_1$ and $X'X'_1$ meet. To construct the image Z'_1 of an arbitrary point Z' of C under the given projective involution, we must join the point O' to the intersection point of the lines Δ and O'_1Z'. Since Δ is the polar line of A, the line $Z'Z'_1$ passes through A.

Now, let us verify that any transformation $X' \mapsto X'_1$ of the conic C under which all lines $X'X'_1$ pass through one point A is projective (obviously, it is involutive). For this, it suffices to note that the intersection points of all pairs of lines $O'X'_1$ and O'_1X' lie on one line, namely, on the polar line of A. This means that the correspondence $X' \mapsto X'_1$ is projective. □

By assumption, the point A does not belong to the conic C. If A lies inside C (Figure A.31 (a)), the involution has no fixed points, and if A lies outside C (Figure A.31 (b)), the involution has two fixed points.

Another description of involutive projective transformations of a line l uses a pencil of conics through four points (see p. 70). These four points must not lie on l. If a conic of the pencil intersects the line l at points X_1 and X_2, then we assign X_2 to X_1 and X_1 to X_2.

THEOREM 10. *The correspondence $X_1 \longleftrightarrow X_2$ obtained by using a pencil of conics through four points is projective.*

Proof. Let the restrictions of two conics of the pencil to the line l have the equations $x^2 + a_1x + b_1 = 0$ and $x^2 + a_2x + b_2 = 0$. Then the restriction of an arbitrary conic from the pencil to l is given by

$$x^2 + a_1x + b_1 + k(x^2 + a_2x + b_2) = 0.$$

This equation has two roots x_1 and x_2, and we have

$$x_1 + x_2 = -\frac{a_1 + ka_2}{1+k}, \qquad x_1x_2 = \frac{b_1 + kb_2}{1+k}.$$

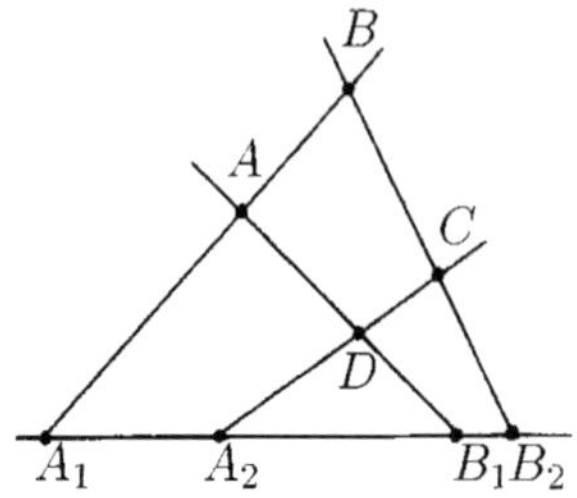

FIGURE A.32

Let p, q, and r be nonzero solutions of the system

$$\begin{cases} pb_1 - qa_1 + r = 0, \\ pb_2 - qa_2 + r = 0. \end{cases}$$

Then $px_1x_2 + q(x_1 + x_2) + r = 0$. This means that x_1 and x_2 are mapped to each other by a linear-fractional transformation, which is obviously involutive. □

Theorem 10 implies that the intersections of conics of the pencil with the line l determine a projective involution. It is easy to show that an arbitrary projective involution can be obtained in this way. Indeed, suppose that a projective involution maps points A_1 and B_1 to points A_2 and B_2, respectively; then it maps A_2 to A_1. Being a projective transformation, the involution is uniquely determined by the images of three points. Therefore, the pencil of conics through the points A, B, C, and D (Figure A.32) determines precisely this involution.

Note that we have not used the fact that B_2 is mapped to B_1; this was possible because of the following assertion.

THEOREM 11. *If a linear-fractional transformation interchanges two different points A_1 and A_2, then it is involutive.*

Proof. Let f be the complexification of the given transformation. Then f has a fixed point B. The transformation $f^2(X) = f(f(X))$ has three fixed points A_1, A_2, and B, which are all different. Therefore, f^2 is the identity transformation. □

Theorem 11 implies that for any two pairs of points on a projective line, there exists a projective involution that interchanges the points in each pair. We say that three pairs of points A and A', B and B', and C and C' on a projective line are in involution if there exists a projective involution interchanging the points in each pair. For the projective line, a pencil of lines through a point on the plane can be considered.

One of the most interesting examples of pairs of points in involution can be obtained by applying Theorem 10. Consider the pairs of intersection points of a line l with three conics of a pencil of conics through four points. The three pairs of points obtained are in involution. In particular, if two or three of the conics under consideration are degenerate, we obtain the following assertions.

1. *The pairs of intersection points of a line l with lines AB and CD, AC and BD, and AD and BC are in involution.*

2. *The pairs of intersection points of a line l with a conic circumscribed about a quadrilateral $ABCD$, with the lines AB and CD, and with the lines AD and BC are in involution.*

Recall that a point b on a line with origin a is said to be the harmonic mean of points x_1 and x_2 if

$$\frac{2}{b-a} = \frac{1}{x_1 - a} + \frac{1}{x_2 - a}.$$

The harmonic mean is invariant with respect to projective transformations, so we can assume that $a = 0$ and $b = \infty$. We then have $x_1 = -x_2$, and hence $[a, b, x_1, x_2] = -1$. A quadruple of points $\{a, b, c, d\}$ is said to be *harmonic* if $[a, b, c, d] = -1$.

THEOREM 12. *If a and b are fixed points of a projective involution f, then the quadruple $\{a, b, x, f(x)\}$ is harmonic for an arbitrary point x.*

Proof. We can assume that $a = 0$ and $b = \infty$. Then $f(x) = \lambda x$; since f is an involution, we have $\lambda = -1$. Clearly,

$$[0, \infty, x, f(x)] = \frac{x}{f(x)} = -1. \quad \square$$

THEOREM 13 (Darboux [**D**]). *Let f be a transformation of the real projective line $\mathbb{R}P^1$ that maps every harmonic quadruple of points to a harmonic quadruple of points. Then f is a projective transformation.*

Proof. Consider the composition $g(f(x))$, where g is the projective transformation that maps the points $f(0)$, $f(1)$, $f(\infty)$ to the points 0, 1, ∞, respectively. As the result, we obtain a transformation φ that leaves the points 0, 1, and ∞ fixed and maps any harmonic quadruple to a harmonic quadruple. Let us prove that this is the identity transformation. The quadruple $\{\infty, (\lambda + \mu)/2, \lambda, \mu\}$ is harmonic; hence

$$\varphi\left(\frac{\lambda + \mu}{2}\right) = \frac{\varphi(\lambda) + \varphi(\mu)}{2}.$$

Therefore, $2\varphi(\lambda) = \varphi(2\lambda)$ and $\varphi(\lambda + \mu) = \varphi(\lambda) + \varphi(\mu)$. Now it is easy to verify that $\varphi(m/n) = m/n$ for any positive integers m and n, $\varphi(0) = 0$, and $\varphi(-\lambda) = -\varphi(\lambda)$ for all $\lambda \in \mathbb{R}$. Thus $\varphi(\lambda) = \lambda$ for $\lambda \in \mathbb{Q}$. It is readily seen that

$$\frac{\lambda^2 - \lambda}{\lambda^2 + \lambda} : \frac{1 - \lambda}{1 + \lambda} = -1,$$

i.e., the quadruple $\{\lambda, -\lambda, \lambda^2, 1\}$ is harmonic. Therefore,

$$\frac{\varphi(\lambda^2) - \varphi(\lambda)}{\varphi(\lambda^2) + \varphi(\lambda)} : \frac{1 - \varphi(\lambda)}{1 + \varphi(\lambda)} = -1,$$

i.e., $\varphi(\lambda^2) = (\varphi(\lambda))^2$. This, in particular, means that the inequalities $x > 0$ and $\varphi(x) > 0$ are equivalent. Taking into account the fact that $\varphi(\lambda) = \lambda$ for $\lambda \in \mathbb{Q}$ and $\varphi(\lambda + \mu) = \varphi(\lambda) + \varphi(\mu)$ for $\lambda, \mu \in \mathbb{R}$, we obtain $\varphi(\lambda) = \lambda$ for $\lambda \in \mathbb{R}$. $\square$

Problems

The projective plane and projective spaces.

A.1. Prove that any line in $\mathbb{C}P^2$ intersects $\mathbb{R}P^2$.

A.2. Prove that the intersection of an arbitrary sphere in $\mathbb{R}^n \subset \mathbb{R}P^n \subset \mathbb{C}P^n$ with the hyperplane at infinity $z_{n+1} = 0$ is the quadric $z_1^2 + \cdots + z_n^2 = 0$.

A.3. The sides BC, CA, and AB of a triangle ABC contain points A_1, B_1, and C_1, respectively; the line segments $[A, A_1]$, $[B, B_1]$, and $[C, C_1]$ intersect at a point M. Prove that if the points A_1, B_1, C_1, M have homogeneous coordinates $(1, 0, 0)$,

$(0, 1, 0)$, $(0, 0, 1)$, $(1, 1, 1)$, respectively, then the homogeneous coordinates of the points A, B, C are $(-1, 1, 1)$, $(1, -1, 1)$, $(1, 1, -1)$.

A.4. Prove that the tangent line to the ellipse $x^2/a^2+y^2/b^2 = 1$ through the point $(1, \pm i, 0)$ passes through one of the foci of the ellipse.

A.5. Given a hyperbola with perpendicular asymptotes which intersects the line at infinity at the points $(x_1, y_1, 0)$ and $(x_2, y_2, 0)$, prove that $x_1x_2 + y_1y_2 = 0$.

A.6. (a) Given three circles, prove that the radical axes of the pairs of these circles meet at one point.

(b) Given three conics with common points A and B, prove that the common chords of these conics meet at a point.[4]

Polar line of a point with respect to a conic.

A.7. Given a conic on the plane and a straightedge, draw

(a) a point lying on the polar line of a given point A with respect to the conic;
(b) the polar line of a given point A with respect to the conic;
(c) the tangent line to the conic from a given point A;
(d) the tangent line to the conic at a given point A lying on the conic.

A.8. Prove that the polar line of a focus of a conic coincides with the directrix of the conic.

A.9. Given polar lines a, b, c, d of collinear points A, B, C, D with respect to a conic, prove that $[a, b, c, d] = [A, B, C, D]$.

A.10. Prove that if the sides of a triangle $A_1B_1C_1$ are polar lines of the vertices of a triangle ABC with respect to a conic, then the lines AA_1, BB_1, CC_1 meet at one point.

Projective duality.

A.11. (a) Prove that if the sides AC, CE, EA of a triangle ACE are tangent to some conic at points B, D, F, respectively, then the lines AD, CF, EB meet at one point.

(b) Formulate the dual assertion.

A.12. A triangle PQR is called *self-polar* with respect to a conic C if each of its sides is the polar line of the opposite vertex (with respect to C).

(a) Prove that a triangle is self-polar with respect to a conic C if and only if there exists a quadrilateral circumscribed about C such that the vertices of PQR coincide with the intersection points of its diagonals and of the pairs of its opposite sides.
(b) Given a plane Π intersecting the axes Ox, Oy, and Oz at points X, Y, and Z, prove that the triangle XYZ is self-polar with respect to the conic C that is the intersection of Π with the cone $z^2 = x^2/a^2 + y^2/b^2$.

A.13. (a) Prove that the polar line of a point a with respect to the conic $xAx^T = 0$ is determined by the equation $aAx^T = 0$.

(b) Prove that the conic dual to the conic $xBx^T = 0$ with respect to the conic $xAx^T = 0$ is given by the equation $xCx^T = 0$, where $C = AB^{-1}A$.

[4]Every pair of conics under consideration has four common points. Two of them are the points A and B. The common chord is a line passing through the two other points.

A.14. Given the circle C of radius R centered at $(a, 0)$ and the conic C^* dual to C with respect to the unit circle $x^2 + y^2 = 1$,

(a) prove that the equation of C^* has the form $x^2 + y^2 = e^2 (x - 1/a)^2$, where $e^2 = a^2/R^2$;

(b) prove that the eccentricity of C^* equals e, the focus is the origin, and the directrix is the polar line of the center of C with respect to the unit circle.

A.15. Two conics meet at points A, B, C, D. The intersection points of the line AB with the line CD, of BC with AD, of AC with BD are P, Q, R, respectively. Prove that the intersection points of the common tangent lines to the conics lie on the lines PQ, QR, PR.

Projective involutions and harmonic quadruples of points.

A.16. Prove that an arbitrary projective involution of a line l is obtained as the section of l by the pencil of circles passing through two given points.

A.17. Given a projective involutive transformation of a pencil of lines through a given point, prove that this pencil contains two orthogonal lines that correspond to each other under the involution.

A.18. Given points A', B', C' lying on lines BC, AC, AB, respectively, prove that the pairs of the orthogonal projections of A and A', of B and B', of C and C' on an arbitrary line are in involution.

A.19. Given lines AB and CD with intersection point E, lines AD and BC with intersection point F, and an arbitrary point O, prove that the pairs of lines OA and OC, OB and OD, OE and OF are in involution.

A.20. The diagonals of a quadrilateral $ABCD$ meet at a point O, and the lines containing its sides meet at points E and F.

(a) Prove that $\{AO, BO, EO, FO\}$ is a harmonic quadruple of lines.

(b) Prove that if the line AC meets EF at a point P, then $\{A, C, O, P\}$ is a harmonic quadruple of points.

A.21. Construct an n-gon whose sides (or their continuations) pass through n given points and whose vertices lie on a given circle.

A.22. Given two different projective involutions f and g of a conic C for which all lines XX' with $X' = f(X)$ or $g(X)$, pass through a point F or G, respectively, prove that $fg = gf$ if and only if F belongs to the polar line of G with respect to C.

A.23. Given a point A on a conic C, consider all chords X_1X_2 of C that are viewed at a right angle from A and prove that

(a) the map $X_1 \mapsto X_2$ is projective;

(b) all chords X_1X_2 pass through one point.

A.24. Circles S_a and S_b intersect the line l passing through their centers at points A_1 and A_2 and at points B_1 and B_2, respectively. The common tangent lines to the circles S_a and S_b intersect l at points C_1 and C_2. Prove that the pairs of points A_1 and A_2, B_1 and B_2, C_1 and C_2 are in involution.

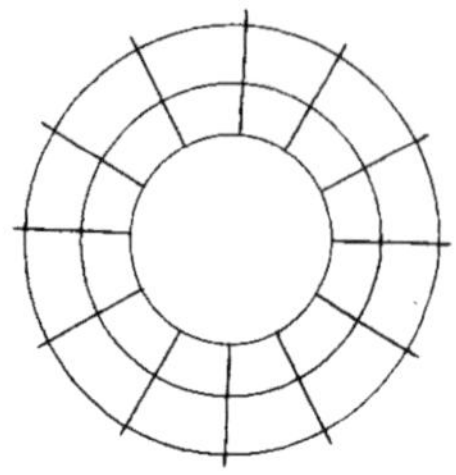

FIGURE A.33

4. Special properties of conics and quadrics

Confocal conics and quadrics. Let a and b be numbers such that $a > b$. Consider the family of conics

$$\frac{x^2}{a-\lambda} + \frac{y^2}{b-\lambda} = 1, \tag{1}$$

depending on the parameter λ. This family includes ellipses (for $b > \lambda$) and hyperbolas (for $a > \lambda > b$); their foci all lie on the Ox axis at the distance of

$$\sqrt{(a-\lambda)-(b-\lambda)} = \sqrt{(a-\lambda-\lambda-b)} = \sqrt{a-b}$$

from the origin. Thus the foci of all these conics coincide; such conics are called *confocal*.

Performing a homothety, we can position the foci at the points ± 1. The family (1) of confocal conics with foci at the points ± 1 is related to an interesting function of a complex variable known as the *Zhukovskii function*; this is the function

$$w = \frac{1}{2}\left(z + \frac{1}{z}\right). \tag{2}$$

Let us see where the rays $z = te^{i\varphi}$ with $t > 1$ and a fixed φ and the circles $z = te^{i\varphi}$ with a fixed $t > 1$, which are orthogonal to these rays (Figure A.33), are mapped. We have

$$z = te^{i\varphi} \Rightarrow w = \frac{1}{2}\left(t + \frac{1}{t}\right)\cos\varphi + \frac{i}{2}\left(t - \frac{1}{t}\right)\sin\varphi.$$

Therefore, the circles (with $t > 1$) transform into the ellipses

$$\frac{x^2}{\cosh^2\alpha} + \frac{y^2}{\sinh^2\alpha} = 1$$

(to write these equations, we have denoted t by e^{α}), and the rays become the hyperbolas

$$\frac{x^2}{\cos^2\varphi} - \frac{y^2}{\sin^2\varphi} = 1.$$

We have obtained a family of confocal conics. The Zhukovskii function conformally maps the exterior of the unit disk onto the exterior of the interval $[-1,1]$. This, in particular, means that through any point (x,y) with $x \neq 0$ and $y \neq 0$, there pass precisely one ellipse and precisely one hyperbola from the family, and they are orthogonal to each other.

Certainly, this can easily be obtained directly, without referring to complex analysis; we shall demonstrate this in connection with the n-dimensional case.

The expressions obtained for confocal ellipses have several interesting corollaries.

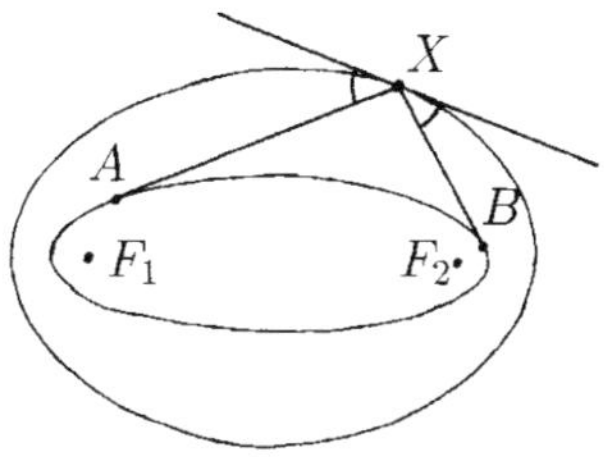

FIGURE A.34

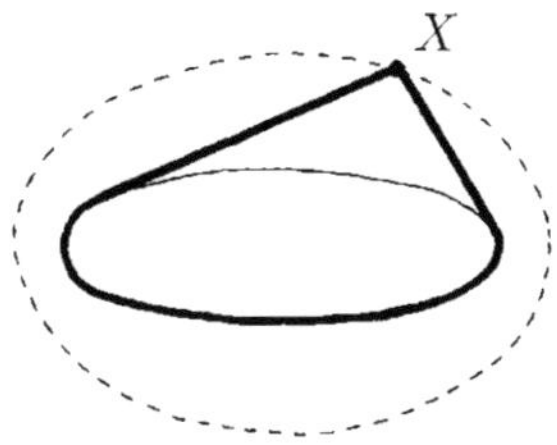

FIGURE A.35

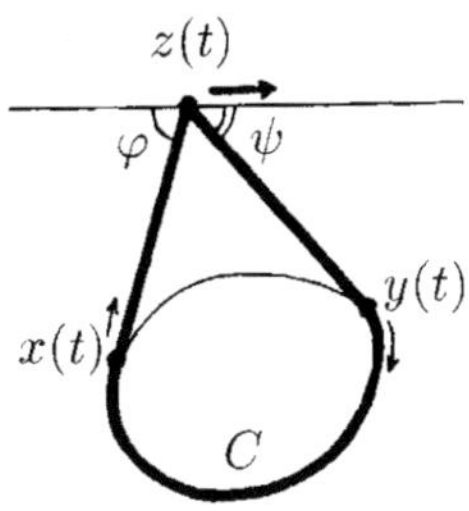

FIGURE A.36

1. *Let Γ_1 and Γ_2 be two confocal ellipses. Then the map of Γ_1 onto Γ_2 along the hyperbolas confocal with these ellipses is the restriction to Γ_1 of an affine transformation.*

Indeed, this map sends the point $(\cosh\alpha_1 \cos\varphi, \sinh\alpha_1 \sin\varphi)$ to the point $(\cosh\alpha_2 \cos\varphi, \sinh\alpha_2 \sin\varphi)$. Obviously, it is affine.

2. *Any curvilinear quadrilateral $w_1w_2w_3w_4$ formed by confocal conics have equal diagonals.*

Indeed, suppose that $w_{1,2}$ are the images of $z_{1,2} = t_{1,2}e^{i\varphi_1}$ and $w_{3,4}$ are the images of $z_{3,4} = t_{2,1}e^{i\varphi_2}$. Simple computations show that $|w_1 - w_3| = |w_2 - w_4|$.

Here is yet another interesting fact.

Let XA and XB be tangent lines to an ellipse with foci F_1 and F_2. Then the segments $[A, X]$ and $[B, X]$ make equal angles with the tangent line at X to an ellipse with the same foci (Figure A.34).

Indeed, according to the assertion of Exercise 4.14, the angle AXF_1 is equal to the angle BXF_2. In addition, the segments $[F_1, X]$ and $[F_2, X]$ make equal angles with the tangent line at the point X (this was proved on p. 66).

The last result has an interesting corollary. Let us surround a rigid ellipse by a belt and pull the belt holding it at some point X (we assume that the belt is longer than the ellipse). *All possible positions of the point X describe an ellipse confocal with the initial one* (Figure A.35).

This is proved by using the following assertion.

LEMMA. *If C is an unmovable convex curve and $l = |xz| + |yz| + \operatorname{arc}(xy)$ is the length of the curve shown in Figure* A.36, *then*

$$\frac{dl}{dt} = \left|\frac{dz}{dt}\right| (\cos\varphi - \cos\psi).$$

Proof. Clearly,

$$\frac{d}{dt}(\operatorname{arc}(xy)) = \left|\frac{dx}{dt}\right| - \left|\frac{dy}{dt}\right|.$$

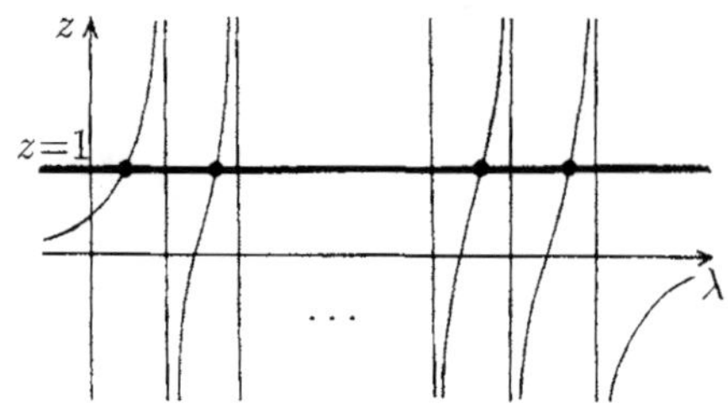

FIGURE A.37

In addition,

$$\frac{d}{dt}(|xz|) = \frac{d}{dt}\sqrt{(x-z, x-z)} = |x-z|^{-1}\left(\frac{dx}{dt} - \frac{dz}{dt}, x-z\right)$$

$$= -\left|\frac{dx}{dt}\right| + \left|\frac{dz}{dt}\right| \cos\varphi.$$

Similarly,

$$\frac{d}{dt}(|yz|) = \left|\frac{dy}{dt}\right| - \left|\frac{dz}{dt}\right| \cos\psi,$$

which yields the required formula. □

Thus, when the point X moves on an ellipse confocal with the initial one, the value l (the length of the belt) remains constant because $\varphi = \psi$ in this situation.

In addition to the families of confocal hyperbolas and ellipses described above, there are important families of confocal parabolas. Consider the simplest orthogonal system of straight lines, namely, vertical and horizontal lines. The map $z \mapsto w = z^2$ takes the vertical lines $a + it$ to the parabolas $x = a^2 - y^2/4a^2$ whose foci coincide with the origin. These parabolas are orthogonal to the images of horizontal lines, which are also parabolas with foci at the origin.

There are very many orthogonal systems on the plane: the application of a schlicht (i.e., univalent) conformal transformation to the exterior of the unit disk with the orthogonal system of circles and rays considered somewhat earlier yields an orthogonal system of curves. In higher-dimensional spaces, orthogonal systems are very rare.

By analogy with systems of confocal conics, we can consider the system of *confocal quadrics*

$$\frac{x_1^2}{a_1 - \lambda} + \cdots + \frac{x_n^2}{a_n - \lambda} = 1. \tag{3}$$

To be definite, we assume that $a_1 > a_2 > \cdots > a_n > 0$.

Similarly to the case $n = 2$, in this case *precisely n quadrics from the family* (3) *pass through any point* $(x_1, \ldots, x_n)$, *where* $x_1 \cdots x_n \neq 0$, *and the quadrics passing through one point are orthogonal* (this means that the hyperplanes tangent to the quadrics at this point are pairwise orthogonal).

Let us prove this assertion. Given a point $(x_1, \ldots, x_n)$, consider the function

$$z(\lambda) = \frac{x_1^2}{a_1 - \lambda} + \cdots + \frac{x_n^2}{a_n - \lambda}.$$

Its graph is shown in Figure A.37. The line $z = 1$ intersects the graph at precisely n points.

Next, the tangent hyperplane to the quadric (3) at the point $(\xi_1, \dots, \xi_n)$ is given by

$$\sum_{i=1}^{n} \frac{x_i \xi_i}{a_i - \lambda} = 1.$$

Clearly, the vector

$$\left(\frac{\xi_1}{a_1 - \lambda}, \dots, \frac{\xi_n}{a_n - \lambda} \right)$$

is orthogonal to this hyperplane.

Suppose that the quadrics with parameters λ and μ pass through the point $(\xi_1, \dots, \xi_n)$, i.e.,

$$\sum_{i=1}^{n} \frac{x_i \xi_i}{a_i - \lambda} = 1 \quad \text{and} \quad \sum_{i=1}^{n} \frac{x_i \xi_i}{a_i - \mu} = 1.$$

Then

$$0 = \sum_{i=1}^{n} \xi_i^2 \left(\frac{1}{a_i - \lambda} - \frac{1}{a_i - \mu} \right) = (\lambda - \mu) \sum_{i=1}^{n} \frac{\xi_i^2}{(a_i - \lambda)(a_i - \mu)}.$$

This means that the vectors

$$\left(\frac{\xi_1}{a_1 - \lambda}, \dots, \frac{\xi_n}{a_n - \lambda} \right) \quad \text{and} \quad \left(\frac{\xi_1}{a_1 - \mu}, \dots, \frac{\xi_n}{a_n - \mu} \right)$$

are orthogonal, and the hyperplanes orthogonal to these vectors are also orthogonal.

To every point $(x_1, \dots, x_n)$ with $x_1 \cdots x_n \neq 0$ one can assign the n-tuple $(\lambda_1, \dots, \lambda_n)$ of numbers such that

$$\frac{x_1^2}{a_1 - \lambda_i} + \cdots + \frac{x_n^2}{a_n - \lambda_i} = 1.$$

These numbers are called the *ellipsoidal coordinates* of the point $(x_1, \dots, x_n)$.

Consider the case of $n = 3$ in more detail. In this case,

$$x_1^2 = \frac{(a_1 - \lambda_1)(a_1 - \lambda_2)(a_1 - \lambda_3)}{(a_1 - a_2)(a_1 - a_3)};$$

the numbers x_2^2 and x_3^2 are computed similarly.

Let us fix two ellipsoids corresponding to λ_3 and λ_3'; λ_1 and λ_2 can be treated as coordinates on these ellipsoids. We map each point on one ellipsoid to the point with the same coordinates λ_1 and λ_2 on the other. This corresponds to carrying a point from one ellipsoid to the other along the trajectory orthogonal to all ellipsoids from family (3). In Cartesian coordinates, this transformation has the form $x \mapsto x'$, where

$$x_1' = \sqrt{\frac{a_1 - \lambda_3'}{a_1 - \lambda_3}}\, x_1,$$

and so on. Surprisingly, the transformation of ellipsoids constructed above, as is seen from the formula, is the *restriction to an ellipsoid of an affine transformation of three-dimensional space.* A similar transformation can be considered not only for ellipsoids but also for other quadrics of the same family, i.e., for one- and two-sheeted hyperboloids. An especially interesting result is obtained for one-sheeted hyperboloids: the rectilinear generators of one hyperboloid transform into the rectilinear generators of the other. Moreover, the map of *the rectilinear generators is an isometry.* The point is that if the endpoints of a segment $[x, y]$ move along

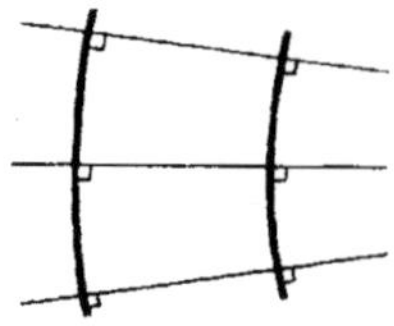

FIGURE A.38

trajectories orthogonal to this segment (Figure A.38), then the length of $[x, y]$ does not change. Indeed, by assumption, we have

$$\left(\frac{dx}{dt}, x-y\right) = \left(\frac{dy}{dt}, x-y\right) = 0.$$

Therefore,

$$\frac{d}{dt}(x-y, x-y) = 2\left(\frac{dx}{dt} - \frac{dy}{dt}, x-y\right) = 0.$$

Hence the length of $[x, y]$ is constant.

Rational parametrizations of conics. Let us take a point (x_0, y_0) on a conic and draw a line $y = y_0 + t(x - x_0)$ through this point. It intersects the conic at one more point. Let us evaluate the coordinate of this other point. For this purpose, we substitute $y_0 + t(x - x_0)$ for y in the equation of the conic; as a result, we obtain an equation of the form $A(t)x^2 + B(t)x + C(t) = 0$, where A, B, and C are polynomials in t. We know one root, x_0, of this equation. By the Viète theorem, the second root x_1 is $-x_0 - B(t)/A(t) = P(t)/A(t)$. In addition,

$$y_1 = y_0 + t(x_1 - x_0) = y_0 + t\left(-2x_0 - \frac{B(t)}{A(t)}\right) = \frac{Q(t)}{A(t)}.$$

Therefore, for a second-order curve, there exist polynomials $A(t)$, $P(t)$, and $Q(t)$ such that the point with coordinates $(P(t)/A(t), Q(t)/A(t))$ moves along the given curve as t varies from $-\infty$ to $+\infty$; adding the point that corresponds to the infinite value of t, we obtain all points of the curve. Such a representation of a curve using the polynomials A, P, and Q is called a *rational parametrization* of the curve.

For example, let us explicitly write the rational parametrization of the circle $x^2 + y^2 = 1$ corresponding to the point $(1, 0)$. The equation of the line through $(1, 0)$ has the form $y = t(x - 1)$; the line $x = 1$ corresponds to the infinite value of t. Substituting $y = t(x-1)$ in the equation $x^2+y^2 = 1$, we obtain $x^2+t^2(x^2-2x+1) = 1$, i.e., $(1+t^2)x^2+(-2t^2)x+(t^2-1) = 0$. The product of the roots of this equation is $(t^2-1)/(t^2+1)$, and one of the roots is 1. Therefore, the root x_1 we are interested in is $(t^2 - 1)/(t^2 + 1)$. Hence $y_1 = t(x_1 - 1) = -2t/(t^2 + 1)$. Thus the unit circle admits the rational parametrization $\big((t^2 - 1)/(t^2 + 1), -2t/(t^2 + 1)\big)$.

The obtained rational parametrization of the circle makes it possible to find all points both coordinates of which are rational. They are precisely those points for which the parameter t is rational. Indeed, if x_1 and y_1 are rational, then so is $t = y_1/(x_1 - 1)$, and if t is rational, then so are

$$x_1 = \frac{t^2 - 1}{t^2 + 1}, \qquad y_1 = \frac{-2t}{t^2 + 1}.$$

Similarly, all rational points of a second-order curve with rational coefficients can be found if at least one rational point (x_0, y_0) on this curve is known.

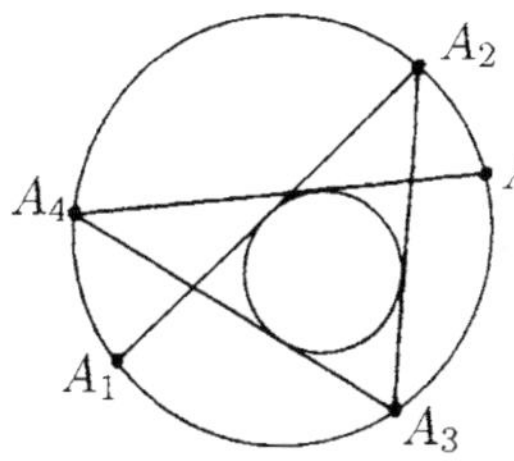

FIGURE A.39

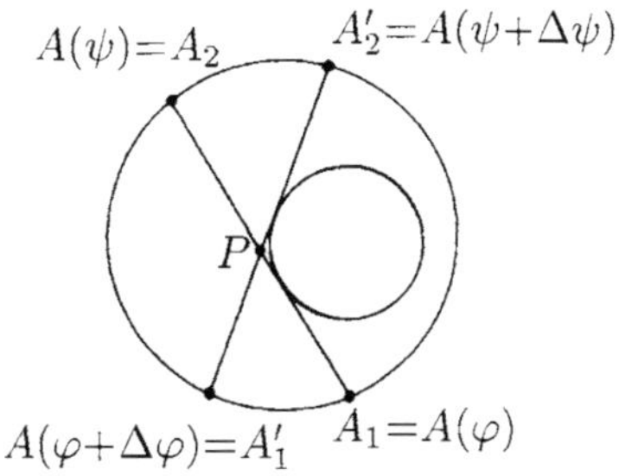

FIGURE A.40

Another application of rational parametrization is related to evaluation of integrals of the form

$$\int R(\sqrt{ax^2+bx+c}, x)\,dx,$$

where R is a rational function in two variables. The point here is that the curve $y^2 = ax^2 + bx + c$ admits the rational parametrization $(x(t), y(t))$. Therefore,

$$\int R(\sqrt{ax^2+bx+c}, x)\,dx = \int R(y(t), x(t))\,dx(t) = \int Q(t)\,dt,$$

where Q is some rational function. To find the required integral, we must compute the integral of this function and substitute the expression of t in terms of x and y, which is a rational function, for t.

Poncelet's theorem and the zigzag theorem. Consider two circles, one of which lies inside the other. Take a point A_1 on the outer circle and draw a chord A_1A_2 tangent to the inner circle. Next, from the point A_2, draw a chord A_2A_3 (different from A_1A_2) tangent to the inner circle, and so on (Figure A.39).

THEOREM 1 (Poncelet). *If $A_n = A_1$, then, for any other point B_1 on the outer circle, a similar construction gives $B_n = B_1$.*

Proof. We can assume that the points on the outer circle are of the form $A(\varphi) = (\cos\varphi, \sin\varphi)$. Consider

$$A_1 = A(\varphi), \qquad A_2 = A(\psi), \qquad A_1' = A(\varphi+\Delta\varphi), \qquad A_2' = A(\psi+\Delta\psi);$$

let P be the intersection point of the chords A_1A_2 and $A_1'A_2'$ (Figure A.40). Then $A_1A_1' : A_1P = A_2A_2' : A_2'P$. For an infinitely small $\Delta\varphi = d\varphi$, we obtain

$$\frac{d\varphi}{l(\varphi)} = \frac{d\psi}{l(\psi)}, \tag{4}$$

where $l(\alpha)$ is the length of a tangent segment from $A(\alpha)$ to the inner circle (there are two such segments, but they have equal lengths).

Consider the function $\psi(\varphi)$ corresponding to the map $A_1 \mapsto A_2$. Relation (4) implies

$$\frac{d}{d\varphi}\left(\int_\varphi^{\psi(\varphi)} \frac{d\alpha}{l(\alpha)}\right) = \frac{d\psi}{d\varphi}\frac{1}{l(\psi)} - \frac{1}{l(\varphi)} = 0, \qquad \text{i.e.,} \quad \int_\varphi^{\psi(\varphi)} \frac{d\alpha}{l(\alpha)} = c$$

is a constant. Let us denote $\psi^2(\varphi) = \psi(\psi(\varphi)), \ldots, \psi^n(\varphi) = \psi(\psi^{n-1}(\varphi))$. Then

$$nc = \int_\varphi^{\psi(\varphi)} \frac{d\alpha}{l(\alpha)} + \int_{\psi(\varphi)}^{\psi^2(\varphi)} \frac{d\alpha}{l(\alpha)} + \cdots + \int_{\psi^{n-1}(\varphi)}^{\psi^n(\varphi)} \frac{d\alpha}{l(\alpha)} = \int_\varphi^{\psi^n(\varphi)} \frac{d\alpha}{l(\alpha)}.$$

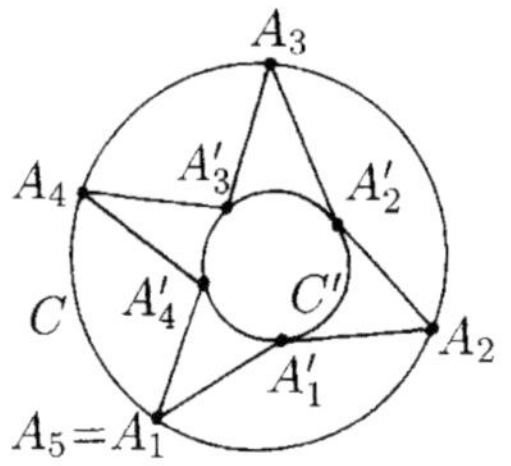

FIGURE A.41

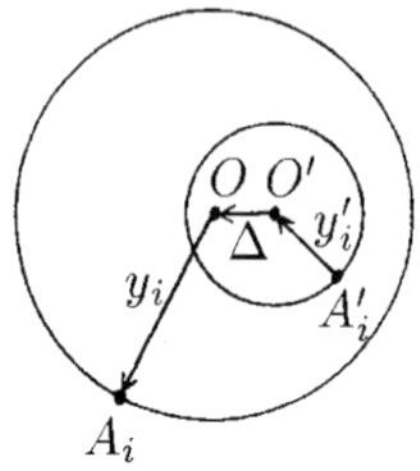

FIGURE A.42

The polygonal line $A_1A_2A_3\ldots$ closes after n steps involving m revolutions if $\psi^n(\varphi) - \varphi = 2\pi m$. Since the function $\int_0^t d\alpha/l(\alpha)$ monotonically increases with t, the condition $\psi^n(\varphi) - \varphi = 2\pi m$ is equivalent to

$$\int_{\varphi}^{\varphi+2\pi m} \frac{d\alpha}{l(\alpha)} = \int_0^{2\pi m} \frac{d\alpha}{l(\alpha)} = nc.$$

This condition is either fulfilled or not for all φ simultaneously. □

In the proof of Poncelet's theorem, it is essential that both outer and inner curves be circles (if the outer curve is not a circle, then the triangles $A_1PA'_1$ and A'_2PA_2 are not similar, and if the inner one is not a circle, then the function $l(\varphi)$ is two-valued rather than single-valued).

The following generalization of Poncelet's theorem can be proved by different methods.

THEOREM 2. *Let C and C' be conics in general position* (i.e., they are nondegenerate and do not touch each other). *Take a point A_1 on C and consider the point $A_2 \in C$ for which the line A_1A_2 is tangent to C'. Next, consider the point $A_3 \in C$ for which the line A_2A_3 is tangent to C', and so on. If $A_n = A_1$, then $B_n = B_1$ for any $B_1 \in C$.*

We shall not prove this theorem because all proofs we know are very complicated.

The statement of the following theorem somewhat resembles Poncelet's theorem.

THEOREM 3 (about zigzags). *Let C and C' be two circles in $\mathbb{R}^3$ such that every point on each of these circles is at distance ρ from exactly two points on the other circle. Consider the polygonal line $A_1A'_1A_2A'_2\ldots$ with side ρ shown in Figure* A.41. *If $A_{n+1} = A_1$ for some point $A_1 \in C$, then $B_{n+1} = B_1$ for any point $B_1 \in C$.*

Proof [**BHH**]. Let O and O' be the centers of the circles C and C', respectively. We put $y_i = \overrightarrow{OA_i}$, $y'_i = \overrightarrow{A'_iO'}$, and $\Delta = \overrightarrow{O'O}$ (Figure A.42). Let n and n' be unit vectors orthogonal to the planes of the circles C and C', respectively.

It is sufficient to prove that an infinitesimal shift dy_1 of the point A_1 leads to a shift dy_{n+1} of the point A_{n+1} such that $|dy_1|/|dy_{n+1}| = 1$. To prove this, we must also consider the shift dy'_1.

Since the vector dy_1 is orthogonal to the vectors y_1 and n, we have

$$dy_1 = \pm(n \times y_1)\frac{|dy_1|}{|y_1|}.$$

Replacing n by $-n$ if necessary, we obtain $dy_1 = (n \times y_1)\frac{|dy_1|}{|y_1|}$; similarly, $dy_1' = (n' \times y_1')\frac{|dy_1'|}{|y_1'|}$.

The vector $y_1' + \Delta + y_1$ has constant length ρ; therefore,

$$(y_1' + \Delta + y_1, dy_1' + dy_1) = 0,$$

i.e.,

$$(y_1' + \Delta + y_1, n' \times y_1')\frac{|dy_1'|}{|y_1'|} + (y_1' + \Delta + y_1, n \times y_1)\frac{|dy_1|}{|y_1|} = 0.$$

Thus

$$\frac{|dy_1'|}{|dy_1|} = -\frac{(\Delta + y_1', n \times y_1)}{(\Delta + y_1, n' \times y_1')}\frac{|y_1'|}{|y_1|}.$$

Since the points A_1 and A_2 play similar roles with respect to A_1', we also have

$$\frac{|dy_2|}{|dy_1'|} = -\frac{(\Delta + y_2, n' \times y_1')}{(\Delta + y_1', n \times y_2)}\frac{|y_2|}{|y_1'|}.$$

Let us multiply these equalities. Taking into account that $|y_1| = |y_2|$, we obtain

$$\frac{|dy_2|}{|dy_1|} = \frac{(\Delta + y_1', n \times y_1)(\Delta + y_2, n' \times y_1')}{(\Delta + y_1, n' \times y_1')(\Delta + y_1', n \times y_2)}. \tag{5}$$

Equality (5) can be simplified by using the following auxiliary assertion.

LEMMA. *We have* $(\Delta + y_1', n \times y_1) = -(\Delta + y_1', n \times y_2)$, $(\Delta + y_2, n' \times y_1') = -(\Delta + y_2, n' \times y_2')$.

Proof. Clearly, $|y_1| = |y_2|$, $|\Delta + y_1' + y_1| = |\Delta + y_1' + y_2|$, and $(n, y_1) = (n, y_2)$. Therefore,

$$(y_1 - y_2, y_1 + y_2) = 0, \qquad (y_1 - y_2, \Delta + y_1') = 0, \qquad (y_1 - y_2, n) = 0.$$

Thus the vectors $y_1 + y_2$, $\Delta + y_1'$, and n lie in one plane Π orthogonal to the vector $y_1 - y_2$. The vector $n \times (y_1 + y_2)$ is orthogonal to the plane Π; hence

$$(\Delta + y_1', n \times (y_1 + y_2)) = 0.$$

The equality $(\Delta + y_2, n' \times (y_1' + y_2')) = 0$ is proved similarly. □

Taking the lemma into account, we can write (5) in the form

$$\frac{|dy_2|}{|dy_1|} = \frac{(\Delta + y_2, n' \times y_2')}{(\Delta + y_1, n' \times y_1')}.$$

Having such a symmetric form equality, we can conclude that

$$\frac{|dy_{n+1}|}{|dy_1|} = \frac{|dy_2|}{|dy_1|} \cdot \frac{|dy_3|}{|dy_2|} \cdot \dots \cdot \frac{|dy_{n+1}|}{|dy_n|} = \frac{(\Delta + y_{n+1}, n' \times y_{n+1}')}{(\Delta + y_1, n' \times y_1')}.$$

Since $y_{n+1} = y_1$ and $y_{n+1}' = y_1'$, we have

$$\frac{|dy_{n+1}|}{|dy_1|} = 1,$$

as required to prove the zigzag theorem. □

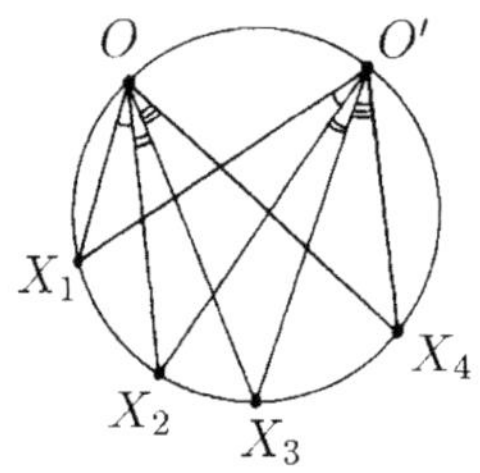

FIGURE A.43

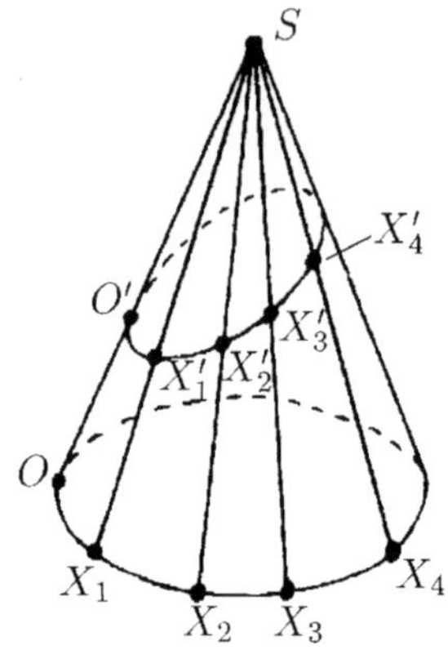

FIGURE A.44

The cross ratio of four points on a conic. For points X_1, X_2, X_3, X_4 on a circle C, the *cross ratio* $[X_1, X_2, X_3, X_4]$ can be defined as follows. We take an arbitrary point O on the circle and put

$$[X_1, X_2, X_3, X_4] = [OX_1, OX_2, OX_3, OX_4].$$

If we take another point O' instead of O, then the angle between the lines OX_i and OX_j will be equal to the angle between $O'X_i$ and $O'X_j$ (Figure A.43); therefore,

$$[O'X_1, O'X_2, O'X_3, O'X_4] = [OX_1, OX_2, OX_3, OX_4].$$

Similarly, the cross ratio of four points on one conic can be defined. The projections preserve the cross ratios of four lines; therefore,

$$[O'X_1', O'X_2', O'X_3', O'X_4'] = [OX_1, OX_2, OX_3, OX_4]$$

(see Figure A.44). This means that the cross ratio of the points X_1', X_2', X_3', X_4' on the conic C' does not depend on the choice of the point O' on this conic.

The independence of the cross ratio of points on a conic of the choice of the point O can be used to obtain yet another proof of Pascal's theorem about hexagons inscribed in conics. (See also the proofs given on pp. 15 and 71.)

THEOREM 4 (Pascal). *If points A, B, C, D, E, F lie on one conic, then the intersection points of the line AB with the line DE, of BC with EF, of CD and FA lie on one line.*

Proof. Consider the projections A_1, C_1, D_1, E_1 and A_2, C_2, D_2, E_2 of the points A, C, D, E from B and F on the lines $l_1 = DE$ and $l_2 = CD$, respectively (Figure A.45). The cross ratios $[A_1, C_1, D_1, E_1]$ and $[A_2, C_2, D_2, E_2]$ equal the cross ratio $[A, C, D, E]$ of the points on the conic. Therefore, the correspondence between the respective points on the lines l_1 and l_2 is projective and $D_1 = D_2$. Hence the lines joining the corresponding points meet at one point (see p. 50), i.e., the lines A_1A_2, C_1C_2, and E_1E_2 meet at one point. In other words, the point $C_1C_2 \cap E_1E_2 = BC \cap EF$ lies on the line joining the points $A_1 = AB \cap DE$ and $A_2 = CD \cap FA$. □

The independence of the cross ratio of points on a conic of the choice of O has the following geometric interpretation. Consider lines l and l' and a conic C. For points O and O' on C, we construct a map $l \to l'$ as follows: for a point Y on l, we take the point X at which the line YO intersects the conic and we send Y to the point Y' at which XO' intersects l' (Figure A.46). The obtained map is projective.

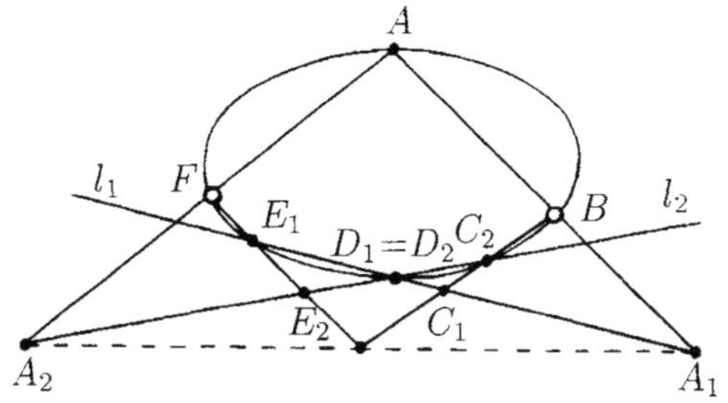

FIGURE A.45

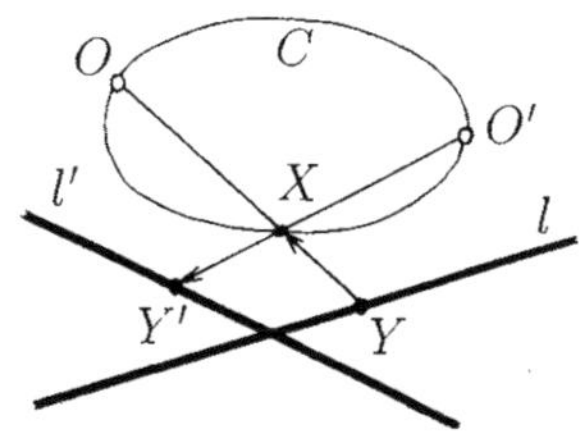

FIGURE A.46

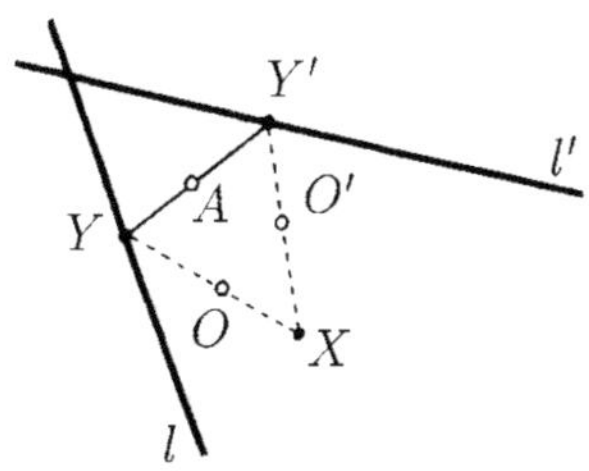

FIGURE A.47

On the other hand, suppose that a projective correspondence $l \to l'$ is given. Let us fix some points O and O' and consider the intersection points X of all lines OY and $O'Y'$ such that Y and Y' are corresponding points. It is natural to expect that the points X lie on a conic passing through O and O'. This is indeed so; it should only be taken into account that in the degenerate case, we may obtain a line rather than a conic (see p. 50).

To show this, consider three intersection points X_1, X_2, and X_3 of three pairs OY_i, $O'Y_i'$ of corresponding lines. Together with O and O', we have five points; in the nondegenerate situation, we can draw exactly one conic C through these points. Suppose that lines OY and $O'Y'$, where Y and Y' are corresponding points, intersect C at points X and X'. Then $[X_1, X_2, X_3, X] = [X_1, X_2, X_3, X']$, i.e., X and X' coincide. Therefore, the intersection point of the lines OY and $O'Y'$ lies on the conic C.

Thus, in the case of a projective correspondence of pencils of lines through points O and O', the intersection points of pairs of corresponding lines lie on a conic passing through O and O'. This property provides us with diverse ways to obtain conics from lines and conics.

EXAMPLE 1. If the lines l and l' and points A, O, O' are fixed, points Y and Y' lie on l and l', respectively, and the line YY' rotates about A, then the intersection point X of the lines OY and $O'Y'$ (Figure A.47) moves along a conic.

EXAMPLE 2. If the line l and points O and O' are fixed and a point Y moves along l, then the intersection point X of the lines obtained by rotating OY and $O'Y$ about O and O' through fixed angles φ and ψ, respectively (Figure A.48), moves along a conic passing through O and O'.

EXAMPLE 3. In Example 2, the point Y can move not only along a line l but also along a conic C.

The cross ratio of points X_1, X_2, X_3, and X_4 on a conic can be defined differently. Suppose that l is an arbitrary tangent line and x_i are the tangent lines at

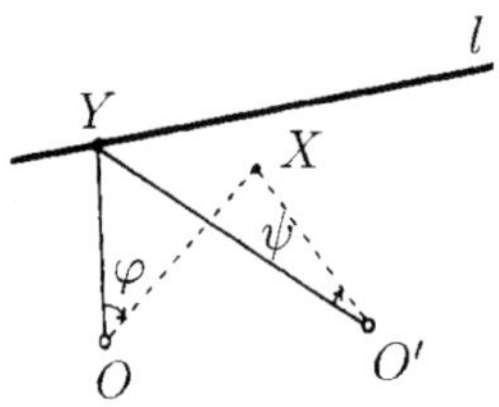

FIGURE A.48

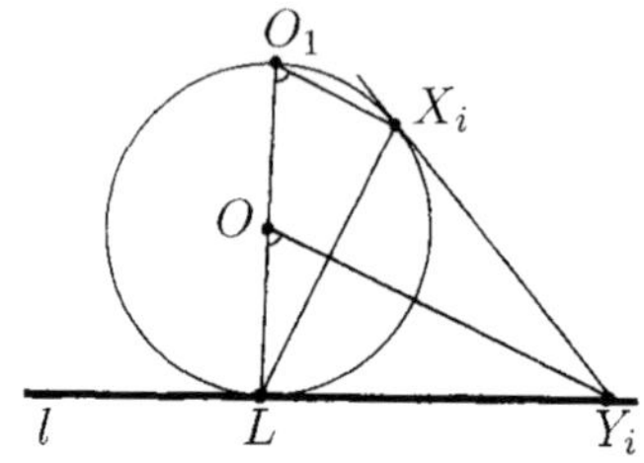

FIGURE A.49

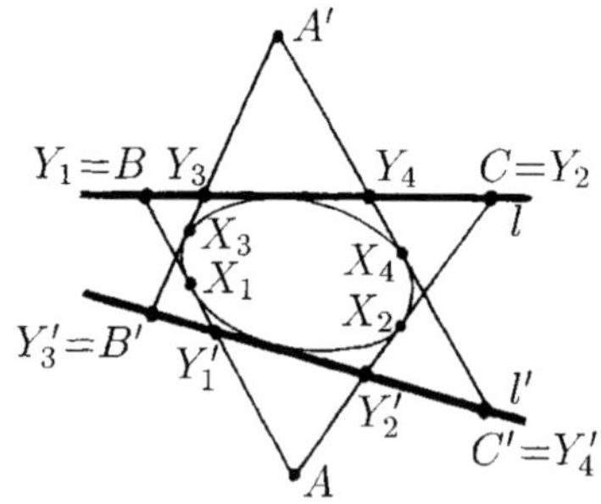

FIGURE A.50

the points X_i to the conic. Let Y_i be the intersection points of x_i with l. Then

$$[X_1, X_2, X_3, X_4] = [Y_1, Y_2, Y_3, Y_4].$$

It is sufficient to prove this for a circle. Let O be the center of the circle. Suppose that L is the point of tangency of l with the circle and O_1 is the point antipodal to L (Figure A.49). The lines O_1X_i and OY_i are perpendicular to LX_i; therefore, they are parallel to each other and, hence, $\angle LOY_i = \angle LO_1X_i$. Thus

$$\sin \angle X_iO_1X_j = \sin \angle Y_iOY_j,$$
$$[O_1X_1, O_1X_2, O_1X_3, O_1X_4] = [OY_1, OY_2, OY_3, OY_4].$$

The independence of the cross ratio $[Y_1, Y_2, Y_3, Y_4]$ of the choice of the tangent line l also has interesting applications. Let us consider an example.

THEOREM 5. *If triangles ABC and $A'B'C'$ are circumscribed about a conic Γ (i.e., their sides or extensions of sides are tangent to Γ), then the points A, B, C, A', B', C' lie on one conic.*

Proof. We choose one side of each of the triangles ABC and $A'B'C'$, say, BC and $B'C'$, as tangent lines l and l' to the conic. Let X_1, X_2, X_3, X_4 be the points of tangency of the other sides of the triangles with the conic (Figure A.50). If the tangent line at X_i intersects the lines l and l' at points Y_i and Y_i', respectively, then the correspondence $Y_i \mapsto Y_i'$ is projective. Therefore, the intersection points of the lines AY_i' and $A'Y_i$ lie on a conic passing through the points A and A'. Clearly,

$$AY_1' \cap A'Y_1 = B,\ AY_2' \cap A'Y_2 = C,\ AY_3' \cap A'Y_3 = B',\ AY_4' \cap A'Y_4 = C'. \quad \square$$

COROLLARY (Poncelet's triangle theorem). *If two conics Γ_1 and Γ_2 have a triangle inscribed in Γ_1 and circumscribed about Γ_2, then there are infinitely many such triangles; moreover, any point of the conic Γ_1 is a vertex of such a triangle.*

Proof. Let a triangle ABC be inscribed in Γ_1 and circumscribed about Γ_2. We draw a chord $A'B'$ of Γ_1 tangent to Γ_2. From points A' and B', we draw tangent lines to Γ_2; they meet at some point C. According to Theorem 5, the points A, B, C, A', B', and C' lie on one conic. On the other hand, through the five points A, B, C, A', and B', passes a unique conic; this conic is Γ_1. Therefore, the point C' lies on Γ_1. □

Problems

A.25. Prove that all conics dual to the conics

$$\frac{x^2}{a-\lambda}+\frac{y^2}{b-\lambda}=z^2$$

with respect to the conic $x^2+y^2+z^2=0$ pass through the points $(\pm 1, \pm i, \sqrt{a-b})$.

A.26. Prove that all confocal conics share four tangent lines passing through the points $(1, \pm i, 0)$.

A.27. Prove that if a line l in $\mathbb{C}P^3$ is not contained in a nondegenerate quadric Q, then there exist precisely two tangent planes to this quadric that pass through l.

5. Additional topics of non-Euclidean geometries

Paving the sphere, the plane, and the Lobachevsky plane by triangles. Consider a curvilinear triangle formed by three arcs of circles or by segments of straight lines in the Euclidean plane. Such a triangle determines a group generated by all reflections (symmetries) in its sides. We are interested in triangles whose reflections in their sides generate pavings, i.e., such that their images under reflections either coincide or have no common interior points. By coincidence, we mean pointwise coincidence, i.e., not only the triangles themselves but also all their respective points must coincide; for example, a regular triangle does not pointwise coincide with its image under the symmetry with respect to an altitude.

Without loss of generality, we can assume that two sides of a curvilinear triangle are line segments (we can achieve this by applying an inversion centered at one of the intersection points of the circles containing the sides of the triangle). The images of such a triangle under reflections in the two rectilinear sides form a paving if and only if the angle between these sides equals π/n, where n is a positive integer. (If the pointwise coincidence of the images is not required, then the angle can also be $2\pi/n$.) Thus the triangle must have angles $(\pi/p, \pi/q, \pi/r)$; we say for short that such a triangle has type (p, q, r).

The problem about reflections of a curvilinear triangle with angles α, β, and γ is actually a problem of spherical, Euclidean, or hyperbolic geometry, depending on the sign of the number $\alpha+\beta+\gamma-\pi$. (If $\alpha+\beta+\gamma-\pi>0$, three additional inequalities of the form $\alpha+\beta<\gamma+\pi$ must hold, but they are automatically true for triangles of type (p, q, r).)

Let us explain what we mean in more detail. Recall that the sides AB and AC of a curvilinear triangle ABC can be assumed to be rectilinear. Let O be the center of the circle containing the arc BC. If the points O and A lie on one side of the line BC, then $\alpha+\beta+\gamma>\pi$, and if they lie on different sides of BC, then $\alpha+\beta+\gamma<\pi$. Therefore, $\alpha+\beta+\gamma=\pi$ only if BC is a line segment. Thus, if $\alpha+\beta+\gamma=\pi$, we obtain the problem of reflections in three Euclidean lines.

Now, consider the case where $\alpha+\beta+\gamma<\pi$, i.e., the center of the circle S containing the arc BC and the vertex A lie on different sides of the line BC. Let us draw the tangent lines AP and AQ to the circle S from the point A (Figure A.51). Let S_1 be the circle of radius $|AP|$ centered at A. Then the sides of the triangle ABC are hyperbolic lines in the Poincaré model in the circle S_1. Clearly, the symmetries about the sides of the curvilinear triangle ABC are symmetries about hyperbolic lines.

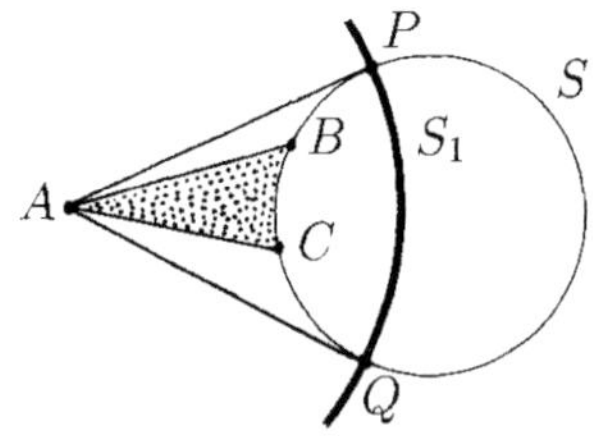

FIGURE A.51

This simple argument has an important corollary. Namely, if $\alpha+\beta+\gamma<\pi$, then all images of the curvilinear triangle ABC under the reflections in its sides lie either inside some disk or in some half-plane.

It remains to consider the case $\alpha+\beta+\gamma>\pi$. This inequality alone does not ensure the existence of a spherical triangle with angles α, β, and γ; additionally, three inequalities of the form $\alpha+\beta<\pi+\gamma$ dual to inequalities of the form $a+b>c$ must hold. But if $\alpha,\beta,\gamma<\pi/2$, then $\alpha+\beta<\pi+\gamma$ automatically. For this reason, we shall assume the existence of a spherical triangle ABC with angles α, β, and γ. The stereographic projection of this triangle on the plane tangent to the sphere at the vertex A is a curvilinear triangle $AB'C'$ with angles α, β, and γ, and the angle A in this triangle is rectilinear. It is easy to verify that if curvilinear triangles $AB'C'$ and AB_1C_1 with equal respective angles have a common rectilinear angle A, then these triangles are homothetic, and the center of the homothety is A. Indeed, if O' and O_1 are the centers of the circles containing the arcs $B'C'$ and B_1C_1, then the triangles $O'B'C'$ and $O_1B_1C_1$ are isosceles triangles with parallel lateral sides, and hence with parallel bases $B'C'$ and B_1C_1. Therefore, changing the radius of the sphere if necessary, we can make the image of the spherical triangle under the stereographic projection to coincide with the plane curvilinear triangle under consideration. Clearly, the symmetries about the sides of the curvilinear triangle correspond to the symmetries about the sides of the spherical triangle. Indeed, the points A and B are symmetric with respect to a spherical line l if and only if an arbitrary spherical circle intersects the spherical line l at a right angle. Therefore, the stereographic projection maps points symmetric with respect to the spherical line l to points symmetric with respect to the image of l.

To find all possible spherical or Euclidean triangles with angles $(\pi/p,\pi/q,\pi/r)$, we must find integer solutions of the inequality $1/p+1/q+1/r>1$ or of the equation $1/p+1/q+1/r=1$, respectively. We can assume that $p\leqslant q\leqslant r$. Then, in both cases, $p\leqslant 3$. A simple review of all possible cases yields spherical triangles of types $(2,2,n)$, $(2,3,3)$, $(2,3,4)$, $(2,3,5)$ and Euclidean triangles of types $(3,3,3)$, $(2,4,4)$, $(2,3,6)$.

It can be verified directly that the images of all these triangles under the reflections in the sides pave the sphere and the Euclidean plane. The Euclidean triangles

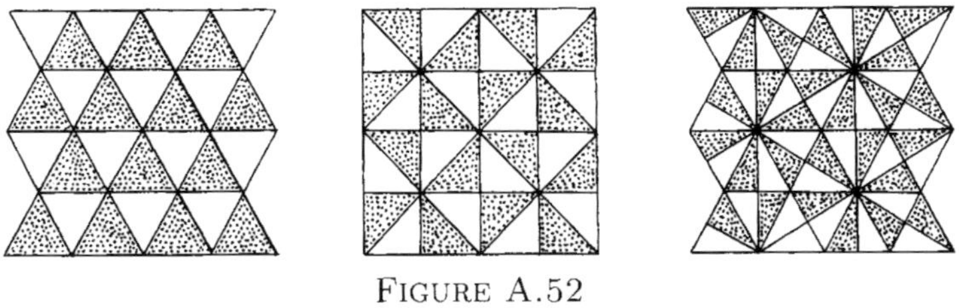

FIGURE A.52

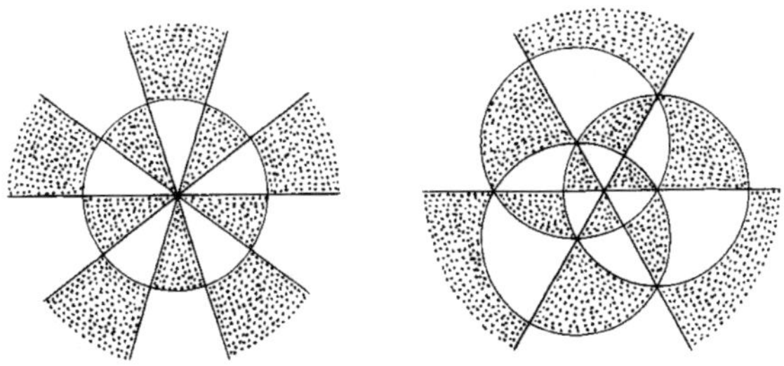

FIGURE A.53

are shown in Figure A.52. For the images of the spherical triangles, similar pictures can be drawn. The first two cases, which are the simplest, are shown in Figure A.53. Note that the required pavings of the sphere can be obtained as follows (except for the first series of triangles). We join the center of each face of a regular polyhedron with the vertices and the midpoints of the sides of this face. Next, we construct the projection of the system of segments obtained (including the edges) on the sphere circumscribed about the polyhedron from the center of this sphere. For a tetrahedron, we obtain a system of triangles of type $(2, 3, 3)$, for a cube and an octahedron, a system of triangles of type $(2, 3, 4)$, and for a dodecahedron and icosahedron, a system of triangles of type $(2, 3, 5)$.

To find all possible hyperbolic triangles with angles $(\pi/p, \pi/q, \pi/r)$, we must find integer solutions of the inequality $1/p+1/q+1/r < 1$. This inequality holds for all sufficiently large numbers p, q, and r, so it has infinitely many solutions. It can be proved that if $1/p + 1/q + 1/r < 1$, then the images of a hyperbolic triangle of type (p, q, r) under the reflections in its sides fill the entire hyperbolic plane without overlaps. This follows from the Poincaré theorem about pavings of the hyperbolic plane by polygons (see p. 191).

Fundamental domains of the modular group. Let G be the group generated by the reflections in the sides of a triangle T with angles $(\pi/p, \pi/q, \pi/r)$. Then the sphere, plane, or Lobachevsky plane can be represented as the union of the triangles gT $(g \in G)$, and the triangles g_1T and g_2T with $g_1 \neq g_2$ have no common interior points. The latter condition is ensured by the requirement of the pointwise coincidence of the images of the triangles. Indeed, if the triangles g_1T and g_2T share an interior point, then $g_1(x) = g_2(x)$ for all $x \in T$, and hence $g_1 = g_2$.

The triangle T is said to be a *fundamental domain* of the group G. This definition admits the following generalization. Let G be a subgroup of the motion group of the sphere, plane, or Lobachevsky plane. We say that D is a *fundamental domain* of the group G if the following conditions hold:

(i) D is a convex polygon (in the case of the Lobachevsky plane, it may have vertices and sides at infinity);

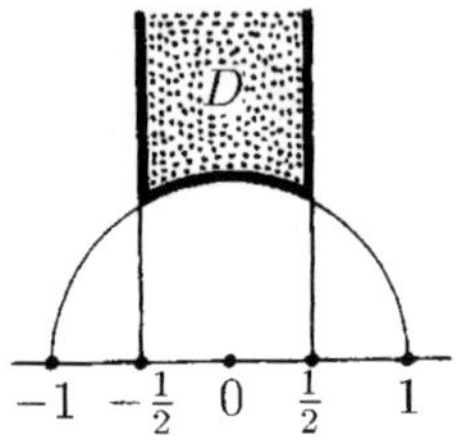

FIGURE A.54

(ii) the polygons gD $(g \in G)$ cover the entire sphere, plane, or Lobachevsky plane;

(iii) the polygons g_1D and g_2D with $g_1 \neq g_2$ have no common interior points.

Note that not all subgroups of the motion group admit fundamental domains. For example, the entire motion group has no fundamental domain.

As an example, let us find a fundamental domain for the subgroup of proper motions in the group generated by the reflections in sides of a triangle T of type (p, q, r). Let s be the symmetry about a side of T. Then the union of T with sT is the required fundamental domain. To prove this, we divide the triangles of the form gT into pairs $\{gT, gsT\}$; this can be done because gsT determines the same pair as gT, since $(gs)sT = gT$.

Consider a more interesting example of a fundamental domain. Recall that the proper motions of the Poincaré upper half-plane model have the form

$$z \mapsto \frac{az+b}{cz+d}, \qquad \text{where} \quad a, b, c, d \in \mathbb{R}, \quad ad - bc = 1.$$

The group of 2×2 matrices with real elements and determinant 1 is denoted by $SL(2, \mathbb{R})$. Any two proportional matrices (and only such matrices) correspond to the same linear-fractional transformation. Therefore, the group of proper motions of the Lobachevsky plane is isomorphic to the quotient group $SL(2, \mathbb{R})/\pm I = PSL(2, \mathbb{R})$; here I is the identity matrix. The group $SL(2, \mathbb{R})$ has the important subgroup $SL(2, \mathbb{Z})$, which consists of the matrices with integer elements. It corresponds to the subgroup $PSL(2, \mathbb{Z})$ in $PSL(2, \mathbb{R})$. The group $PSL(2, \mathbb{Z})$ is called the *modular group*.

THEOREM 1. *The triangle D with angles $(0, \pi/3, \pi/3)$ shown in Figure* A.54 *is a fundamental domain of the modular group.*

Proof. Consider the elements $S(z) = -1/z$ and $T(z) = z + 1$ in the group $G = PSL(2, \mathbb{Z})$. They generate a subgroup G'. We start by proving that D is a fundamental domain of the group G'; then we shall prove that $G' = G$, i.e., that the elements S and T generate the entire group G.

First, we check that the triangles $g'D$ with $g' \in G'$ cover the entire Lobachevsky plane, i.e., that $\mathrm{Im}(z) > 0$ implies $g'z \in D$ for some $g' \in G'$.

LEMMA 1. *If* $\mathrm{Im}(z) > 0$, *then* $\mathrm{Im}(gz)$ *with* $g \in G$ *takes only finitely many values exceeding* $\mathrm{Im}(z)$.

Proof. Clearly,

$$\mathrm{Im}(gz) = \mathrm{Im}\left(\frac{az+b}{cz+d}\right) = \mathrm{Im}\,\frac{adz + bc\bar{z}}{|cz+d|^2} = \frac{\mathrm{Im}(z)}{|cz+d|^2}.$$

Therefore, $\operatorname{Im}(gz) \geqslant \operatorname{Im}(z)$ only if $|cz + d| \leqslant 1$. The last inequality only holds for a finite number of integers (c, d), and the value $\operatorname{Im}(gz)$ is determined uniquely for every such pair. □

The group G' is contained in G; therefore, for any point z in the upper half-plane, we can select an element $g' \in G'$ for which $\operatorname{Im}(g'z)$ is maximal. The transformation $T(z) = z + 1$ does not change the imaginary part of the number z; hence we have $|\operatorname{Re}(w)| \leqslant 1/2$ for some $w = T^k g'z$, where $k \in \mathbb{Z}$, and the value $\operatorname{Im}(w)$ is still maximal. In particular,

$$\operatorname{Im}(w) \geqslant \operatorname{Im}\left(-\frac{1}{w}\right) = \frac{\operatorname{Im}(w)}{|w|^2}.$$

Therefore, $|w| \geqslant 1$, and hence $w \in D$.

LEMMA 2. *If z is an interior point of the domain D and $gz \in D$ for $g \in G$, then g is the identity transformation.*

Proof. Let $g(z) = (az + b)/(cz + d)$. First, consider the case $c = 0$. Then $ad = 1$, i.e., $g(z) = z \pm b$. If $b \neq 0$, then $g(D)$ intersects D only for $g(z) = z \pm 1$ and in this case the intersection is contained in the set $|\operatorname{Re}(z)| = 1/2$, which does not contain interior points of D.

Now, suppose that $c \neq 0$. Then

$$g(z) = \frac{a}{c} - \frac{1}{c(cz + d)};$$

hence

$$\left|g(z) - \frac{a}{c}\right| \cdot \left|z + \frac{d}{c}\right| = \frac{1}{c^2}. \tag{1}$$

The numbers a/c and d/c are real; therefore, the imaginary parts of $g(z) - a/c$ and $z + d/c$ are equal to the imaginary parts of $g(z)$ and z, and, since the imaginary part of no point in D is smaller than $\sqrt{3}/2$, the absolute values of $g(z) - a/c$ and $z + d/c$ are not smaller than $\sqrt{3}/2$. Therefore, $|c| \leqslant 2/\sqrt{3}$, and c is a nonzero integer. Thus $c = \pm 1$; hence relation (1) can be written as

$$|g(z) \mp a| \cdot |z \pm d| = 1.$$

This contradicts the inequalities $|g(z) \mp a| \geqslant 1$ and $|z \pm d| > 1$, which hold for any integers a and d whenever $g(z) \in D$ and z is an interior point of D. □

Lemma 2 implies, in particular, that the sets $g_1'D$ and $g_2'D$ have no common interior points for two different elements g_1' and g_2' of the group G'. Thus D is a fundamental domain of G'.

Now, it is easy to prove that $G = G'$. Indeed, let g be an arbitrary element in the group G. Take an arbitrary interior point z in the domain D. The point gz lies in the upper half-plane; hence there exists an element $g' \in G'$ for which $g'(gz) \in D$. The motion $g'g \in G$ maps the interior point z of D to some point of D. Therefore, according to Lemma 2, the transformation $g'g$ is the identity transformation, i.e., $g = (g')^{-1} \in G'$.

This completes the proof of the theorem. □

Therefore, a fundamental domain of the group $PSL(2, \mathbb{Z})$ is a triangle with angles $(0, \pi/3, \pi/3)$. But it is more convenient to treat this domain as a quadrilateral $ABCD$ with angles 0, $\pi/3$, π, and $\pi/3$ (Figure A.55). The point is that the sides of

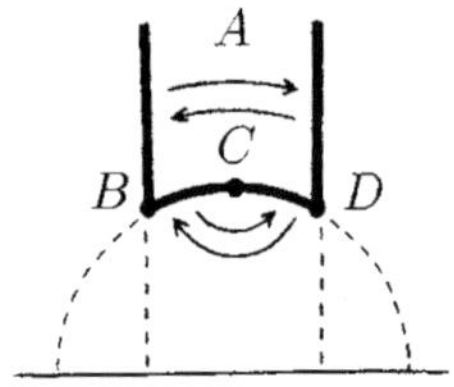

FIGURE A.55

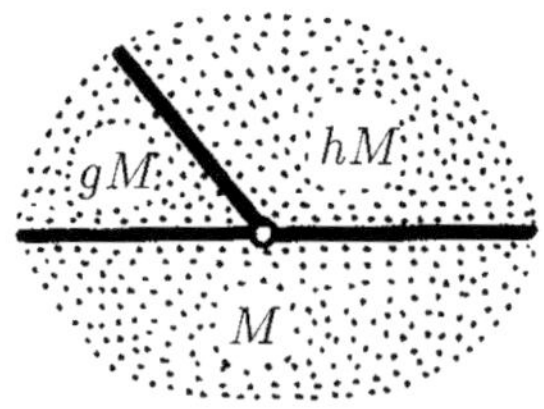

FIGURE A.56

this quadrilateral are divided into pairs so that the sides in each pair are transformed into each other by some elements of the group $PSL(2, \mathbb{Z})$.

Poincaré's theorem about fundamental polygons. The examples of the sphere, the plane, and the Lobachevsky plane paved by triangles show that pavings of the Lobachevsky plane have the most complex structure. Nevertheless, all convex polygons in the Lobachevsky plane that are fundamental domains for the groups of proper motions can be described. We emphasize that, as opposed to pavings by triangles, we only consider proper motions. This, however, does not prevent us from simultaneously obtaining results about pavings by triangles. The point is that if we glue together triangles T and sT, where s is the symmetry about a side of the triangle T, then we obtain a fundamental domain D of a group that only includes proper motions. On the other hand, from pavings of the Lobachevsky plane by the images of the figure D, it is easy to obtain a paving by the images of the triangle T.

Let M be a convex polygon on the Lobachevsky plane. We only consider the case where M is compact, i.e., it has no vertices on the absolute. Suppose that M is a fundamental domain for some group G of proper motions. Consider the polygons of the form gM (where $g \in G$) that have nonempty intersections with M; there are finitely many such polygons. The intersections of the convex sets M and gM have no interior points and are therefore either points or line segments. We call all segments of the form $M \cap gM$ the *sides* of the fundamental domain M and all points of triple intersections $M \cap gM \cap hM$ the *vertices* of the fundamental domain. The endpoints of each side of the fundamental domain are vertices of the fundamental domain. Note that a vertex of the fundamental domain M may or may not be a vertex of the polygon M (Figure A.56).

To each side a of the fundamental domain M, a polygon $g(M)$ is attached; thus $a = g(b)$, where b is a side of the fundamental domain M. Clearly, the side $g^{-1}(a)$ of $g^{-1}(M)$ is attached to the side b; so the sides of the fundamental domain M are divided into pairs of sides that are mapped to each other by motions $g \in G$. These motions g are said to *pair the sides*.

We do not eliminate the case $a = g(a)$. It is easy to verify that a proper motion $g \neq \mathrm{id}$ maps a segment to itself only if g is the symmetry about the midpoint of the segment. Therefore, we can get rid of the situation where $a = g(a)$ by adding the midpoint of a to the set of vertices. We stress that vertices of this type are only added to make the sides in the pairs different; no other complications arise if we do not add them.

Each motion g pairing the sides maps a pair of vertices A_1, A_2 to another pair of vertices $g(A_1), g(A_2)$. Let us join the vertices A_i and $g(A_i)$ by an oriented edge; the same edge with opposite orientation corresponds to the motion g^{-1}. We draw such edges for all pairing motions. As a result, we obtain a graph. To each vertex of the fundamental domain, precisely two pairing motions correspond; therefore, each vertex of the graph is incident to precisely two edges. Thus the obtained graph consists of several cycles. The vertices belonging to one cycle are said to be *equivalent*.

Consider a cycle $A_1 \dots A_n$. Its edges can be oriented so that $g_i(A_i) = A_{i+1}$. Let us show that the transformation $g = g_n g_{n-1} \cdots g_1$ is the rotation about A_1 through the angle $\alpha_1 + \cdots + \alpha_n$, where α_i is the angle at the vertex A_i. Clearly,

$$g(A_1) = g_n g_{n-1} \cdots g_2(A_2) = \cdots = A_1.$$

Let a be the side of the fundamental domain that is mapped to another side by the motion g_1. The proper motion

$$f = R_{A_1}^{\alpha_1} g_n \cdots R_{A_3}^{\alpha_3} g_2 R_{A_2}^{\alpha_2} g_1$$

takes the point A_1 and the line a to themselves; therefore, $f = \mathrm{id}$. It is easy to verify that for any motion h,

$$hR_A^{\alpha} = R_{h(A)}^{\varepsilon\alpha} h, \tag{2}$$

where $\varepsilon = 1$ if the motion h is proper, and $\varepsilon = -1$ otherwise. Using this equality (for proper motions), we obtain

$$f = R_{A_1}^{\alpha_1} g_n \cdots R_{A_3}^{\alpha_3} g_2 g_1 R_{A_1}^{\alpha_2} = \cdots = g R_{A_1}^{\alpha_1 + \cdots + \alpha_n}.$$

Therefore, $g = R_{A_1}^{-(\alpha_1 + \cdots + \alpha_n)}$.

The polygons $g_n(M), g_n g_{n-1}(M), \dots, g_n g_{n-1} \cdots g_1(M)$ have the common vertex A_1, and the angles at this vertex equal $\alpha_n, \alpha_{n-1}, \dots, \alpha_1$, respectively. Let us apply powers of the transformation $g = g_n g_{n-1} \cdots g_1$ to these polygons. A necessary and sufficient condition for the result to be a paving of a neighborhood of A_1 is the fulfillment of the equality $\alpha_1 + \cdots + \alpha_n = 2\pi/k$ for some $k \in \mathbb{N}$.

We have obtained some necessary conditions for a convex polygon on the Lobachevsky plane to be a fundamental domain. It turns out that these conditions are also sufficient.

POINCARÉ'S THEOREM. *Let a compact convex hyperbolic polygon M have the following properties:*

(i) *the sides of the polygon are divided into pairs $\{a, b\}$, and for each such pair of sides, there exists a proper motion g pairing them, i.e., $b = g(a) = M \cap g(M)$;*
(ii) *the sum of angles at equivalent vertices is $2\pi/k$, where $k \in \mathbb{N}$.*

Then M is a fundamental domain of the group G generated by the motions pairing the sides of M.

Proof. First, we prove that if the polygons $h_1 M$ and $h_2 M$ (where $h_1, h_2 \in G$) share an interior point, then $h_1 = h_2$. Let us apply the transformation h_1^{-1} to both polygons. After that, we can assume that $h_1 = \mathrm{id}$. The transformation

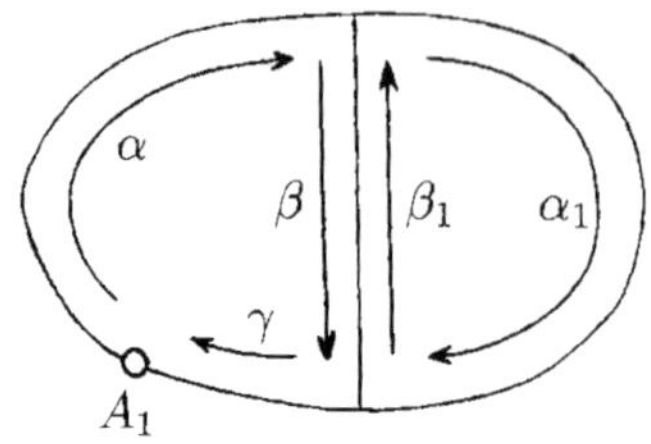

FIGURE A.57

$g = h_2$ can be decomposed into transformations pairing the sides, i.e., represented as $g = g_k g_{k-1} \cdots g_1$. Suppose that $g \neq \mathrm{id}$ and the length k of the decomposition is minimal. Let A be a common interior point of the polygons M and $g(M)$. The decomposition $g = g_k \cdots g_1$ determines a closed polygonal line $A_1 \dots A_k$, where we have $A_1 = A$ and $A_{s+1} = g_s(A_s)$ for $s \geqslant 1$. The points A_s and A_{s+1} cannot coincide because they are interior points of two convex polygons lying on different sides of a line (the extension of the common side). The minimality of the decomposition length implies that the polygonal line $A_1 \dots A_n$ has no self-intersections. Indeed, a self-intersecting polygonal line determines a self-intersecting sequence of polygons, which contains a shorter closed chain.

If the polygonal line $A_1 \dots A_k$ winds around some vertex of M, the transformation g is by assumption the rotation about this vertex through the angle 2π, and we then have $g = \mathrm{id}$. Suppose that the polygonal line is the union of two polygonal lines, and the common sides of these lines are passed in opposite directions (Figure A.57). The equalities $\alpha\beta\gamma = 1$, $\beta_1\alpha_1 = 1$, and $\beta_1 = \beta^{-1}$ imply $\alpha\alpha_1\gamma = 1$. Therefore, if the closed polygonal line is obtained as a union of several loops around vertices of polygons, then we also have $g = \mathrm{id}$. Clearly, an arbitrary closed non-self-intersecting polygonal line corresponding to a decomposition of g can be represented in such a way; therefore, $g = \mathrm{id}$.

It remains to show that the images of the polygon M fill the entire Lobachevsky plane. Let X be an arbitrary point. The vertices of the polygons gM $(g \in G)$ form a countable set; therefore, we can choose a point A inside the polygon M so that the segment $[X, A]$ contains no vertices of the polygons gM. It suffices to show that the segment $[X, A]$ intersects only finitely many polygons gM.

The distance between two sides of a convex polygon is zero only if these sides are neighbors. Therefore, the distance between any two points of nonneighboring sides of the polygon M is longer than some fixed positive number. Hence the segment $[X, A]$ contains only finitely many segments that join nonneighboring sides of the polygons gM. We say that a segment *subtends* a vertex of a polygon if its endpoints lie on sides incident to this vertex. Suppose that a pair of neighboring segments subtend nonequivalent vertices of polygons g_1M and g_2M (Figure A.58). Then the segment formed by the two segments under consideration cannot subtend a vertex in the polygon $g_1M \cup g_2M$ (only equivalent vertices can be situated at one point); hence it joins points of two nonneighboring sides. Therefore, the length of the obtained segment exceeds some fixed positive number. Thus, discarding a finite number of segments, we can assume that all segments subtend vertices of the same equivalence class.

Suppose that the segment $[X, A]$ contains infinitely many such segments. If the vertex subtended by these segments does not lie on the absolute, then the line XA must wind around this vertex like a spiral (Figure A.59), which is impossible. □

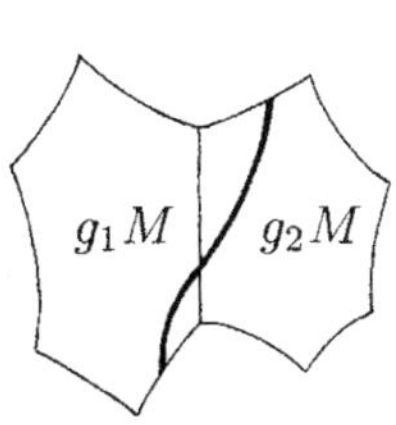

FIGURE A.58

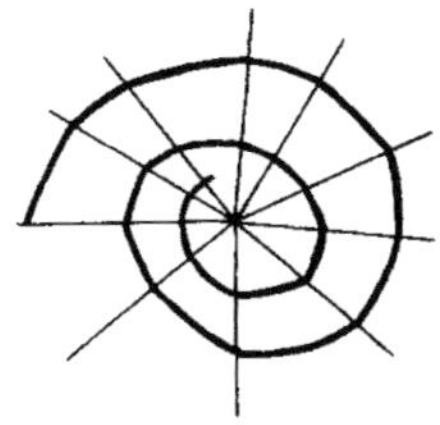

FIGURE A.59

The proof of Poincaré's theorem shows that all relations between pairing motions follow from the relation that arise in going around the vertices of the polygon. The relations that arise in going around vertices belonging to one cycle are equivalent.

The Lobachevsky space. All models of the Lobachevsky plane that have been considered above can be generalized to n-dimensional space; the isometry of these models is proved without substantial changes.

For the Klein model and the Poincaré model in the n-ball $x_1^2 + \cdots + x_n^2 < 1$, the distance between points A and B can be defined as the distance between the points A and B in the model of the Lobachevsky plane in the section of the ball by the plane through the center of the ball and the points A and B.

For the Poincaré model in the half-space

$$H^n = \{ (x_1, \dots, x_n) \in \mathbb{R}^n \mid x_n > 0 \},$$

the distance between points A and B is defined as the distance between these points in the Poincaré model in the half-plane passing through the points A and B and perpendicular to the boundary of the half-space H^n (i.e., to the *absolute*).

For the model on a sheet of the pseudosphere

$$[x, x] = -c^2, \qquad \text{where} \quad [x, y] = x_1y_1 + \cdots + x_ny_n - x_{n+1}y_{n+1},$$

the distance d between points x and y is defined by

$$\cosh^2\left(\frac{d}{c}\right) = \frac{[x, y]^2}{[x, x][y, y]}.$$

In what follows, we only deal with the Poincaré models. In these models, the hyperbolic subspaces are parts of spheres orthogonal to the absolute (the Euclidean subspaces are treated as spheres of infinite radius). The subspaces of dimension $n-1$ in the space of dimension n are called *hyperplanes.* To define the symmetry about a hyperplane in the Lobachevsky space, it suffices to define the symmetry about the sphere $\mathbb{S}^{n-1}$ in $\mathbb{R}^n$. We say that points A and B are *symmetric* with respect to the sphere $\mathbb{S}^{n-1}$ in $\mathbb{R}^n$ if any $n-1$-sphere through the points A and B is orthogonal to $\mathbb{S}^{n-1}$. Symmetry about a sphere coincides with the inversion with respect to this sphere. The inversions preserve the cross ratios; therefore, the symmetries about hyperplanes are motions of the Lobachevsky space.

THEOREM 2. *An arbitrary motion of the Lobachevsky n-space can be represented as a composition of no more than $n+1$ symmetries about hyperplanes.*

Proof. We have already proved a similar theorem for the Lobachevsky plane, so we only discribe the part of the proof that is not carried over automatically. Namely, we show that any motion with $n+1$ fixed points not lying in one hyperplane is the identity map. Let $A_1, \dots, A_{n+1}$ be the fixed points of the motion under consideration. Clearly, the points $A_1, \dots, A_m$ cannot lie in one $(m-2)$-dimensional subspace. The strict triangle inequality implies that if X and Y are two different fixed points, then all points of the line XY are fixed. In particular, all points of the line A_1A_2 are fixed. Through an arbitrary point in the plane $A_1A_2A_3$, we can draw a line that intersects two of the three fixed lines A_1A_2, A_2A_3, A_1A_3 at two different points. This means that all points of the plane $A_1A_2A_3$ are fixed. Similarly, all points of the subspace $A_1A_2A_3A_4$ are fixed, and so on. □

THEOREM 3. *Every motion of the Lobachevsky space is uniquely determined by its restriction to the absolute.*

Proof. Consider the Poincaré model in the half-space. Among the restrictions to the absolute of all hyperplanes through a point X, we can select a sphere of minimal radius (its center coincides with the orthogonal projection of X on the absolute). Clearly, the point X is uniquely reconstructed from the sphere selected. A motion g maps the family of hyperplanes under consideration to the family of hyperplanes through the point $g(x)$. In the corresponding family of spheres on the absolute, we can again select a sphere of minimal radius, from which the point $g(x)$ can be reconstructed. This means that we can find the point $g(x)$ solely from the transformations of the spheres on the absolute, and these transformations are determined by the restriction of g to the absolute. □

THEOREM 4. *The group of proper motions of the Lobachevsky 3-space is isomorphic to $PSL(2, \mathbb{C})$.*

Proof. According to Theorems 2 and 3, the group of motions of the Lobachevsky 3-space is isomorphic to the group of transformations of the plane generated by the symmetries about lines and circles. We treat the plane as the complex plane $\mathbb{C}$. The symmetries about the real axis and the unit circle transform z into $\bar{z}$ and $-1/\bar{z}$, respectively. The compositions of symmetries about two parallel lines, two lines through the origin, and two circles centered at the origin transform z into $z + a$, $e^{i\alpha}z$, and kz $(k > 0)$, respectively. Therefore, an arbitrary map of the form

$$z \mapsto \frac{az+b}{cz+d}, \quad z \mapsto \frac{a\bar{z}+b}{c\bar{z}+d}, \quad \text{where } \begin{pmatrix} a & b \\ c & d \end{pmatrix} \in GL(2, \mathbb{C}),$$

can be represented as a composition of symmetries. Clearly, any composition of symmetries about lines and circles is of such a form.

The proper motions correspond to the transformations $z \mapsto (az+b)/(cz+d)$; thus, since proportional matrices determine the same transformation, the group of proper motions of the Lobachevsky 3-space is isomorphic to $SL(2, \mathbb{C})/\pm I$. □

By analogy with the Lobachevsky plane, we can define elliptic, parabolic, and hyperbolic pencils of planes in the Lobachevsky 3-space. This enables us to prove the following assertions.

(i) The group $PSL(2, \mathbb{C})$ contains (as subgroups) the groups of proper motions of the sphere, the plane, and the Lobachevsky plane.

(ii) The sphere, the plane, and the Lobachevsky plane can be isometrically embedded in the Lobachevsky 3-space.

An *elliptic* pencil (in Lobachevsky 3-space) is a family of planes passing through a fixed point; a *parabolic* pencil is a family of planes through a fixed point on the absolute; and a *hyperbolic* pencil is a family of planes orthogonal to a fixed plane.

THEOREM 5. *The set of all compositions of symmetries about pairs of planes from one pencil is a subgroup in $PSL(2,\mathbb{C})$. This subgroup is isomorphic to the group of proper motions of the sphere, the plane, or the Lobachevsky plane depending on whether the pencil is elliptic, parabolic, or hyperbolic.*

Proof. Suppose that the pencil is elliptic. Consider the Poincaré ball model such that the planes of the pencil pass through the center of the ball. The symmetries about such planes generate the motion group of a sphere concentric with the absolute.

For a parabolic pencil, we consider the Poincaré half-space model and assume that the planes of the pencil are Euclidean half-planes orthogonal to the absolute (all such planes pass through the point ∞ on the absolute). The symmetries about such planes transform each Euclidean plane parallel to the absolute into itself. The restrictions of these symmetries to the planes considered generate the group of Euclidean motions of these planes.

For a hyperbolic pencil, we consider the Poincaré half-space model and assume that all planes of the pencil are orthogonal to some hyperbolic plane Π passing through the point ∞. Let l be the intersection line of Π with the absolute. The symmetries about the planes of the pencil under consideration are inversions with respect to spheres centered on the line l; these motions transform any Euclidean half-plane Π' bounded by l into itself. The restrictions of these transformations to Π' are symmetries about circles orthogonal to the line l. These transformations generate the group of motions of the Poincaré upper half-plane model. □

On a surface F in the Lobachevsky space, the *induced metric* can be defined; the length element on F (i.e., the distance between infinitely close points) in this metric is the distance between the corresponding points in the Lobachevsky space.

Consider the surface consisting of the images of a fixed point under the action of the proper motions corresponding to the planes of some pencil. Such a surface is called a *sphere*, a *horosphere*, or a *hypersphere* depending on whether the pencil is elliptic, parabolic, or hyperbolic. In the situations considered above, the spheres are Euclidean spheres centered at the center of the model, the horospheres are Euclidean planes parallel to the absolute, and the hyperspheres are Euclidean planes containing the line l.

For the Poincaré model in the half-space $H^3 = \{\,(x,y,z) \mid z > 0\,\}$, the length element has the form ds/z, where $ds^2 = dx^2 + dy^2 + dz^2$. Therefore, the squared length element on the horosphere $z = a$ has the form $a^{-2}(dx^2 + dy^2)$. Thus, in the coordinates (x, y), the distance between points on an horosphere is proportional to the Euclidean distance.

On the hypersphere $y = kz$, the squared length element has the form

$$\frac{dx^2 + (1+k^2)dz^2}{z^2} = (1+k^2)\frac{du^2 + dv^2}{v^2},$$

where $u = x/\sqrt{1+k^2}$ and $v = z$. Thus, in the coordinates (u, v), the distance between points on a hypersphere is proportional to the distance on the Lobachevsky plane.

In the Poincaré unit ball model, the length element has the form $2\,ds/(1-R^2)$, where ds is the Euclidean length element and R is the distance to the center of the ball. Therefore, the distance between points on the hyperbolic sphere $x^2+y^2+z^2 = R^2$ is $2/(1-R^2)$ times longer than the distance between points on the Euclidean sphere $x^2 + y^2 + z^2 = R^2$. Thus the hyperbolic sphere under consideration is isometric to the Euclidean sphere of radius $2R/(1 - R^2)$.

The quaternion model. The motions of the Lobachevsky 3-space can be represented by quaternions (introduced by Hamilton). Recall that the *quaternions* are the algebra over $\mathbb{R}$ with generators 1, i, j, k and relations

$$i^2 = j^2 = k^2 = -1, \qquad ij = -ji = k, \qquad jk = -kj = i, \qquad ki = -ik = j;$$

this algebra is associative. We put

$$H^3 = \{\, z + tj \mid z \in \mathbb{C},\ t > 0\,\}.$$

The action of a matrix $\gamma = \left(\begin{smallmatrix} a & b \\ c & d \end{smallmatrix}\right) \in SL(2, \mathbb{C})$ on the points of the absolute $\mathbb{C}$ is given by the formula $\gamma z = (az + b)/(cz + d)$. We must extend this action to the quaternions of the form $q = z + tj$, where $t > 0$. This can be done in two equivalent ways:

$$\gamma q = (aq + b)(cq + d)^{-1} = (qc + d)^{-1}(qa + b).$$

First, let us verify that both formulas are indeed equivalent, i.e.,

$$(qc + d)(aq + b) = (qa + b)(cq + d).$$

The numbers a, b, c, and d pairwise commute; therefore, the required equality is equivalent to $(ad - bc)q = q(ad - bc)$, which is obvious because $ad - bc = 1$.

Now we show that $\gamma q = z' + t'j$, where $t' > 0$. Taking into account that $j\overline{(cz + d)} = (cz + d)j$ and $ad - bc = 1$, we obtain

$$(aq + b)\overline{(cq + d)} = (az + b)\overline{(cz + d)} + act^2 + tj;$$

hence

$$\gamma q = \frac{(aq + b)\overline{(cq + d)}}{|cq + d|^2} = \frac{(az + b)\overline{(cz + d)} + act^2 + tj}{|cq + d|^2}.$$

It remains to verify that the transformation γ is a motion of the Lobachevsky space. It suffices to show that γ preserves the cross ratios. Clearly,

$$\begin{aligned} \gamma q - \gamma q' &= (aq + b)(cq + d)^{-1} - (q'c + d)^{-1}(q'a + b) \\ &= (q'c + d)^{-1}[(q'c + d)(aq + b) - (q'a + b)(cq + d)](cq + d)^{-1} \\ &= (q'c + d)^{-1}(q - q')(cq + d)^{-1}. \end{aligned}$$

In addition, $q'c + d = (z'c + d) + t\bar{c}j$ and $cq' + d = (cz' + d) + tcj$. Therefore, $|q'c + d| = |cq' + d|$, and hence

$$|\gamma q - \gamma q'| = \frac{|q - q'|}{|cq + d|\,|cq' + d|}.$$

This formula ensures that γ preserves the cross ratio $\frac{|CA|}{|CB|} : \frac{|DA|}{|DB|}$.

About the axiomatic approach to Euclidean and non-Euclidean geometries. Among the reasons for including the principles of geometry in this book, there are two of memorial character; the first is the 200th birthday of Nikolai Ivanovich Lobachevsky (1792–1856), one of the greatest Russian scientists, called the Copernicus of geometry by the famous English mathematician *William Clifford* (1845–1879). The geometry course on which this book is based was first taught at the Independent University of Moscow in 1992, the year of Lobachevsky's anniversary. It was therefore pertinent to make the listeners familiar with the main work of

Lobachevsky's life. The second reason is related to Andrei Nikolaevich Kolmogorov, as we have already mentioned in the Introduction.

There were also other reasons for touching upon the fundamentals of non-Euclidean geometries. The exposition gave us the occasion to discuss a number of fundamental ideas and concepts whose development required so much efforts from our great predecessors. We mean the idea of a deductive construction of geometry; the concept of a geometric model (we have already touched upon this idea in discussing Euclidean situation); the idea of the uniformity of non-Euclidean geometries and the uniform approach to the notion of geometry itself which is based on the concept of "motion;" finally, we would like to lift the veil hiding the nonlinear, "curved," world of geometric images.

The geometry of Euclid lives on a plane board. The geometry of Riemann is almost the geometry of the surface of a ball (we must only "glue together" the antipodal points of the sphere). Unfortunately, the geometry of Lobachevsky cannot be globally realized in three-dimensional space. But locally, it can be realized on a special surface called the *pseudosphere*.

The pseudosphere is the surface of revolution obtained by rotating the curve called the *tractrix*, or the *pursuit curve* (the segments of the tangents to this curve cut out by the curve and its asymptote have constant length).

If we cut a piece of the sphere, then we can move it along the surface of the sphere so that it will never lose contact with the sphere. The same is true of the pseudosphere: a piece of the pseudosphere can freely slide along the pseudosphere without being folded. On the sphere and pseudosphere, Riemann and Lobachevsky geometries, respectively, are locally realized. Constancy of "curvature" is characteristic of both of these geometries, but there is only one step to geometric objects curved arbitrarily (such geometries, however, are beyond the scope of this book).

So far, we have got acquainted with many important notions of geometry, such as subspaces, volumes, affine varieties, cones, and convex sets, and established relations between algebraic objects—linear equalities and inequalities—and their geometric images in the Cartesian model.

But the basic idea that made it possible to translate algebra in the geometric language and geometry into the language of algebra was the idea of a model. Now we shall use this idea once more in combination with another idea that also has a long history. Let us repeat some points. We have already discussed the deductive construction of science. The notion of deduction or proof proper must lean on some basic facts, which cannot be derived without tautology. They are called *axioms*. These are, so to speak, the "atoms" of science. The "molecules," theorems, are formed from axioms with the use of logical connectives, and collections of theorems, in their turn, form "compounds," scientific theories.

Among the first mathematicians who tried to develop geometry using the deductive method was Euclid. The deductive construction of geometry was completed by *David Hilbert* (1862–1943). His work *Foundations of Geometry* (1899) is undoubtedly one of the most outstanding creations of the human mind (this work, however, led to excessive enthusiasm for the deductive method, which had many negative after-effects).

But what does the phrase "completion of Euclid's project" mean? To explain this, we describe the notion of "mathematical structure" introduced by Nicolas Bourbaki. Let us turn to his fundamental paper *L'Architecture des mathématiques* ([**Bo2**]).

"Now,—he writes,—it should be explained what is meant by a mathematical structure in the general case. The common feature of various notions united under this generic name is that they apply to sets of elements of indefinite nature. To define a structure means first to specify one or several relations between these elements ... ; then, it is postulated that the given relation or relations satisfy certain conditions, which are the axioms of the structure under consideration. To construct an axiomatic theory of a structure means to deduce logical consequences from the axioms of this structure forgetting all other assumptions about the elements under consideration (in particular, any conjectures concerning their nature)."

What is the plane from this "structural" point of view? What structure is the "Euclidean plane" (and, simultaneously, what is the "non-Euclidean plane") according to Kolmogorov?

The plane (either Euclidean or Lobachevsky—they differ in only one axiom) is a triple (X, L, d), where X is a set of "points," L is a family of subsets called "straight lines," and d is a map assigning a nonnegative real number $d(x, y)$ (the "distance" from x to y) to each pair of points x and y in X. Next follow the axioms.

We state the axioms of Euclidean (and non-Euclidean) geometry following A. N. Kolmogorov (see the school textbook by A. N. Kolmogorov, A. F. Semenovich, and P. S. Cherkasov, *Geometry 6–8*, Moscow: Prosveshchenie, 1982, pp. 373–376). They are divided into five groups.

I. *Membership axioms.*

I_1. *Each straight line is a set of points.*

I_2. *One and only one straight line passes through any two distinct points.*

I_3. *There exists at least one straight line; to each straight line, at least one point belongs; there exists a point outside an arbitrary straight line.*

II. *Distance axioms.*

A distance d is a function $d\colon X \times X \to \mathbb{R}_+$ satisfying the following axioms:

II_1. $d(x_1, x_2) = 0$ *if and only if* $x_1 = x_2$;

II_2. $d(x_1, x_2) = d(x_2, x_1)$ *for any* x_1 *and* x_2;

II_3. *The triangle inequality* $d(x_1, x_3) \leqslant d(x_1, x_2) + d(x_2, x_3)$ *holds.*

These axioms turn X into a *metric space*; this is one of the fundamental notions in geometry and functional analysis.

III. *Order axioms* (axiomatization of the basic betweenness relation).

DEFINITION 1. Let points x, y, and z lie on one straight line. We say that the point y *is between* x and z if $d(x, y) + d(y, z) = d(x, z)$. The set of points lying between x and z is called the *closed interval* $[x, z]$.

III_1. *Three points* x_1, x_2, *and* x_3 *lie on one straight line if and only if one of them lies between the other two.*

III_2. *Any point* ξ *on a straight line* l *divides the set of all points on* l *different from* ξ *into two nonempty subsets so that* ξ *lies between any two points belonging to distinct subsets.*

DEFINITION 2. *A ray from* ξ is one of the subsets specified in the preceding axiom together with the point ξ itself.

III_3. *For any* $a \geqslant 0$, *on a ray from* ξ *there exists exactly one point for which the distance to* ξ *is equal to* a.

DEFINITION 3. A set A of points on the plane is said to be *convex* if for any x and y in A, the entire interval $[x, y]$ lies in A.

III_4. *Any straight line divides the set of all points on the plane that do not belong to this line into two nonempty convex subsets.*

IV. *Motion axiom.*

DEFINITION 4. The set of points belonging to one of the subsets into which a straight line l divides X, together with the points on the line l itself, is called a *half-plane.* The line l is said to be the boundary line for the half-plane. If a line containing a ray is a boundary for some half-plane, then we say that this half-plane is *adjacent to the ray.*

DEFINITION 5. A one-to-one isometric (i.e., such that the distance between images equals the distance between the preimages) map of the plane onto itself is called a *displacement* (or a motion, or an isometry).

IV. *For any pair of rays and half-planes Π_1 and Π_2 adjacent to them, there exists a unique displacement that maps the first ray onto the second and the half-plane Π_1 onto the half-plane Π_2.*

IV′. *For any two intervals $[x_1, x_2]$ and $[y_1, y_2]$ of equal lengths, there exist exactly two displacements that map x_1 to y_1 and x_2 to y_2.*

V. *Parallelism axiom* (this axiom distinguishes between Euclidean and non-Euclidean geometries).

In Euclidean plane geometry: *Through a point outside a straight line, passes one and only one straight line parallel to the given one.*

In Lobachevsky plane geometry: *Through a point outside a straight line, pass at least two lines parallel to the given one.*

As we have mentioned, it was D. Hilbert who had finally completed Euclid's project. In his *Foundations of Geometry* (1899), Hilbert constructed a consistent (noncontradictory) and complete system of axioms and proved the independence of the basic groups of axioms in his axiomatics. Later, similar axiomatics based on the description of different structures were constructed by Schur, Weyl (who, apparently, first suggested to use the notion of vector space as the foundation for an axiomatics of the Euclid space), Birkhoff, and many other mathematicians, in particular, by A. N. Kolmogorov in his geometry course.

Let us elucidate the meaning of all these terms—consistency, completeness, and independence.

Consistency. A system of axioms is called *consistent* if there exists a model in which all axioms turn into valid assertions.

We described several geometry models, including the Cartesian model for Euclidean and affine geometries and the Klein and Poincaré models for Lobachevsky geometry.

We leave to the reader the justification of the following two statements:

— Kolmogorov's axiomatics of the Euclidean plane is consistent (in the Cartesian model);

— Kolmogorov's axiomatics of the Lobachevsky plane is consistent (in the Poincaré and Klein models).

Completeness means that the system of axioms provides unambiguity of derivations. A statement is called *independent* of some axioms if there exist two models such that this statement is an axiom in the first, its negation is an axiom in the second, and the given axioms are axioms in both.

A brief excursion into the history of non-Euclidean geometry. During more than two centuries, the attention of geometers was attracted by the special role played by Euclid's fifth postulate in the *Elements*. First, it has a somewhat unnaturally cumbersome formulation: *If a straight line intersects two straight lines so as to make the interior angles on one side of it together less than two right angles, the two straight lines will intersect, if indefinitely produced, on the side on which are the angles which are together less than two right angles.* (The other four postulates taken together are shorter: (i) *It is possible to draw a straight line through any two points*; (ii) *An interval can be unboundedly extended*; (iii) *It is possible to describe a circle from any center with an arbitrary span*; (iv) *All right angles are equal to each other.*)

Secondly, Euclid did not use his postulate from the very beginning; for the first time, he used it only in the 28th proposition. All this led to the temptation to try to deduce the fifth postulate from the other four. These attempts resulted in establishing many assertions equivalent to the fifth postulate, such as (a) *There exists a triangle similar but not equal to a given one*; (b) *There exists at least one triangle whose sum of angles equals two right angles*; (c) *Through a point outside a straight line, exactly one straight line parallel to the given one can be drawn*; (d) *A perpendicular line always intersects an oblique line*; and (e) *About any triangle, a circle can be circumscribed.* (We suggest that the reader prove that none of these assertions holds in the Poincaré model.)

Gauss was the first who questioned the possibility of deducing the fifth postulate from the other four. He had advanced greatly in his doubts, but published nothing. He wrote (in 1817): "I become increasingly convinced that the necessity of our geometry cannot be proved, at least by a human mind for a human mind."

The first publication (1829) concerning a new geometry where one of the axioms is the negation of the fifth postulate was due to the great Russian mathematician *Nikolai Ivanovich Lobachevsky* (1792–1856). Lobachevsky developed the new geometry very deeply; he had written formulas relating sides and angles in triangles and found that they precisely corresponded to the formulas of spherical geometry. Already in his first work *On the principles of geometry*, he wrote: "These equations [the relations between angles and sides in triangles] change into ... [equations] of spherical Trigonometry as soon as, instead of sides a, b, and c, we put $a\sqrt{-1}$, $b\sqrt{-1}$, and $c\sqrt{-1}$, but in ordinary Geometry and in spherical Trigonometry, the material aspects of lines alone enter everywhere; therefore, ordinary Geometry, Trigonometry, and this new Geometry will always agree with each other." The discussion concerned the similarity between the law of cosines, the law of sines, and some others. (The reader may compare, say, the laws of sines and cosines in spherical and Lobachevsky geometries.)

Simultaneously with Lobachevsky, the idea of non-Euclidean geometry had occurred to the great Hungarian mathematician *Janos Bolyai* (1802–1860). After graduating from a military engineering academy in Vienna, he served in the Temesvar (this is the Hungarian name of Timisoara) fortress. There, he got interested in the problem of parallel lines and, in 1823, wrote to his father (who was an eminent mathematician and a friend of Gauss): "I have not reached the goal, but ... I have created a whole world from nothing." In 1832, J. Bolyai published an appendix to his father's book in mathematics. This is how his famous *Appendix Containing a Science about a Space Absolutely True Independently of Whether the XI Axiom of Euclid is True or False (Which by no Means Can Be Decided A Priori)*, known as merely *Appendix*, had appeared. Gauss read this work and wrote in one letter: "I consider the young geometer von Bolyai a genius of first rank." To Bolyai's father, however, he wrote with much more restraint and mentioned that he knew many facts himself. When Gauss afterwards looked through *Geometric Investigations* of Lobachevsky, he advised J. Bolyai to read it. J. Bolyai did so and could not believe the truth. He wrote: "Gauss, this colossus, on taking possession of such treasures, could not accept the fact that somebody had left him behind in this question and, since he was not able to hide this, he himself elaborated the idea and published it under the name of Lobachevsky."

However, neither Gauss, nor Lobachevsky, nor Bolyai were bound to prove the independence of Euclid's fifth postulate of the other axioms. Let us mention in this relation one interesting fact.

In 1838, the well-known *Crelle's Journal* published some papers by Minding, where formulas similar to those of the Pythagorean theorem and the laws of cosines and sines for the geometry of the pseudosphere were given. For reasons unknown, Lobachevsky, who regularly read *Crelle's Journal*, never took (according to the card index) any of the three issues with Minding's papers. Otherwise, a road to proving the consistency of his geometry would have opened up for him. But the way it was, the proof was due to Beltrami, who established the equivalence of the formulas of Minding and Lobachevsky and thereby showed that Lobachevsky geometry is realized on the pseudosphere. This happened 12 years after Lobachevsky's death. And then, the Cayley–Klein and Poincaré models were constructed, the Erlangen Program emerged, and Lobachevsky geometry became so trivial that it is no longer included in university mathematical courses.

Conic sections in spherical and Lobachevsky geometries. In Euclidean geometry, a conic section (conic) is defined as a section of the cone $ax^2+by^2+cz^2=0$ by a Euclidean plane. In spherical geometry, instead of the Euclidean plane, we take the sphere $x^2+y^2+z^2=1$, and in Lobachevsky geometry, the hyperboloid $x^2+y^2=z^2-1$.

First, let us examine conic sections in spherical geometry. Recall that there is a duality on the sphere, namely, each point $\pm(x_0, y_0, z_0)$ on the sphere $\mathbb{S}^2$ is assigned the section of $\mathbb{S}^2$ by the plane $x_0x+y_0y+z_0z=0$. This allows us to define the dual cone to the cone $ax^2+by^2+cz^2=0$ as follows. The tangent plane to the initial cone at a point (x_0, y_0, z_0) is given by the equation $ax_0x+by_0y+cz_0z=0$. This plane is dual to the point (ax_0, by_0, cz_0); the dual cone consists of all such points, i.e., of the points dual to the tangent planes to the initial cone. To write the equation of the dual cone, we must find numbers A, B, and C such that the equality $ax_0^2+by_0^2+cz_0^2=0$ is equivalent to $A(ax_0)^2+B(by_0)^2+C(cz_0)^2=0$; the numbers $A=a^{-1}$, $B=b^{-1}$, and $C=c^{-1}$ suit this purpose. Thus the cone dual to $ax^2+by^2+cz^2=0$ is

$$\frac{x^2}{a}+\frac{y^2}{b}+\frac{z^2}{c}=0.$$

The coordinate axes are the symmetry axes of the cone $ax^2+by^2+cz^2=0$, and the coordinate planes are its planes of symmetry. The intersection points of the coordinate axes with the sphere $x^2 + y^2 + z^2 = 1$ are called the *centers* of the conic section (conic) $ax^2 + by^2 + cz^2 = 0$.

Let O_1 be a center of the conic section. Through O_1, we draw a spherical line, i.e., a great circle. The conic divides it into four arcs. The midpoints of two of these arcs are the point O_1 and the antipodal point O_2, and the midpoints of the other two arcs lie in the plane dual to O_1 and O_2.

Now we consider the dual cone and see what the dual assertion looks like. Let l be the line through O_1 and O_2. Consider the intersection points A and B of l with the conic such that the midpoint of the arc AB coincides with O_1. We denote the dual objects by $l^\perp$, $O^\perp$, $A^\perp$, and $B^\perp$. The dual assertion can be stated as follows. Suppose that tangent lines $A^\perp$ and $B^\perp$ to the conic intersect at a point $l^\perp$ lying in the plane $O^\perp$. Then one bisector of the angle between $A^\perp$ and $B^\perp$ coincides with the section of the sphere by the plane $O^\perp$, and the other bisector is the line joining the points $l^\perp$ and O_1. (The former assertion is obvious, because $O^\perp$ is a plane of symmetry.)

The spherical conics, like the plane ones, have foci, which possess the corresponding focal properties. It is convenient to approach this notion gradually. Consider the complexification of the conic $ax^2 + by^2 + cz^2 = 0$ on the sphere $x^2 + y^2 + z^2 = 1$ (it is obtained by allowing x, y, and z to take complex values). The *foci* of this conic are the intersection points of the common tangent lines to the conic under consideration and the conic $x^2 + y^2 + z^2 = 0$. The tangent line to $x^2+y^2+z^2 = 0$ at (x_1, y_1, z_1) is given by $x_1x+y_1y+z_1z = 0$, and the tangent line to $ax^2+by^2+cz^2 = 0$ at (x_2, y_2, z_2), by $ax_2x+by_2y+cz_2z = 0$. Both these equations must determine the same line. This means that $ax_2 : by_2 : cz_2 = x_1 : y_1 : z_1$. The required point (x_1, y_1, z_1) can be found as the solution of the system of equations

$$x_1^2 + y_1^2 + z_1^2 = 0, \qquad \frac{x_1^2}{a} + \frac{y_1^2}{b} + \frac{z_1^2}{c} = 0,$$

i.e., as the intersection point of the conic $x^2 + y^2 + z^2 = 0$ with the dual conic $x^2/a + y^2/b + z^2/c = 0$.

Thus the foci of a conic $ax^2 + by^2 + cz^2 = 0$ can be defined as the points dual to the common chords of the dual conic $x^2/a + y^2/b + z^2/c = 0$ and the conic $x^2 + y^2 + z^2 = 0$. In the real case, these common chords have a very simple geometric meaning. Namely, they correspond to the circular sections of the cone $x^2/a + y^2/b + z^2/c = 0$. By this we mean the following. Suppose that a common chord of the two conics under consideration has the equation $px+qy+rz = 0$. Then the section of the cone specified above by the plane $px + qy + rz = s$, where $s \neq 0$, is a circle. Indeed, the section is a circle if and only if the system of equations

$$px + qy + rz = s, \qquad \frac{x^2}{a} + \frac{y^2}{b} + \frac{z^2}{c} = 0$$

is equivalent to the system

$$px + qy + rz = s, \qquad (x - x_0)^2 + (y - y_0)^2 + (z - z_0)^2 = R^2.$$

Let us pass from $\mathbb{R}^3$ to $\mathbb{R}P^3$, i.e., consider the system of equations

$$px + qy + rz = sw, \qquad (x - x_0w)^2 + (y - y_0w)^2 + (z - z_0w)^2 = R^2w^2.$$

Sections of a cone by parallel planes are similar; therefore, it suffices to consider the case of $w = 0$ (which corresponds to the section of the cone by the plane at infinity). The system

$$px + qy + rz = 0, \qquad \frac{x^2}{a} + \frac{y^2}{b} + \frac{z^2}{c} = 0$$

must then be equivalent to the system

$$px + qy + rz = 0, \qquad x^2 + y^2 + z^2 = 0.$$

This means that $px + qy + rz = 0$ is a common chord of the two conics under consideration.

The conics $x^2 + y^2 + z^2 = 0$ and $ax^2 + by^2 + cz^2 = 0$ have six common chords; if a, b, and c are real and pairwise different, precisely two of these chords are real. Indeed, we can assume that $a > b > c$. We put $x_1 = x$, $x_2 = y$, $x_3 = z$, $a_1 = a$, $a_2 = b$, and $a_3 = c$. Then the common chords have equations of the form

$$(a_i - a_j)x_i^2 + (a_k - a_j)x_k^2 = 0.$$

Such an equation determines real lines only if the numbers $a_i - a_j$ and $a_k - a_j$ have different signs. But this is only possible if $a_j = b$. In this case, we obtain the pair of lines

$$(a - b)x^2 = (b - c)z^2. \tag{3}$$

Consider conics with fixed foci on the unit sphere $x^2 + y^2 + z^2 = 1$. Suppose that the foci are determined by vectors with coordinates, say, $(1, 0, \pm\alpha)$. The pair of planes orthogonal to these vectors has the equation $x^2 = \alpha^2 z^2$. This equation must coincide with an equation of the form (3) for the dual cone, i.e., with (3) where a is replaced by a^{-1}, etc. Therefore, the conics with the specified foci have the form $ax^2 + by^2 + cz^2 = 0$, where

$$a^{-1} - b^{-1} = 1, \qquad b^{-1} - c^{-1} = \alpha^2.$$

Setting $b = \lambda^{-1}$, we obtain $a = (1 + \lambda)^{-1}$ and $c = (\lambda - \alpha^2)^{-1}$, i.e., the conics have equations of the form

$$\frac{x^2}{\lambda + 1} + \frac{y^2}{\lambda} + \frac{z^2}{\lambda - \alpha^2} = 0.$$

Now we can prove some properties of the conics on the sphere similar to those of the plane conics. We start by proving an analog of the assertion that the product of the distances from the foci to a tangent line does not depend on the choice of the tangent line.

THEOREM 6. *Let F_1 and F_2 be the foci of a conic on the sphere (centered at O). Suppose that the conic is determined by a cone C, Π is a tangent plane to C, and α_1 and α_2 are the angles between the plane Π and the lines OF_1 and OF_2, respectively. Then the value $\sin\alpha_1 \sin\alpha_2$ does not depend on the choice of the tangent plane.*

Proof. We can assume that the sphere is centered at the origin and the foci have coordinates $(1, 0, \pm\alpha)$. Consider a vector normal to the plane Π; let it have coordinates (x_0, y_0, z_0). This vector belongs to the cone dual to C; therefore, its coordinates satisfy the relation

$$(\lambda + 1)x_0^2 + \lambda y_0^2 + (\lambda - \alpha^2)z_0^2 = 0, \tag{4}$$

where λ is a number.

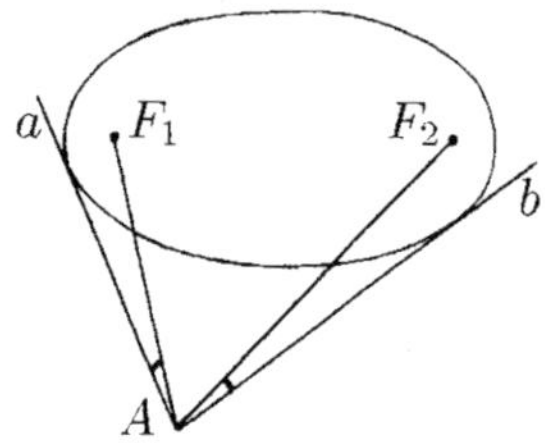

FIGURE A.60

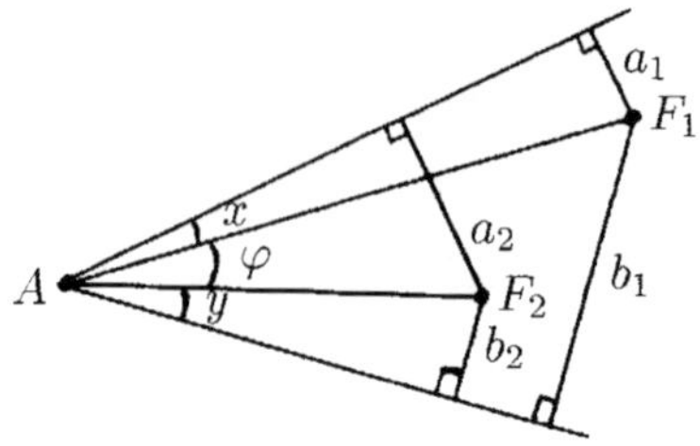

FIGURE A.61

Let β_i be the angle between the line OF_i and the normal to the plane Π. Then $\beta_i = 90° - \alpha_i$. Thus we must prove that $\cos\beta_1 \cos\beta_2 = \text{const}$. It is easy to find the cosines of the angles between the vectors $(1, 0, \pm\alpha)$ and (x_0, y_0, z_0), obtaining

$$\cos\beta_1 \cos\beta_2 = \frac{x_0^2 - \alpha z_0^2}{(x_0^2 + y_0^2 + z_0^2)(1 + \alpha^2)}.$$

Relation (4) gives

$$\cos\beta_1 \cos\beta_2 = \frac{-\lambda}{1 + \alpha^2} = \text{const.} \quad \square$$

THEOREM 7. *Let a and b be tangent lines from a point A to a spherical conic with foci F_1 and F_2* (Figure A.60). *Then the angle between the lines a and AF_1 equals the angle between b and AF_2.*

Proof. Let a_1 and a_2 be the distances from F_1 and F_2 to the line a, and let b_1 and b_2 be the distances from F_1 and F_2 to the line b. According to Theorem 6,

$$\sin a_1 \sin a_2 = \sin b_1 \sin b_2$$

(we assume that the radius of the sphere is 1).

Next, suppose that c_1 and c_2 are the spherical distances from the point A to the points F_1 and F_2 and x, y, and φ are the angles shown in Figure A.61. Applying the spherical law of sines to the four right triangles shown in the figure, we obtain

$$\sin a_1 = \sin c_1 \sin x, \qquad \sin b_1 = \sin c_1 \sin(y + \varphi),$$
$$\sin a_2 = \sin c_2 \sin(x + \varphi), \qquad \sin b_2 = \sin c_2 \sin y.$$

Therefore,

$$\frac{\sin a_1 \sin a_2}{\sin b_1 \sin b_2} = \frac{\sin x \sin(x + \varphi)}{\sin y \sin(y + \varphi)}.$$

Taking into account that $\sin a_1 \sin a_2 = \sin b_1 \sin b_2$, we obtain $\sin x \sin(x + \varphi) = \sin y \sin(y + \varphi)$, i.e.,

$$\cos(2x + \varphi) - \cos\varphi = \cos(2y + \varphi) - \cos\varphi.$$

The equality $x + y + \varphi = \pi$ cannot hold; hence $x = y$. $\square$

COROLLARY. *If a (spherical) line l is tangent to a spherical conic with foci F_1 and F_2 at a point A, then the lines AF_1 and AF_2 make equal angles with l.*

To prove this, we must make the point A mentioned in the statement of Theorem 7 approach the conic.

THEOREM 8. *For all points on a spherical conic, the sum of their distances to the foci of the conic is constant.*

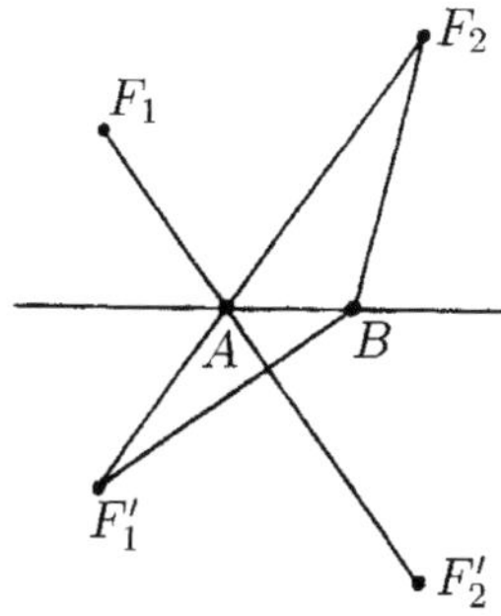

FIGURE A.62

Proof. Consider the tangent line l to the conic at a point A. Let F_1' and F_2' be the points symmetric to the foci F_1 and F_2 with respect to l (Figure A.62). The corollary of Theorem 7 shows that the point A lies on the segment $[F_1', F_2]$, i.e., the sum $|AF_1| + |AF_2|$ under consideration is equal to the length of this segment. It suffices to prove that if B is a point on l close to A, then the difference $|BF_1'| + |BF_2| - (|AF_1'| + |AF_2|)$ is of the order $|AB|^2$. Indeed, if this is so, then the function $|AF_1| + |AF_2|$ has zero derivative.

Let us draw a perpendicular $[B, A_1]$ from the point B to the side $F_1'F_2$. It divides the triangle $F_1'BF_2$ into two right triangles. It is sufficient to prove that the differences $|BF_1'| - |A_1F_1'|$ and $|BF_2| - |A_1F_2|$ are of the order $|A_1B|^2$. Let c be the hypotenuse of a spherical right triangle with legs a and x. Then $\cos c \,/\, \cos a = \cos x = 1 - x^2/2 + \cdots$. Therefore, for small x, the value $c - a$ is of the order of x^2. □

Much of what is said about spherical conics remains valid in hyperbolic geometry, but some changes must be made. Instead of the conic $x^2 + y^2 + z^2 = 0$, the conic $x^2 + y^2 - z^2 = 0$ must be considered, and the cone dual to $ax^2 + by^2 + cz^2 = 0$ is

$$\frac{x^2}{a} + \frac{y^2}{b} - \frac{z^2}{c} = 0.$$

As opposed to spherical geometry, the intersections of cones of the form $ax^2 + by^2 + cz^2 = 0$ with the one-sheeted hyperboloid $x^2 + y^2 - z^2 = 1$ have diverse structures. This means that there are hyperbolic conics of many different types.

In hyperbolic geometry, the foci of a conic $ax^2 + by^2 + cz^2 = 0$ are, by definition, the intersection points of the common tangent lines to this conic and to the conic $x^2 + y^2 - z^2 = 0$. It is easy to verify that the foci are dual to the common chords of the conics $x^2 + y^2 - z^2 = 0$ and $x^2/a + y^2/b - z^2/c = 0$.

Parabolic mirrors in Lobachevsky geometry. In Euclidean geometry, the parabolas have the property that incident light beams parallel to the axis of a parabola, being reflected off the parabola, converge in the focus of the parabola. In Lobachevsky geometry, there are lines with a similar property; these are the curves specified in the Poincaré upper half-plane model by the equations

$$\frac{x^2}{a^2 + \lambda} + \frac{y^2}{b^2 + \lambda} = 1, \tag{5}$$

where $0 < a^2 < -\lambda < b^2$. Incident light beams parallel to the imaginary axis, being reflected off a mirror (5), converge in the point $(0, c)$, where $c^2 = b^2 - a^2$.

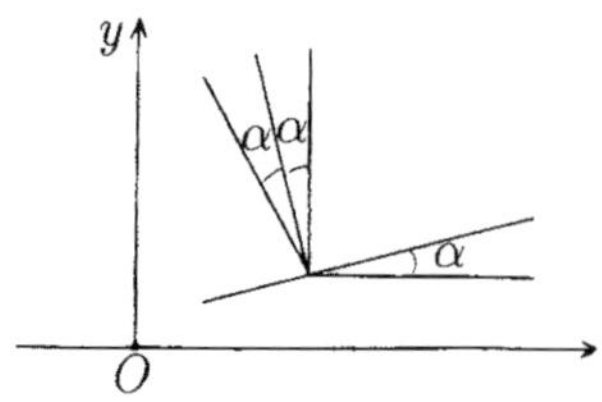

FIGURE A.63

Recall that in Euclidean geometry, the curves (5) are confocal hyperbolas with foci $(0, \pm c)$.

A proof of this property of the mirrors (5) appears in [**P**]; we reproduce it here.

In Euclidean geometry, an incident light beam parallel to the Oy axis, being reflected off the line $y = (\tan\alpha)x + d_1$, moves along the line $y = -(\cot 2\alpha)x + d_2$ (Figure A.63). Therefore, being reflected off a curve $y = f(x)$ at a point (x_0, y_0), a beam parallel to the Oy axis moves along a straight line with slope $m = -\cot 2\alpha$, where $\tan\alpha = f'(x_0)$, i.e.,

$$m = \frac{(f'(x_0))^2 - 1}{2f'(x_0)}. \tag{6}$$

This property remains valid in the Poincaré upper half-plane model. Namely, being reflected off a curve $y = f(x)$ at a point (x_0, y_0), an incident light beam parallel to the imaginary axis moves along a hyperbolic line whose slope m at the point (x_0, y_0) is defined by (6). This hyperbolic line is a Euclidean semicircle. The Euclidean normal to this semicircle at the point (x_0, y_0) has the slope $-1/m$, and therefore its equation is

$$y - y_0 = -(x - x_0)/m.$$

The center of the semicircle is the point $(x_1, 0)$ at which the normal intersects the real axis. The equation of the normal shows that $x_1 = my_0 + x_0$. The squared radius of the semicircle equals $(x_0 - x_1)^2 + y_0^2 = (m^2 + 1)y_0^2$.

Thus, being reflected off a curve $y = f(x)$ at a point (x_0, y_0), the light beam moves along the hyperbolic line

$$(x - x_1)^2 + y^2 = (m^2 + 1)y_0^2.$$

This hyperbolic line intersects the imaginary axis at the point $(0, c)$, where c is found from the equalities $x_1^2 + c^2 = (m^2 + 1)y_0^2$ and $x_1 = my_0 + x_0$, i.e.,

$$c^2 = y_0^2 - 2mx_0y_0 - x_0^2. \tag{7}$$

It remains to prove that $c^2 = b^2 - a^2$ for all the mirrors (5).

Let us denote $f'(x) = y'$ for short. Then (7) can be written as

$$c^2 = y^2 - \frac{(y')^2 - 1}{y'}xy - x^2, \quad \text{i.e.,} \quad \frac{c^2}{xy} = \frac{1}{y'} - \frac{x}{y} - y' + \frac{y}{x}.$$

Differentiating (5) with respect to x, we obtain

$$y' = -\frac{x}{y} \cdot \frac{b^2 + \lambda}{a^2 + \lambda}.$$

Therefore,

$$\frac{1}{y'} - \frac{x}{y} - y' + \frac{y}{x} = \frac{b^2 - a^2}{xy}\left(\frac{x^2}{a^2 + \lambda} + \frac{y^2}{b^2 + \lambda}\right) = \frac{b^2 - a^2}{xy}.$$

Hence $c^2 = b^2 - a^2$, as required.

The curves (5) with $-\lambda < a^2 < b^2$ are ellipses with foci $(0, \pm c)$. Incident light beams parallel to the imaginary axis, being reflected off such mirrors (in the Poincaré model), move along hyperbolic rays whose extensions pass through the point $(0, c)$.

The volume of a simplex with vertices on the absolute. On the Lobachevsky plane, the area of a triangle with angles α, β, and γ equals $\pi-\alpha-\beta-\gamma$ (we assume that the imaginary radius c is 1). In particular, the area of an arbitrary triangle with vertices on the absolute equals π.

All the triangles with vertices on the absolute are congruent. But already in the three-dimensional Lobachevsky space, not all simplices (tetrahedra) with vertices on the absolute are congruent. The volume of a 3-simplex with vertices on the absolute is expressed in terms of the angles between its faces through the use of the *Lobachevsky function*

$$\mathrm{L}(\theta) = -\int_0^\theta \ln|2\sin u|\, du.$$

The formula for the volume of a tetrahedron was obtained by Lobachevsky. Our proof of this formula follows [**M**].

Prior to obtaining an expression for the volume of a tetrahedron, we prove the basic properties of the Lobachevsky function.

LEMMA. *The function* $\mathrm{L}(\theta)$ *is odd, has period* π, *and satisfies the relation*

$$\mathrm{L}(2\theta) = 2\,\mathrm{L}(\theta) + \mathrm{L}\left(\theta + \frac{\pi}{2}\right).$$

Proof. As $u \to 0$, the integrand tends to $-\infty$, but at a rate lower than $-1/u^2$. Thus the function $\mathrm{L}(\theta)$ is defined by a convergent integral. Obviously, $\mathrm{L}(\theta)$ is an odd function and $\mathrm{L}(0) = 0$.

The equality $\sin 2\theta = 2\sin\theta\sin(\theta+\pi/2)$ shows that

$$\mathrm{L}'(2\theta) = 2\,\mathrm{L}'(\theta) + 2\,\mathrm{L}'\left(\theta + \frac{\pi}{2}\right);$$

therefore,

$$\mathrm{L}(2\theta) = 2\,\mathrm{L}(\theta) + 2\,\mathrm{L}\left(\theta + \frac{\pi}{2}\right) + C.$$

Let us substitute $\theta = 0$ and $\theta = \pi/2$ in this equality and subtract one of the resulting equalities from the other. The result is $\mathrm{L}(\pi) = \mathrm{L}(0) = 0$. The graph of the integrand is symmetric with respect to the line $u = \pi/2$; therefore, the equality $\mathrm{L}(\pi) = 0$ implies $\mathrm{L}(\pi/2) = 0$. Thus

$$C = \mathrm{L}(0) - 2\,\mathrm{L}(0) - 2\,\mathrm{L}(\pi/2) = 0.$$

The function $\mathrm{L}'(\theta) = -\ln|2\sin\theta|$ has period π, and $\mathrm{L}(\pi) = \mathrm{L}(0) = 0$; hence $\mathrm{L}(\theta)$ also has period π. □

Let α, β, and γ be the dihedral angles at the edges of a tetrahedron incident to one vertex. If the vertices of the tetrahedron lie on the absolute, then $\alpha+\beta+\gamma = \pi$ and the dihedral angles at opposite edges are α, β, γ (see Problem A30).

THEOREM 9. (a) *The volume of a tetrahedron* Δ *with vertices on the absolute equals* $\mathrm{L}(\alpha) + \mathrm{L}(\beta) + \mathrm{L}(\gamma)$.

(b) *A tetrahedron with vertices on the absolute has maximum volume if and only if the tetrahedron is regular, i.e.,* $\alpha = \beta = \gamma = \pi/3$.

Proof. (a) Consider the Poincaré model in the half-space $H^3 = \{(x, y, z) | z > 0\}$. The squared length element in this model is $(dx^2 + dy^2 + dz^2)/z^2$, and the volume element is $(dx\,dy\,dz)/z^3$.

Let us position the tetrahedron Δ so that one of its faces lies on the sphere $x^2 + y^2 + z^2 = 1$ and the opposite vertex is at a point at infinity. The orthogonal projection of the tetrahedron Δ on the absolute is a triangle ABC with angles α, β, γ inscribed in the circle $x^2 + y^2 = 1$.

First, consider the case in which all the angles α, β, γ are acute. We can assume that the Oy axis is parallel to the side BC and the ray Ox intersects the side BC. Then the part of Δ above the triangle OBC is determined by the inequalities

$$z \geqslant \sqrt{1 - x^2 - y^2}, \quad -x \tan\alpha \leqslant y \leqslant -x \tan\alpha, \quad 0 \leqslant x \leqslant \cos\alpha.$$

Its volume equals

$$\int_0^{\cos\alpha} \left(\int_{-x\tan\alpha}^{x\tan\alpha} \left(\int_{\sqrt{1-x^2-y^2}}^{\infty} \frac{dz}{z^3} \right) dy \right) dx$$
$$= \int_0^{\cos\alpha} \left(\int_{-x\tan\alpha}^{x\tan\alpha} \frac{dy}{2(1 - x^2 - y^2)} \right) dx.$$

Recall that

$$\int \frac{dy}{A^2 - y^2} = \frac{1}{2A} \ln \frac{A + y}{A - y}.$$

Therefore,

$$\int_{-x\tan\alpha}^{x\tan\alpha} \frac{dy}{2(1 - x^2 - y^2)} = \frac{1}{4A} \ln \frac{A + y}{A - y} \Bigg|_{-x\tan\alpha}^{x\tan\alpha}$$
$$= \frac{1}{2A} \ln \frac{A + x\tan\alpha}{A - x\tan\alpha} = \frac{1}{2A} \ln \frac{A\cos\alpha + x\sin\alpha}{A\cos\alpha - x\sin\alpha},$$

where $A = \sqrt{1 - x^2}$.

To evaluate the integral

$$\int_0^{\cos\alpha} \frac{1}{2A} \ln \frac{A\cos\alpha + x\sin\alpha}{A\cos\alpha - x\sin\alpha}\, dx,$$

we substitute $x = \cos\theta$. Then $A = \sqrt{1 - x^2} \sin\theta$ and $dx = -A d\theta$, and we obtain the integral

$$-\int_{\pi/2}^{\alpha} \frac{1}{2A} \ln \frac{\sin\theta\cos\alpha + \cos\theta\sin\alpha}{\sin\theta\cos\alpha - \cos\theta\sin\alpha} A\, d\theta$$
$$= \frac{1}{2} \int_{\alpha}^{\pi/2} \ln \frac{2\sin(\theta + \alpha)}{2\sin(\theta - \alpha)}\, d\theta$$
$$= \frac{1}{2} \int_{2\alpha}^{\frac{\pi}{2}+\alpha} \ln(2\sin u)\, du - \frac{1}{2} \int_0^{\frac{\pi}{2}-\alpha} \ln(2\sin u)\, du$$
$$= \frac{1}{2} \left(-\mathrm{L}\left(\frac{\pi}{2} + \alpha\right) + \mathrm{L}(2\alpha) + \mathrm{L}\left(\frac{\pi}{2} - \alpha\right) - \mathrm{L}(0) \right).$$

According to the lemma, $\mathrm{L}(2\alpha) = 2\,\mathrm{L}(\alpha) + \mathrm{L}(\alpha + \pi/2)$. Therefore, the expression within the parentheses equals

$$2\,\mathrm{L}(\alpha) + \mathrm{L}\left(\frac{\pi}{2} + \alpha\right) + \mathrm{L}\left(\frac{\pi}{2} - \alpha\right) - \mathrm{L}(0).$$

The same lemma implies that the function $\mathrm{L}(\theta)$ is odd and has period π. Hence $\mathrm{L}(0) = 0$ and

$$\mathrm{L}\left(\frac{\pi}{2} - \alpha\right) = -\mathrm{L}\left(\alpha - \frac{\pi}{2}\right) = -\mathrm{L}\left(\alpha - \frac{\pi}{2} + \pi\right) = -\mathrm{L}\left(\frac{\pi}{2} + \alpha\right).$$

Thus the volume of the part of Δ above the triangle OBC is equal to $\mathrm{L}(\alpha)$. Therefore, the volume of the entire tetrahedron Δ is $\mathrm{L}(\alpha) + \mathrm{L}(\beta) + \mathrm{L}(\gamma)$.

If, say, $\gamma \geqslant \pi/2$, then the same argument shows that the volume of Δ equals

$$\mathrm{L}(\alpha) + \mathrm{L}(\beta) - \mathrm{L}(\pi - \gamma) = \mathrm{L}(\alpha) + \mathrm{L}(\beta) + \mathrm{L}(\gamma).$$

(b) We need to find the maximum of the function $\mathrm{L}(\alpha) + \mathrm{L}(\beta) + \mathrm{L}(\gamma)$ for nonnegative α, β, γ related by $\alpha + \beta + \gamma = \pi$. The maximum is attained at an interior point of the domain under consideration (if one of the angles α, β, and γ is zero, then the volume of the tetrahedron equals 0). Therefore, at the point of maximum, we have $\mathrm{L}'(\alpha) = \mathrm{L}'(\beta) = \mathrm{L}'(\gamma)$. Indeed, the function

$$f(x) = \mathrm{L}(\alpha + x) + \mathrm{L}(\beta - x)$$

has a local maximum at $x = 0$, and hence

$$0 = f'(0) = \mathrm{L}'(\alpha) - \mathrm{L}'(\beta).$$

By assumption, $\mathrm{L}'(\theta) = -\ln|2\sin\theta|$. If $0 \leqslant \theta \leqslant \pi$, then $\mathrm{L}'(\theta) = -\ln(2\sin\theta)$. Therefore, $\sin\alpha = \sin\beta = \sin\gamma$. This means that the sides of a triangle with angles α, β, and γ are pairwise equal, i.e., $\alpha = \beta = \gamma = \pi/3$. □

In the Lobachevsky n-space with $n > 3$, any regular simplex with vertices on the absolute is also of maximum volume (see [**HM**]).

Problems

A.28. Prove that the restriction of a motion of three-dimensional Lobachevsky space to the absolute has one or two fixed points.

A.29. Let A, B, C, and D be pairwise distinct points on the absolute of three-dimensional Lobachevsky space. Prove that there exists a motion that interchanges A with B and C with D.

A.30. Given a tetrahedron with vertices on the absolute, prove that

(a) its opposite dihedral angles are equal;

(b) the sum of its dihedral angles at each vertex equals π.

A.31. Prove that the sum of the dihedral angles of an n-hedral angle with vertex on the absolute is equal to $(n-2)\pi$.

A.32. Prove that the volume of a tetrahedron with vertices on the absolute is finite.

A.33. The dihedral angle of a cube in the Lobachevsky space is α, and the angle between the sides of each face is β. Prove that $\sin\alpha/2\cos\beta/2 = 1/2$.

A.34. Given a motion f of the Lobachevsky space such that the distance between the points X and $f(X)$ is the same for all X, prove that f is the identity transformation.

A.35. Given a transformation f of the Lobachevsky space which increases the distance between any two points by the same factor k, i.e., $d(f(X), f(Y)) = kd(X,Y)$ for any X and Y, prove that $k = 1$, i.e., f is an isometry.

A.36. Let $z_1, \ldots, z_{n+2}$ be points in n-dimensional hyperbolic space. Consider the matrix $A = (a_{ij})$ with elements $a_{ij} = \cosh(d_{ij})$, where each d_{ij} is the distance between z_i and z_j. Prove that $\det A = 0$.

Solutions, Hints, and Answers

Chapter 1

1.1. Write the equations and find the intersection point of the perpendiculars to the sides of the triangle through their midpoints.

1.2. The vectors (a_1, a_2, a_3) and (b_1, b_2, b_3) are orthogonal to the given planes.

1.3. The coordinates are

$$\left(\begin{vmatrix} a_2 & a_3 \\ b_2 & b_3 \end{vmatrix}, -\begin{vmatrix} a_1 & a_3 \\ b_1 & b_3 \end{vmatrix}, \begin{vmatrix} a_1 & a_2 \\ b_1 & b_2 \end{vmatrix} \right).$$

1.4. The planes $\alpha e_1 + \beta e_2$ and $\gamma e_3 + \delta e_4$.

1.5. The planes $\alpha e_1 + \beta e_2 + e_3$ and $\gamma e_1 + \delta e_4$.

1.6. Move the point A first perpendicularly to $\mathbb{R}^3$, next parallel to $\mathbb{R}^3$, and then return it to $\mathbb{R}^3$.

1.7. Move one circle first perpendicularly to $\mathbb{R}^3$, next parallel to $\mathbb{R}^3$, and then return it to $\mathbb{R}^3$.

1.8. If $e_1, \dots, e_n$ is an orthonormal basis, $a_k = \sum a_k^i e_i$, and $x = \sum x^i e_i$, then the sum of the squared projections of the vectors $a_1, \dots, a_n$ to the line parallel to the vector x equals

$$\sum_k \frac{(a_k, x)^2}{|x|^2} = \frac{\sum_{i,k}(a_k^i, x^i)^2 + 2\sum_k \sum_{i<j} a_k^i a_k^j x^i x^j}{x_1^2 + \cdots + x_n^2}.$$

This sum is constant if and only if $\sum_k (a_i^k)^2 = \lambda^2$ and $\sum_k a_k^i a_k^j = 0$. Let $b_k^i = \lambda^{-1} a_k^i$. Then the matrix $B = (b_k^i)$ is orthogonal, and hence $(b_p, b_q) = \sum_i b_p^i b_q^i = \delta_{pq}$.

1.9. (a) No, this is not possible. Cut the cube into eight equal small cubes with edge $d/2$. The distance between any two points lying in a small cube does not exceed $d\sqrt{3}/2$, i.e., is less than d. Hence, each of the eight small cubes contains at most one given point.

(b) Yes, this is possible. Place 16 points at the vertices of the cube and one more point at the center of the cube.

1.10. We can assume that the length of each vector $a_i = (a_{i1}, \dots, a_{in})$ equals 1. Suppose that these vectors are linearly dependent. Then there exist numbers $y_1, \dots, y_n$ such that the sum of their squares is 1 and $\sum_j y_j a_{ij} = 0$ (we use the fact that the linear dependence of the rows of a square matrix is equivalent to the linear dependence of its columns). We then have

$$a_{ii}^2 y_i^2 = \Big(\sum_{j\neq i} y_j a_{ij}\Big)^2 \leqslant \Big(\sum_{j\neq i} y_j^2\Big)\Big(\sum_{j\neq i} a_{ij}^2\Big) = (1 - y_i^2)(1 - a_{ii}^2).$$

Therefore, $y_i^2 + a_{ii}^2 \leqslant 1$, i.e., $a_{ii} \leqslant \sqrt{1 - y_i^2}$. Hence

$$\sum a_{ii} \leqslant \sum \sqrt{1 - y_i^2} \leqslant \sqrt{n} \cdot \sqrt{\sum (1 - y_i^2)} = \sqrt{n(n-1)}.$$

Thus, if the vectors $a_1, \dots, a_n$ are linearly dependent, then $\sum \cos \alpha_i = \sum a_{ii} \leqslant \sqrt{n(n-1)}$. Therefore, the inequality $\sum \cos \alpha_i > \sqrt{n(n-1)}$ implies the linear independence of these vectors.

Remark. For linearly dependent vectors $a_i = ne_i - (e_1 + \dots + e_n)$, we have the equality $\sum \cos \alpha_i = \sqrt{n(n-1)}$.

1.11. Let P be a common point of the initial balls. Consider $\max\{XB_i/PA_I\} = f(X)$ (the maximum is over i). We have $f(X) \to \infty$ as $X \to \infty$; therefore, f has a minimum point Q. It is sufficient to prove that $f(Q) \leqslant 1$ (then $QB_i \leqslant PA_i \leqslant R_i$). The graph of the function $y = f(X)$ is the boundary of the intersection of cones with vertices at $B_1, \dots, B_m$. The point where the minimum is attained belongs to the boundaries of at least two cones with vertices B_i and B_j. Considering the restriction of the function f to the line B_iB_j, it is easy to see that $f(Q) \leqslant 1$.

1.12. (a) Let a_3' be the projection of the vector a_3 to the plane Π generated by a_1 and a_2. Then $(a_3', a_i) = (a_3, a_i) \geqslant 0$ for $i = 1, 2$; therefore, we can choose axes of a Cartesian coordinate system on the plane Π so that the coordinates of the vectors a_1, a_2, and a_3' will be nonnegative. The third axis is perpendicular to the plane Π and directed so that the third coordinate of the vector a_3 is nonnegative.

(b) In the space $\mathbb{R}^3 \subset \mathbb{R}^n$, consider the vectors $a_1 = (1, 0, 0)$, $a_2 = (0, 1, 0)$, $a_3 = (1, 0, 1)$, $a_4 = (0, 1, 1)$, and $a_5 = (1, 1, -1)$. It is easy to verify that $(a_i, a_j) \geqslant 0$. Let us show that there exists no Cartesian coordinate system in $\mathbb{R}^n$ with respect to which the coordinates of $a_1, \dots, a_5$ are nonnegative. First, we prove that in such a system, all coordinates of the vector $a_6 = (0, 0, 1)$ are also nonnegative. Suppose that some coordinate of a_6 is negative. Then the same coordinate of a_1 is positive because all the coordinates of the vector $a_3 = a_1 + a_6$ are nonnegative. Similarly, we prove that the same coordinate of the vector a_2 is positive because all the coordinates of the vector $a_4 = a_2 + a_6$ are nonnegative. But a_1 and a_2 cannot have the same positive coordinates. Indeed, all coordinates of these vectors are nonnegative, and their scalar product is zero. Therefore, the vector a_6 has no nonnegative coordinates.

Now it is easy to obtain a contradiction. On the one hand, all coordinates of the vectors a_5 and a_6 are nonnegative; therefore, their scalar product is nonnegative. On the other hand, in $\mathbb{R}^3$ the vectors a_5 and a_6 have coordinates $(1, 1, -1)$ and $(0, 0, 1)$, respectively; therefore, their scalar product equals -1.

1.13. The determinant under consideration is equal to the oriented volume of the parallelepiped spanning the vectors $(x_1, y_1, 1)$, $(x_2, y_2, 1)$, $(x_3, y_3, 1)$. The volume is zero if and only if these vectors are linearly dependent, i.e., the points under consideration are collinear.

1.14. Use the observation that the volume of a parallelepiped does not exceed the product of its edge lengths.

1.15. Let A, B, C be the endpoints of the given vectors with common initial point O, and let H be the projection of A to the plane OBC. Then $V = |AH|bc \sin \alpha$. To evaluate $|AH|$, consider the projections M and N of the point A to the lines OB and OC, respectively. The points M and N lie on the circle with diameter $[O, H]$; therefore, $|OH| = |MN|/\sin \alpha$. An expression for the side MN of the triangle

OMN can be written by using the law of cosines, and $|AH|$ is then found from the relation $|AH|^2 = a^2 - |OH|^2$.

1.16. *First solution.* Let S_i be the area of the ith face. Then we must have $S_1 = S_2 \cos\varphi_{12} + S_3 \cos\varphi_{13} + S_4 \cos\varphi_{14}$. Similar expressions can be written for S_2, S_3, and S_4. These expressions can be treated as a system of linear equations with respect to S_1, S_2, S_3, and S_4. This system has a nonzero solution; therefore, its determinant is zero, and this determinant coincides with the determinant under consideration.

Second solution. Let n_i be the outer normal vector to the ith face. Then $(n_i, n_j) = -\cos\varphi_{ij}$ for $i \neq j$. Clearly, $\det \|(n_i, n_j)\|$ is a Gram determinant of order 4 in 3-space. Hence it is zero.

1.17. (a) Clearly,

$$V_n = \int_0^h \left(\frac{t}{h}\right)^{n-1} V_{n-1}\, dt = \frac{1}{n}\frac{h^n}{h^{n-1}} V_{n-1}.$$

(b) Let V_k be the volume of the simplex $A_1 \dots A_k A_{n+1}$. Then $V_k = hV_{k-1}/k$. On the other hand, $|V(a_1, \dots, a_k)| = h|V(a_1, \dots, a_{k-1})|$. Clearly, $V_1 = |V(a_1)|$. By induction, we obtain

$$V_k = \frac{1}{k!}|V(a_1, \dots, a_k)|.$$

1.18. Let W_n be the volume of the parallelepiped generated by the edges incident to the vertex A_1 in the simplex $A_1 \dots A_{n+1}$. Then $V_n = \frac{1}{n!}W_n$. Therefore, we must prove that $W_n^2 = \big((-1)^{n+1}/2^n\big)D$, where D is the determinant under consideration.

Consider an auxiliary determinant

$$G = \begin{vmatrix} (a_1, a_1) & \dots & (a_1, a_{n+1}) & 1 \\ \dots & \dots & \dots & \dots \\ (a_{n+1}, 1) & \dots & (a_{n+1}, a_{n+1}) & 1 \\ 1 & \dots & 1 & 0 \end{vmatrix},$$

where $a_i = \overrightarrow{OA_i}$ and O is the origin. The proof is in two steps. First, we prove that $W_n^2 = -G$, and then, that $G = \frac{(-1)^n}{2^n}D$.

The determinant G can be represented as the product of two determinants:

$$G = \begin{vmatrix} a_1^1 & \dots & a_1^n & 1 & 0 \\ \dots & \dots & \dots & \dots & \dots \\ a_{n+1}^1 & \dots & a_{n+1}^1 & 1 & 0 \\ 0 & \dots & 0 & 0 & 1 \end{vmatrix} \cdot \begin{vmatrix} a_1^1 & \dots & a_{n+1}^1 & 0 \\ \dots & \dots & \dots & \dots \\ a_1^n & \dots & a_{n+1}^n & 0 \\ 0 & \dots & 0 & 1 \\ 1 & \dots & 1 & 0 \end{vmatrix}.$$

Taking into account that

$$\begin{vmatrix} a_1^1 & \dots & a_1^n & 1 \\ \dots & \dots & \dots & \dots \\ a_{n+1}^1 & \dots & a_{n+1}^n & 1 \end{vmatrix} = \begin{vmatrix} a_1^1 & \dots & a_1^n & 1 \\ a_2^1 - a_1^1 & \dots & a_2^n - a_1^n & 0 \\ \dots & \dots & \dots & \dots \\ a_{n+1}^1 - a_1^1 & \dots & a_{n+1}^n - a_{n+1}^n & 0 \end{vmatrix} = \pm W_n,$$

we obtain $G = -W_n^2$.

Next, let us prove the equality $G = \big((-1)^n/2^n\big)D$. According to the law of cosines, we have

$$d_{ij}^2 = a_i^2 + a_j^2 + 2(a_i, a_j);$$

therefore, each element (a_i, a_j) in G can be replaced by $\frac{1}{2}(a_i^2 + a_j^2 - d_{ij}^2)$; for the diagonal elements, $d_{ii} = 0$. For each $k = 1, 2, \dots, n+1$, we subtract the last column of the determinant G multiplied by $a_k^2/2$ from the kth column. Then we perform the same operation for the rows. As the result, we obtain the determinant

$$\begin{vmatrix} 0 & -\frac{1}{2}d_{12}^2 & \dots & -\frac{1}{2}d_{1,n+1}^2 & 1 \\ -\frac{1}{2}d_{21}^2 & 0 & \dots & -\frac{1}{2}d_{2,n+1}^2 & 1 \\ \dots & \dots & \dots & \dots & \dots \\ -\frac{1}{2}d_{n+1,1}^2 & -\frac{1}{2}d_{n+1,2}^2 & \dots & 0 & 1 \\ 1 & 1 & \dots & 1 & 0 \end{vmatrix}.$$

Let us multiply the last row and the last column by $-1/2$ (the determinant will be multiplied by $1/4$). Next, we multiply each element by -2 (the determinant will be multiplied by $(-2)^{n+2}$). As the result, we obtain the determinant D. Thus $D = ((-2)^{n+2}/4)G$, i.e., $G = ((-1)^n/2^n)D$.

1.19. Consider the degenerate $(n+1)$-dimensional simplex $A_1 \dots A_{n+2}$, where A_{n+2} is the center of the sphere circumscribed about the simplex $A_1 \dots A_{n+1}$. Since this simplex is degenerate, its volume is zero. Let us write this condition in the determinant form given by the preceding problem. The determinant has the additional row and column $d_{i,n+2}^2 = d_{n+2,i}^2 = R^2$. We add the last row to the next to last row in this determinant and do the same for columns. Expanding the obtained determinant in the next to last row, we obtain $-2R^2D + \Delta = 0$.

1.20. Let the point A_i have coordinates $(a_i^1, \dots, a_i^n) = a_i$. The points $A_1, \dots, A_{n+2}$ lie on one sphere or in one hyperplane if and only if there exists a nonzero set of numbers $x, y, z_1, \dots, z_n$ such that

$$x|a_i|^2 + y + \sum_{j=1}^{n} z_j a_i^j = 0 \quad \text{for } i = 1, 2, \dots, n+2.$$

We obtained a system of linear equations in x, y, z_j. It has a nonzero solution if and only if its determinant

$$\begin{vmatrix} |a_1|^2 & 1 & a_1^1 & \dots & a_1^n \\ \dots & \dots & \dots & \dots & \dots \\ |a_{n+2}|^2 & 1 & a_{n+2}^1 & \dots & a_{n+2}^n \end{vmatrix}$$

vanishes or, equivalently, the determinant

$$\begin{vmatrix} 1 & \dots & 1 \\ |a_1|^2 & \dots & |a_{n+2}|^2 \\ -2a_1^1 & \dots & -2a_{n+2}^1 \\ \dots & \dots & \dots \\ -2a_1^n & \dots & -2a_{n+2}^n \end{vmatrix}$$

is zero. It is easy to verify that the product of the former determinant by the latter equals the determinant mentioned in the statement of the problem.

1.21. Problem (a) is a special case of problem (b); hence it suffices to solve problem (b) directly. Let d_{ij} be the distance between the centers O_i and O_j of the given spheres. The angle between these spheres equals φ if and only if

$$d_{ij}^2 = r_i^2 + r_j^2 - 2r_i r_j \cos\varphi.$$

These conditions can be applied to obtain an algebraic relation for $r_1, \dots, r_{n+1}$. Indeed, in a suitable coordinate system, the first $n - i$ coordinates of the point O_i

are zero. The $n(n+1)/2$ conditions allow us to eliminate all the coordinates and obtain the required relation for r_i.

Next, let us show that for fixed $r_1, \dots, r_n$, the resulting relation is a quadratic equation in r_{n+1}. Indeed, the distances d_{ij} are fixed for $i, j \leqslant n$; therefore, the configuration of the centers of the first n spheres is determined uniquely. Thus the coordinates of these centers can be assumed fixed. We can treat the relations

$$d_{i,n+1}^2 = r_{n+1}^2 + r_i^2 - 2r_{n+1}r_i \cos\varphi \quad \text{with } i = 1, \dots, n$$

as equations whose unknowns are r_{n+1} and the coordinates of O_{n+1}. Consider the equivalent system of equations obtained by replacing the second, ..., nth equations by their respective differences with the first equation; the first equation remains unchanged. The resulting $n-1$ equations are linear. They allow us to express the coordinates of the point O_{n+1} in terms of the numbers r_{n+1} (and the fixed values). The expressions are linear with respect to r_{n+1}; substituting them in the first equation, we obtain a quadratic equation for r_{n+1}.

Consider a special case where $r_1 = \dots = r_n = x^{-1}$ and $r_{n+1} = y^{-1}$. The squared pairwise distances between the points $O_1, \dots, O_n$ are $d^2 = 2(1-\cos\varphi)y^{-2}$, while the squared distances from these points to O_{n+1} are $l^2 = x^{-2}y^{-2}(x^2+y^2 - xy\cos\varphi)$. Hence $O_1 \dots O_n$ is a regular simplex centered at O_{n+1}. The edge length of this simplex equals d, and the radius of the circumscribed sphere is l. Thus

$$n(x^2+y^2-2xy\cos\varphi) = (n-1)(1-\cos\varphi)x^2. \tag{1}$$

Let $s_k = r_1^{-k} + \dots + r_{n+1}^{-k}$. For $r_1, \dots, r_{n+1}$, we have an algebraic relation $P(s_1, \dots, s_{n+1}) = 0$. The substitution of $r_1 = \dots = r_n = x^{-1}$ and $r_{n+1} = y^{-1}$ in this relation yields (1); no degeneration occurs because in the general case we also have a quadratic equation in r_{n+1} for fixed $r_1, \dots, r_n$. Therefore, P depends only on s_1 and s_2. Since the polynomial P is homogeneous with respect to r_i^{-1}, we have $P(s_1, s_2) = \lambda s_1^2 + \mu s_2$.

In the case under consideration, $s_1 = x + ny$ and $s_2 = x^2 + ny^2$. The polynomial P coincides with (1) at $\lambda = -\cos\varphi$ and $\mu = 1 + n\cos\varphi$. Therefore, in the general case, we also have

$$s_1^2 = \left(n + \frac{1}{\cos\varphi}\right) s_2.$$

1.22. For $n = 1$, the assertion is obviously valid: we need only consider one determinant; it equals $2d_{12}^2$, and the sign in front of the determinant is plus.

In the general case, we prove the required assertion by induction on n. Let us describe the step from n to $n+1$. By the induction hypothesis, there exist simplices $A_1A_3 \dots A_{n+2}$ and $A_2A_3 \dots A_{n+2}$ in $\mathbb{R}^n$ with edges of lengths $d_{ij} = |A_iA_j|$. Consider a hyperplane $\mathbb{R}^{n-1}$ in $\mathbb{R}^n$ that contains the simplex $A_3 \dots A_{n+2}$. The positions of the points A_1 and A_2 are determined up to symmetry about this hyperplane. Let $\rho_{\min}$ and $\rho_{\max}$ be the distances between these two points in the cases where they are on one side and on different sides of the hyperplane, respectively. Our immediate goal is to show that a simplex $A_1A_2 \dots A_{n+2}$ with edges of lengths $d_{ij} = |A_iA_j|$ exists if and only if $\rho_{\min} < d_{12} < \rho_{\max}$. Let us embed $\mathbb{R}^{n-1} \subset \mathbb{R}^n$ in $\mathbb{R}^{n+1}$. We can assume that these subspaces have the equations $x_{n+1} = 0$ and $x_n = x_{n+1} = 0$. Consider the rotation through an angle φ about $\mathbb{R}^{n-1}$ in $\mathbb{R}^{n+1}$ (this transformation does not change the coordinates $x_1, \dots, x_{n-1}$, and acts on x_n, x_{n+1} as the rotation through the angle φ). Let us apply this transformation to the point A_1. As φ varies from 0 to 2π, the point A_1 rotates along a circle in the two-dimensional plane orthogonal to $\mathbb{R}^{n-1}$, the distance from A_1 to A_2 varies from $\rho_{\min}$ to $\rho_{\max}$, and the distances from A_1 to $A_3, \dots, A_{n+2}$ do not change. Therefore, the required simplex exists if $\rho_{\min} < d_{12} < \rho_{\max}$, and it does not exist otherwise. The cases $d_{12} = \rho_{\min}$ and $d_{12} = \rho_{\max}$ correspond to degenerate simplices of zero volume.

It remains to show that the inequality

$$(-1)^{n+2} \begin{vmatrix} 0 & x & \dots & d^2_{1,n+2} & 1 \\ x & 0 & \dots & d_{2,n+2} & 1 \\ \dots & \dots & \dots & \dots & \dots \\ d^2_{n+2,1} & d^2_{n+2,2} & \dots & 0 & 1 \\ 1 & 1 & \dots & 1 & 0 \end{vmatrix} > 0$$

is equivalent to $\rho_{\min} < x < \rho_{\max}$. Clearly, the left-hand side of this inequality has the form $ax^2 + bx + c$, where the numbers a, b, and c do not depend on x and $a < 0$. Indeed, the coefficient a is equal to the product of the number $-(-1)^{n+2}$ with the determinant for the points $A_3, \dots, A_{n+2}$, whose sign is $(-1)^n$ by the induction hypothesis. The roots of the polynomial $ax^2 + bx + c$ correspond to the degeneracies of the simplex $A_1A_2 \dots A_{n+2}$, i.e., they are $\rho_{\min}$ and $\rho_{\max}$. Since $a < 0$, the inequality $ax^2 + bx + c > 0$ is equivalent to $\rho_{\min} < x < \rho_{\max}$.

1.23. Let S be the $(n-1)$-dimensional volume of the face of a regular n-simplex. Let us choose one face of the simplex and project the other n faces to this face. The volume of the projection of each face equals $S \cos\varphi$; therefore, $nS\cos\varphi = S$, i.e., $\cos\varphi = 1/n$.

1.24. The volume of an arbitrary parallelepiped generated by vectors with integer coordinates is an integer. Clearly, the squared length of an edge is also an integer. Thus if a is the edge length of the cube, then the numbers a^{2n+1} and a^2 are integers. Therefore, $a = a^{2n+1}/(a^2)^n$ is rational; since its square is an integer, a is also an integer.

1.25. Let O be a point inside a 4-simplex. Through the point O and each edge of the simplex, we draw a two-dimensional plane. We obtain a finite number of planes (namely, 10 planes). In 4-space, two two-dimensional planes in general position passing through the point O have no other common points. Hence there exists a plane passing through the point O and sharing no other points with all the planes drawn. This plane passes through an interior point of the simplex and does not intersect its edges.

1.26. In n-space, consider the plane generated by vectors a and b and passing through the origin. Its intersection with the cube $|x_k| \leqslant 1$, $k = 1, \dots, n$, is

$$\{\, xa + yb \mid |xa_k + yb_k| \leqslant 1,\ k = 1, \dots, n \,\}.$$

It is sufficient to choose the vectors a and b so that the following conditions hold:

(i) $|a| = |b|$ and $(a, b) = 0$;

(ii) the inequalities $|xa_k + yb_k| \leqslant 1$, where $k = 1, \dots, n$ and x, y are the coordinates on the plane, determine a regular $2n$-gon.

Both conditions hold for $a_k = \sin k\alpha$ and $b_k = \cos k\alpha$, where $\alpha = \pi/n$. Indeed,

$$(a, b) = \frac{1}{2}\sum_{k=1}^{n} \sin 2k\alpha, \quad |a|^2 = \frac{n}{2} - \frac{1}{2}\sum_{k=1}^{n} \cos 2k\alpha, \quad |b|^2 = \frac{n}{2} + \frac{1}{2}\sum_{k=1}^{n} \cos 2k\alpha,$$

and all these sums of sines and cosines are zero.

1.27. (a) Let $ABCDA_1B_1C_1D_1$ be a cube. Consider its sections perpendicular to the diagonal AC_1. The sections not suitable for our purposes are those passing through points of the tetrahedra A_1ABD and $CC_1B_1D_1$. Next, consider the sections perpendicular to the diagonal A_1C. The unsuitable set narrows down to two smaller tetrahedra and the pair of segments $[B, D]$ and $[B_1, D_1]$. Finally, consider the sections perpendicular to the diagonals BD_1 and B_1D. This narrows the bad set to the vertices of the cube and the centers of the faces; the interior points of the cube are not included in this set.

(b) Let us draw a two-dimensional plane Π through the center of the cube such that its intersection with the cube is a $2n$-gon (see Problem 1.26). The plane Π intersects all $(n-1)$-faces of the cube at interior points; therefore, the 3-space generated by Π and the given point A has all the required properties.

1.28. (a) First, suppose that the altitudes of the simplex intersect at a point A_0. For $e_i = \overrightarrow{A_1A_i}$ and $x = \overrightarrow{A_1A_0}$, we have

$$(x, e_i - e_j) = 0 \quad \text{and} \quad (x - e_i, e_j) = 0.$$

Therefore, $(x, e_i) = \text{const}$ and $(e_i, e_j) = (x, e_i)$.

Now suppose that $(e_i, e_j) = \text{const}$. Since the simplex is nondegenerate, the vectors $e_2, \dots, e_{n+1}$ are linearly independent. Let $\varepsilon_2, \dots, \varepsilon_{n+1}$ be the dual basis, i.e., $(\varepsilon_i, \varepsilon_j) = \delta_{ij}$. Now consider $x = c(\varepsilon_2 + \dots + \varepsilon_n)$, where $c = (e_i, e_j)$. It is easy to verify that

$$(x, e_i - e_j) = 0, \qquad (x - e_i, e_j) = 0.$$

Hence, if A_0 is the endpoint of a vector x with initial point A_1, then A_0 is the orthocenter of the simplex.

(b) The edges of the simplex are orthogonal if and only if $(e_i, e_j - e_k) = 0$ and $(e_i - e_j, e_k - e_l) = 0$, and these equalities are equivalent to $(e_i, e_j) = \text{const}$.

(c) First, suppose that $A_2 \dots A_{n+1}$ is an orthocentric simplex and the altitude from A_1 passes through the orthocenter H of this simplex. We must prove that $A_1A_k \perp A_iA_j$ if $i, j, k \geqslant 2$ are pairwise different. Clearly, $A_2H \perp A_iA_j$ and $A_kH \perp A_iA_j$. Therefore, the plane A_1A_kH is perpendicular to the line A_iA_j, and hence $A_1A_k \perp A_iA_j$.

Now suppose that the simplex $A_1 \dots A_n$ is orthocentric. Then the edges A_iA_j and A_kA_l are orthogonal for any pairwise different i, j, k, l. This means, in particular, that the simplex $A_2 \dots A_n$ is orthocentric. Let H be its orthocenter. It is required to prove that the edge A_iA_j with $i, j \geqslant 2$ is orthogonal to the line A_1H. We have $A_kH \perp A_iA_j$ and $A_1A_k \perp A_iA_j$, and hence $A_1H \perp A_iA_j$.

(d) Let a, b, c, d be the vectors of the sides of the spatial quadrilateral with vertices at the four vertices of the simplex. The equality $a^2 + c^2 = b^2 + d^2$ is equivalent to the orthogonality of the diagonals of the quadrilateral, i.e., to the equality $(a + b, b + c) = 0$. Indeed, since $d = -(a + b + c)$, we can write the first equality in the form

$$a^2 + c^2 = b^2 + a^2 + b^2 + c^2 + 2(a, b) + 2(b, c) + 2(a, c).$$

This is equivalent to $(a + b, b + c) = 0$.

(e) First, suppose that $|A_iA_j|^2 = a_i + a_j$. It is easy to verify that then relation (2) on p. 30 holds, and therefore the simplex is orthocentric.

Now, suppose that the simplex is orthocentric. Then by (2) the number

$$a_i = \frac{1}{2}(|A_iA_j| + |A_iA_k| - |A_jA_k|)$$

does not depend on j and k. It is also clear that $|A_iA_j|^2 = a_i + a_j$.

(f) If i, j, k are pairwise different and nonzero, then $A_0A_i \perp A_jA_k$. Therefore, (2) holds, and there exist numbers $a_0, \dots, a_{n+1}$ satisfying the required relations.

We prove the equality $a_0^{-1} + \cdots + a_{n+1}^{-1} = 0$ by induction on n. The argument starts as follows: for a point A_0 in the segment $[A_1, A_2]$, we denote $|A_1A_0| = b$ and $|A_0A_2| = c$; then $a_0 = -bc$, $a_1 = b(b+c)$, and $a_2 = c(b+c)$.

Let us describe the induction step. Consider the orthocenters A_0' and A_0 of the simplices $A_1 \dots A_n$ and $A_1 \dots A_{n+1}$, respectively. Setting $|A_0'A_0| = x$ and $|A_0A_{n+1}| = y$, we can write

$$|A_iA_0|^2 = |A_0'A_i|^2 + |A_0'A_0|^2 = a_i + a_0' + x^2,$$
$$|A_iA_{n+1}|^2 = |A_0'A_i|^2 + |A_0'A_{n+1}|^2 = a_i + a_0' + (x+y)^2.$$

These equalities imply that $a_0 = a_0' + x^2$ and $a_{n+1} = a_0' + (x+y)^2$. In addition, $y^2 = a_0 + a_{n+1}$; therefore, $a_0' = -x(x+y)$.

By the induction hypothesis, $a_1^{-1} + \cdots + a_n^{-1} = -(a_0')^{-1}$. Therefore,

$$a_0^{-1} + \cdots + a_{n+1}^{-1} = \frac{1}{-xy} - \frac{1}{-x(x+y)} + \frac{1}{y(x+y)} = 0.$$

1.29. The relation $\overrightarrow{OA}_1 + \cdots + \overrightarrow{OA}_{n+1} = (n-1)\overrightarrow{OX}$ uniquely determines the point X. Hence it is sufficient to verify that $(\overrightarrow{A_2A}_3, \overrightarrow{A_1X}) = 0$. Clearly,

$$\begin{aligned}(n-1)\overrightarrow{A_1X} &= (n-1)\overrightarrow{A_1O} + (n-1)\overrightarrow{OX} = (n-2)\overrightarrow{A_1O} + \overrightarrow{OA}_2 + \cdots + \overrightarrow{OA}_{n+1} \\ &= \overrightarrow{OA}_2 + \overrightarrow{OA}_3 + \overrightarrow{A_1A}_4 + \cdots + \overrightarrow{A_1A}_{n+1}.\end{aligned}$$

The equality $(\overrightarrow{A_2O} + \overrightarrow{OA}_3, \overrightarrow{OA}_2 + \overrightarrow{OA}_3) = 0$ follows from $|OA_2| = |OA_3|$, and the equalities $(\overrightarrow{A_2A}_3, \overrightarrow{A_1A}_4) = \cdots = (\overrightarrow{A_2A}_3, \overrightarrow{A_1A}_{n+1}) = 0$ hold because the simplex is orthocentric.

Chapter 2

2.4. The case of a parallelogram is easy to handle; hence, we assume that the given convex quadrilateral $ABCD$ has no parallel sides. To be definite, we assume that the ray AB intersects the ray DC and the ray BC intersects the ray AD. Suppose that $\overrightarrow{AB} = \vec{a}$, $\overrightarrow{BC} = \vec{b}$, $\overrightarrow{CD} = p\vec{a} + q\vec{b}$, $\overrightarrow{DA} = u\vec{a} + v\vec{b}$. Then $p < 0$, $q > 0$, $u < 0$, $v < 0$.

Consider an affine transformation which maps the vectors $\vec{a}$ and $\vec{b}$ to orthogonal vectors $\vec{a}'$ and $\vec{b}'$ of lengths λ and μ, respectively. The image of the given quadrilateral under this transformation has two right opposite angles provided that the scalar product

$$(p\vec{a} + q\vec{b}, u\vec{a} + v\vec{b}) = pu\lambda^2 + qv\mu^2$$

vanishes. Since $pu > 0$ and $qv < 0$, this can always be achieved by a suitable choice of the numbers λ and μ.

Note that for any affine transformation, the image of the angle at the vertex C is larger than the image of the angle at the vertex A; these angles cannot be made equal.

2.5. Let A_1, B_1, C_1, D_1, E_1, F_1 be the midpoints of the sides AB, BC, CD, DE, EF, FA, respectively. The diagonals AD and BE are equal if and only if the line A_1D_1 is perpendicular to the lines AB and DE. Let O be the intersection point of the lines A_1D_1 and B_1E_1. We must construct an affine transformation that maps the angles A_1 and B_1 of the quadrilateral A_1BB_1O to right angles. This can be done by applying the preceding problem. The intersection points of the extensions of the sides of the quadrilateral A_1BB_1O are positioned as required because the hexagon is convex.

2.6. Suppose that there exists an affine transformation that maps the vectors $\vec{a}$, $\vec{b}$, $\vec{c}$ to vectors $\vec{a}'$, $\vec{b}'$, $\vec{c}'$ of equal lengths. The equality $\alpha\vec{a}' + \beta\vec{b}' + \gamma\vec{c}' = 0$ implies that there exists a triangle with sides $|\alpha|$, $|\beta|$, $|\gamma|$.

Now suppose that there exists a triangle with sides $|\alpha|$, $|\beta|$, $|\gamma|$. Then $\alpha\vec{a}'+\beta\vec{b}'+\gamma\vec{c}' = 0$ for some vectors $\vec{a}'$, $\vec{b}'$, $\vec{c}'$ of unit length. Consider the affine transformation that maps the vectors $\vec{a}$ and $\vec{b}$ to $\vec{a}'$ and $\vec{b}'$, respectively. The equalities $\alpha\vec{a}+\beta\vec{b}+\gamma\vec{c} = 0$ and $\alpha\vec{a}' + \beta\vec{b}' + \gamma\vec{c}' = 0$ imply that this affine transformation maps $\vec{c}$ to $\vec{c}'$ (we assume that $\gamma \neq 0$).

2.10. Let $y_i = a_{i1}x_1 + \dots + a_{in}x_n$. In space with coordinates $y_1, \dots, y_n$, the inequalities $|y_i| \leqslant b_i$ specify a rectangular parallelepiped of volume $2^n b_1 \cdots b_n$. The figure under consideration is the image of this parallelepiped under the map A^{-1} because the equality $y = Ax$ implies $x = A^{-1}y$. Therefore, the volume of this figure is $|\det A^{-1}| 2^n b_1 \cdots b_n = 2^n b_1 \cdots b_n / |\Delta|$.

2.11. (a) and (b) Let $\lambda_1 A_1 + \lambda_2 A_2$ and $\lambda_1 B_1 + \lambda_2 B_2$ be points in the figure $\lambda_1 M_1 + \lambda_2 M_2$ (here A_i and B_i are points in the polygon M_i). Then the figure $\lambda_1 M_1 + \lambda_2 M_2$ contains the parallelogram with vertices $\lambda_1 A_1 + \lambda_2 A_2$, $\lambda_1 A_1 + \lambda_2 B_2$, $\lambda_1 B_1 + \lambda_2 B_2$, $\lambda_1 B_1 + \lambda_2 A_2$. The figure $\lambda_1 M_1 + \lambda_2 M_2$ is convex because it contains the diagonal of this parallelogram.

Suppose that the polygons M_1 and M_2 lie on one side of some line l. Let us move this line parallel to itself; at some moment it touches the polygon M_1, and at some other moment it touches M_2. Let a_1 and a_2 be the lengths of the intersection segments of l with M_1 and M_2, respectively, at the moment when l first touches the polygons ($a_i = 0$ if l is not parallel to the sides of the polygon M_i). Then l intersects $\lambda_1 M_1 + \lambda_2 M_2$ at the moment of first touch in a segment of length $\lambda_1 a_1 + \lambda_2 a_2$. The number $\lambda_1 a_1 + \lambda_2 a_2$ is nonzero if and only if the numbers a_1 and a_2 are not both zero.

(c) Take a point O_i inside each polygon M_i, where $i = 1, 2$. Let us cut every polygon M_i into triangles with vertex O_i and the polygon $\lambda_1 M_1 + \lambda_2 M_2$ into triangles with vertex $O = \lambda_1 O_1 + \lambda_2 O_2$. The pair of triangles with bases a_1 and a_2 and altitudes h_1 and h_2 corresponds to the triangle with base $\lambda_1 a_1 + \lambda_2 a_2$ and altitude $\lambda_1 h_1 + \lambda_2 h_2$. It remains to note that

$$(\lambda_1 a_1 + \lambda_2 a_2)(\lambda_1 h_1 + \lambda_2 h_2) = \lambda_1^2 a_1 h_1 + \lambda_1 \lambda_2 (a_1 h_2 + a_2 h_1) + \lambda_2^2 a_2 h_2.$$

Chapter 3

3.1. The required equality is equivalent to

$$\left(\frac{c-a}{c-b} \cdot \frac{d-b}{d-a}\right) \cdot \left(\frac{d-a}{d-b} \cdot \frac{e-b}{e-a}\right) \cdot \left(\frac{e-a}{e-b} \cdot \frac{c-b}{c-a}\right) = 1.$$

3.2. The line $l(t)$ intersects the line $y = 1$ at the point $(x, 1)$, where x is related to t by the formula $a_1x + b_1 = t(a_2x + b_2)$. Hence the cross ratio of the lines $l(t_i)$ is equal to the cross ratio of the numbers w_i, where $t_i = (a_1w_i + b_1)/(a_2w_i + b_2)$. It remains to note that the cross ratio of the numbers w_i is equal to the cross ratio of the numbers t_i.

3.3. The cross ratios of the intersection points on the lines l_1 and l_2 equal the cross ratio of the four planes passing through the line l_3 and the lines a, b, c, and d, respectively.

3.4. Let $A_1, B_1, \dots$ be the intersection points of l with the planes $BCD, ACD, \dots,$ and let $A_2, B_2, \dots$ be the intersection points of the line CD with the planes passing through the line l and the vertices $A, B, \dots$. Then $C_2 = C$ and $D_2 = D$; in addition, the lines C_1C_2 and D_1D_2 intersect AB at the points C_3 and D_3 which are the intersection points of the line AB with the planes passing through l and the vertices C and D. Clearly, the points A_1, A_2, and $A_3 = A$ are collinear, and the points B_1, B_2, and $B_3 = B$ are also collinear. Therefore, according to the preceding problem,

$$[A_1, B_1, C_1, D_1] = [A_3, B_3, C_3, D_3].$$

3.5. Consider two lines l_1 and l_2 that intersect the given hyperplanes (which have a common subspace V^{n-2} of dimension $n - 2$). Let W^3 be the three-dimensional subspace containing the lines l_1 and l_2. If W^3 is in general position, then the intersection of W^3 with V^{n-2} is a straight line, and the intersections of W^3 with the given hyperplanes are planes. Now, we can apply the fact that the cross ratio of planes in a three-dimensional space does not depend on the choice of the line intersecting the planes.

3.6. The square of the transformation under consideration equals

$$\frac{(a^2 + bc)x + b(a + d)}{c(a + d)x + (bc + d^2)}.$$

It is the identity map if and only if $b(a+d) = c(a+d) = 0$ and $a^2 + bc = d^2 + bc$. If $a + d \neq 0$, then $b = c = 0$ and $a^2 = d^2$, i.e., $a = d \neq 0$. Such coefficients correspond to the identity transformation, while the initial transformation is not the identity by assumption.

3.7. Any linear transformation leaves the point ∞ fixed; therefore, a linear-fractional transformation with fixed points a and b cannot be linear. Thus we can assume that the transformation under consideration has the form $x \mapsto (px+q)/(x+r)$. The points a and b are fixed points of this transformation if $(p - r)a = a^2 - q$ and $(p - r)b = b^2 - q$. These relations imply that $p - r = a + b$ and $q = -ab$, i.e., $p = \lambda$, $q = -ab$, and $r = \lambda - a - b$.

3.8. Over the field $\mathbb{C}$, f has a fixed point x_0. Hence the transformation $f(f(x))$ has three different fixed points (a, $f(a)$ and x_0) and is therefore the identity.

3.9. Suppose that the given projective transformation f takes the point ∞ to some point $x_0 \neq \infty$. Then the composition of f with the involution $x \mapsto (x_0x+b)/(x-x_0)$ leaves the point ∞ fixed. Therefore, this transformation is either a translation $x \mapsto x + t$ (which can be represented as the composition of the involutions $x \mapsto -x$ and $x \mapsto t-x$) or a homothety $x \mapsto \lambda x$ (which can be represented as the composition of the involutions $x \mapsto x^{-1}$ and $x \mapsto \lambda x^{-1}$).

3.10. Consider the projection parallel to the line l to a plane transversal to the lines l_1 and l_2 (i.e., intersecting these lines at precisely one point).

3.11. Suppose that M and M' are the intersection points of A_1O' with A_2O and of A_1O with A_2O', respectively, N and N' are the intersection points of A_1O' with A_3O and of A_1O with A_3O', and P and P' are the intersection points of A_2O with A_3O' and of A_2O' with A_3O. The lines $O'A_1$, $O'A_2$, $O'A_3$, and $O'O$ intersect A_1O and C_1O at the points $\{A_1, M', N', O\}$ and $\{N, P, A_3, O\}$, respectively. Let us join the points of the first quadruple to the point M and the points of the second quadruple to the point P. The lines MO and PO coincide; therefore, the three remaining intersection points of the respective lines are collinear. Since $MA_1 \cap PN = N$ and $MN' \cap PA_3 = N'$, the point $MM' \cap PP'$ lies on the line NN'.

3.12. The intersection points of the lines O_1A_1 and O_2A_1 lie on the fixed straight line l_1; therefore, $O_1A_1 \mapsto O_2A_1$ is a projective map of the projective pencils of lines through the points O_1 and O_2. A similar argument for the maps $O_2A_2 \mapsto O_3A_2$ etc. shows that $O_1A_n \mapsto O_nA_n$ is a projective correspondence with fixed line $l = O_1O_n$; therefore, the intersection points of the respective lines in the pencils of lines through O_1 and O_n are collinear.

Chapter 4

4.1. Apply the theorem about common chords of conics to the conics Γ, Γ_1, and Γ_2, where Γ is the ellipse under consideration, Γ_1 is the pair of lines AB and CD, and Γ_2 is the pair of lines BC and AD.

4.2. The directions of the axes of the conics depend only on the quadratic terms of the equations of the conics; for this reason, we shall consider only these terms. We can assume that the equation of one conic is $ax^2 + by^2 + \dots = 0$. If a linear combination of this equation and the equation $a_1x^2 + b_1y^2 + c_1xy + \dots = 0$ of the second conic is $x^2 + y^2 + \dots = 0$, then $c_1 = 0$, and the axes of the conics are then perpendicular. Conversely, suppose that $c_1 = 0$. Put $\lambda = -(a-b)/(a_1 - b_1)$ (the case $a_1 = b_1$ corresponds to the circle). Then $a + \lambda a_1 = b + \lambda b_1$. It remains to note that if $a + \lambda a_1 = b + \lambda b_1 = 0$, then the conics under consideration have at most two common points, because among linear combinations of their equations there is a linear equation.

4.3. The condition for a conic from a given pencil to be a parabola can be written as $\det(\lambda A + \mu B) = 0$, where A and B are square matrices of order 2. In this form, the condition is a quadratic equation.

4.4. A conic passing through the points A, B, C, and D has the equation $F = 0$, where $F = \lambda l_{AB} \cdot l_{CD} + l_{BC} \cdot l_{AD}$. The center of this conic is given by the system of equations $F_x = 0$, $F_y = 0$; each of these equations is linear with respect to x, y, and λ. Substituting the expression for λ found from one equation in the other equation, we obtain a second-order equation relating x to y.

4.5. (a) Let C' and D' be the points symmetric to C and D with respect to the midpoint M of the segment $[A, B]$. Then the points A, B, C, D, C', and D' lie on a conic with center M; therefore, $M \in \Gamma$.

Let O be the intersection point of the lines AB and CD. The point O is the center of the degenerate conic formed by the pair of lines AB and CD. Hence $O \in \Gamma$.

(b) The midpoints of the sides of the rectangle $ABCD$ form a parallelogram whose center coincides with the center of mass of the points A, B, C, and D. This parallelogram is inscribed in the conic Γ; therefore, its center coincides with the center of the conic.

(c) Follows from (b).

(d) In the pencil under consideration, take a conic which is neither a circle nor a parabola. The assertion of Problem 4.2 implies that the axes of all other conics are perpendicular to the axes of the conic under consideration. (The axes of the conics are perpendicular to each other; therefore, the axes of all other conics are parallel to the axes of the conic under consideration.)

The pencil includes an ellipse and hyperbolas of two different types (depending on which point, B or D, belongs to the branch containing A). Therefore, there are two parabolas among the conics of the pencil, and their axes are perpendicular to each other. The centers of these two parabolas are the points at infinity of two mutually perpendicular directions.

4.6. For $Y = 4ay$ and $X = 4ax$, the equations $4ay = x^2$ and $Y = X^2$ are equivalent.

4.7. Substituting the equation $y = x^2$ of a parabola into the equation $(x - a)^2 + (y - b)^2 = R^2$ of a circle, we obtain a fourth-order equation in which x^3 occurs with zero coefficient. The sum of roots of this equation is zero.

4.8. Add up the equations $x^2 + py = 0$ and $(y - y_0)^2 + ax + b = 0$.

4.9. The points P_1 and P_2 have polar coordinates (r_1, φ) and $(r_2, \varphi + \pi)$, where

$$r_1 = \frac{2a}{1 - \cos\varphi} \quad \text{and} \quad r_2 = \frac{2a}{1 - \cos(\varphi + \pi)} = \frac{2a}{1 + \cos\varphi}.$$

4.10. The tangent line at the point $(2t_i, t_i^2)$ has the equation $y = xt_i - t_i^2$.

4.11. Let A_1 and B_1 be the projections of A and B to the directrix. Then AO and BO are perpendicular lines to the segments $[A_1, F]$ and $[B_1, F]$ through their midpoints. Hence $|A_1O| = |FO| = |B_1O|$; therefore, $\angle A_1B_1O = \angle B_1A_1O = \varphi$. Thus

$$\angle OFA = \angle OFB = 90^\circ + \varphi, \qquad \angle AFB = 180^\circ - 2\varphi = \angle A_1OB_1 = 2\angle AOB.$$

4.12. We use the notation of Problem 4.11. According to the assertion of Problem 4.11, $\angle AFB = \angle A_1OB_1 = 2\angle AOB$. So, $\angle AOB = 90^\circ$ if and only if $\angle AFB = 180^\circ$ and $\angle A_1OB_1 = 180^\circ$.

4.13. The point $z = \cos\varphi + i\sin\varphi$ is transformed into

$$w = \frac{-\cos\varphi + i\sin\varphi}{1 - \cos\varphi}.$$

4.14. Yes. Put $x = (x_1 + y_1)^2$ and $y = x_1 + y_1$. It is easy to verify that x_1 and y_1 can be expressed in terms of x and y.

4.15. (a) The projection of the focus F to a tangent line to the parabola lies on the tangent line perpendicular to the axis. Therefore, the projections A', B', and C' of F to the lines BC, CA, and AB lie on one straight line. This means that the point F lies on the circle circumscribed about the triangle ABC. Indeed, $\angle AFC' = \angle AB'C' = \angle A'B'C = \angle A'FC$, and hence $\angle CFA = \angle A'FC' = 180^\circ - \angle B$.

(b) The tangent lines to the parabola $x^2 = 4y$ at the points $(2t_i, t_i^2)$ have the equations $y = t_ix - t_i^2$. They intersect at the points $(t_i + t_j, t_it_j)$. It is easy to verify that the orthocenter of the triangle with vertices at these three points is $(t_1 + t_2 + t_3 + t_1t_2t_3, -1)$.

(c) We can assume that the parabola is given by the equation $x^2 = 4y$. Then the points α, β, γ have coordinates $(2t_i, t_i^2)$, where $i = 1, 2, 3$. It is easy to verify that

$$S_{\alpha\beta\gamma} = \frac{1}{2}\begin{vmatrix} 2t_1 & t_1^2 & 1 \\ 2t_2 & t_2^2 & 1 \\ 2t_3 & t_3^2 & 1 \end{vmatrix} \quad \text{and} \quad S_{ABC} = \frac{1}{2}\begin{vmatrix} t_2 + t_3 & t_2t_3 & 1 \\ t_3 + t_1 & t_3t_1 & 1 \\ t_1 + t_2 & t_1t_2 & 1 \end{vmatrix}.$$

(d) There exists an affine transformation that maps the axis of the parabola and the line AC to a pair of perpendicular lines. Therefore, we can assume that the points α, β, γ have coordinates $(2t_1, t_1^2)$, $(0, 0)$, $(2t_3, t_3^2)$ with $t_1 < 0$ and $t_3 > 0$. Then we have

$$S_{\alpha\beta C} = -\frac{1}{2}t_1^3, \quad S_{\beta\gamma A} = \frac{1}{2}t_3^3, \quad \text{and} \quad S_{\alpha\beta B} = \frac{1}{2}\begin{vmatrix} 2t_3 & t_3^2 & 1 \\ 2t_1 & t_1^2 & 1 \\ t_1 + t_3 & t_1t_3 & 1 \end{vmatrix} = \frac{(t_3 - t_1)^3}{2}.$$

4.16. The tangent lines to the parabola $x^2 = 4y$ at the points $A = (2t_1, t_1^2)$ and $B = (2t_2, t_2^2)$ intersect at the point $O = (t_1 + t_2, t_1t_2)$. In the case under consideration, $t_1t_2 = -2$. Now, it is easy to verify that $(0, 0)$ is the orthocenter of the triangle AOB.

4.17. Consider the family of all parabolas with focus at the given point F and axis parallel to the light beams. More precisely, the axis of each parabola is directed so that the light beam under consideration, being reflected off the parabola, passes through the point F.

Suppose that a light beam is reflected at a point M of the curve C and passes through the point F. Then the tangent line to C at the point M coincides with the tangent line at M to the parabola passing through M. This property can hold for all points of the curve C only if it coincides with a parabola from the family under consideration.

4.18. Any parallelogram is the image of a square under an affine transformation, and any triangle is the image of a regular triangle.

4.19. (a) An ellipse is the image of a circle under an affine transformation, and conjugate diameters of the ellipse are the images of perpendicular diameters of the circle. It is also clear that any affine transformation preserves the ratios of areas of figures.

(b) Clearly we can assume that the points O, A, B have the coordinates $(0, 0)$, $(a\cos\varphi, b\sin\varphi)$, $(-a\sin\varphi, b\cos\varphi)$, respectively.

4.20. Let O be the center of the ellipse, and let P_1 and P_2 be the projections of its foci F_1 and F_2 to the tangent line to the ellipse at point A. Then $\angle P_1AF_1 = \angle P_2AF_2 = \varphi$. Put $x = |F_1A|$ and $y = |F_2A|$. The value $x + y = c$ does not depend on the point A; hence,

$$|P_1O|^2 = |P_2O|^2 = \left(\frac{x+y}{2}\cos\varphi\right)^2 + \left(\frac{x+y}{2}\sin\varphi\right)^2 = \frac{c^2}{4}.$$

In addition, $|F_1F_2|^2 = x^2 + y^2 + 2xy\cos 2\varphi = c^2 - 4xy\sin^2\varphi$; therefore, the value $xy\sin^2\varphi = d_1d_2$ is constant.

4.21. Let the points G_1 and G_2 be symmetric to F_1 and F_2 with respect to the lines OA and OB, respectively. The points F_1, B, and G_2 are collinear, and

$$|F_1G_2| = |F_1B| + |BG_2| = |F_1B| + |BF_2|.$$

The triangles G_2F_1O and G_1F_2O have equal sides. Therefore,

$$\angle G_1OF_1 = \angle G_2OF_2, \qquad \angle AF_1O = \angle AG_1O = \angle BF_1O.$$

4.22. Consider the affine transformation that maps the ellipse to a circle. It maps the given parallelogram to a rhombus. The diagonals of the rhombus contain a pair of perpendicular diameters of the circle.

4.23. Consider the map $z \mapsto az + b\bar{z}$. In the coordinates (x, y), where $x + iy = z$, this map is linear, and its determinant equals $|a|^2 - |b|^2$. The image of the circle $|z| = 1$ under a nondegenerate linear transformation is an ellipse and under a degenerate one, a segment or a point.

4.24. Consider a vertex O of the given rectangle and tangent lines OA and OB to the ellipse. In the case under consideration the triangle F_1OG_2 constructed in Problem 4.21 is right. Therefore,

$$|F_1O|^2 + |F_2O|^2 = |F_1G_2|^2 = \text{const}.$$

If M is the center of the ellipse, then

$$|OM|^2 = \frac{1}{4}(2|F_1O|^2 + 2|F_2O|^2 - |F_1F_2|^2)$$

is constant.

4.25. For the ellipse under consideration, the vertices of the circumscribed rectangle with sides parallel to the axes of the ellipse lie on the circle under consideration. The assertion of Problem 4.24 implies that the vertices of all other rectangles circumscribed about the ellipse also lie on the circle under consideration. Therefore, the chord PQ is a side of a rectangle circumscribed about the ellipse. According to Problem 4.22, the diagonals of this rectangle contain conjugate diameters of the ellipse.

4.26. (a) The points A and B have the coordinates

$$(a\cos\varphi, b\sin\varphi) \quad \text{and} \quad (a\sin\varphi, -b\cos\varphi).$$

The points P and Q have the coordinates

$$((a+b)\sin\varphi, -(a+b)\cos\varphi) \quad \text{and} \quad ((a-b)\sin\varphi, (a-b)\cos\varphi).$$

(b) The required construction is essentially described in (a).

4.27. The point Q lies on the circle inscribed in the triangle AF_1F_2, where F_1 and F_2 are the foci of the ellipse, and Q is the midpoint of the arc F_1F_2. If R is the radius of the circumscribed circle and α and β are the angles at the vertices F_1 and F_2, then

$$|AQ| = 2R\cos\frac{\alpha-\beta}{2} \quad \text{and} \quad |AP| = 2R\sin^2\frac{\alpha+\beta}{2}\cos\frac{\alpha-\beta}{2}.$$

Therefore,

$$\begin{aligned}|AP|\cdot|AQ| &= \left(2R\sin\frac{\alpha+\beta}{2}\cos\frac{\alpha-\beta}{2}\right)^2 \\ &= (R\sin\alpha + R\cos\alpha)^2 = \left(\frac{|AF_1|+|AF_2|}{2}\right)^2.\end{aligned}$$

4.28. Let $ABCD$ be a rhombus inscribed in an ellipse with center O. Then the radius r of the circle inscribed in the rhombus is equal to the altitude of the right triangle AOB, i.e.,

$$\frac{1}{r^2} = \frac{1}{|OA|^2} + \frac{1}{|OB|^2}.$$

For the ellipse $x^2/a^2 + y^2/b^2 = 1$, the lines OA and OB have the equations $y = kx$ and $y = -x/k$, and the points A and B have coordinates (x_0, y_0) and (x_1, y_1), where

$$x_0^2\left(\frac{1}{a^2} + \frac{k^2}{b^2}\right) = 1 \quad \text{and} \quad x_1^2\left(\frac{1}{a^2} + \frac{1}{k^2b^2}\right) = 1.$$

Therefore,

$$\frac{1}{|OA|^2} + \frac{1}{|OB|^2} = \frac{\frac{1}{a^2} + \frac{k^2}{b^2}}{1+k^2} + \frac{\frac{1}{a^2} + \frac{1}{k^2b^2}}{1+\frac{1}{k^2}} = \frac{1}{a^2} + \frac{1}{b^2};$$

thus the radius r does not depend on the position of the rhombus.

4.29. The conjugate diameters can be represented as the diagonals of a parallelogram $ABCD$ circumscribed about the ellipse; let O be the center of the ellipse. Suppose that the bisectors of the angles AOB, BOC, COD, DOA intersect the sides of the parallelogram at points A_1, B_1, C_1, D_1, respectively, and the rays OA_1, OB_1, OC_1, OD_1 intersect the ellipse at points A_2, B_2, C_2, D_2. Then the points A_2, B_2, C_2, D_2 are the centers of the circles under consideration.

It is sufficient to prove that $A_2B_2C_2D_2$ is a parallelogram with sides parallel to the lines AC and BD. Indeed, the diagonals of this parallelogram are perpendicular; hence it is a rhombus. According to the assertion of Problem 4.28, the radius of the circle inscribed in the rhombus depends on the ellipse only and does not depend on the position of the rhombus. Since the lines A_2B_2 and AC are parallel, the radius of the circle inscribed in the rhombus equals the distance from the point A_2 to the line AC; i.e., it is equal to the radius of the circle under consideration centered at A_2.

Let us show, say, that $A_2B_2 \parallel AC$. First, note that $A_1B_1 \parallel AC$ because

$$|AA_1| : |A_1B| = |AO| : |BO| = |CO| : |BO| = |CB_1| : |BB_1|.$$

Let us apply an affine transformation that maps the conjugate diameters under consideration to diameters of the circle. The images of the lines A_1O and B_1O under such a transformation are symmetric with respect to the line OB; therefore, the image of the line A_2B_2 is parallel to the image of the line AC.

4.30. (a) Let us denote $\angle PF_1Q = 2\alpha$, $\angle PF_2Q = 2\beta$, $\angle POF_1 = p$, $\angle F_1OF_2 = q$. According to Problem 4.21, we have

$$\angle PF_1O = \alpha, \qquad \angle PF_2O = \beta, \qquad \angle F_2OQ = p;$$

the last equality implies $\angle POQ = 2p + q$.

The segments $[P, F_1]$ and $[P, F_2]$ make equal angles with the tangent line PO; therefore,

$$\alpha + p = \angle F_2PO = \pi - \beta - (p+q),$$

i.e.,

$$\angle POQ = 2p + q = \pi - (\alpha + \beta) = \pi - \frac{1}{2}(\angle PF_1Q + \angle PF_2Q).$$

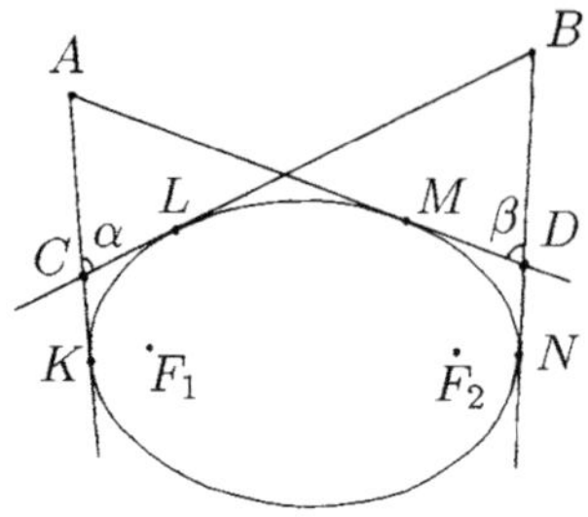

FIGURE S.1

(b) Let us denote the points of tangency as in Figure S.1. According to (a),

$$\alpha = \frac{1}{2}(\angle KF_1L + \angle KF_2L), \qquad \beta = \frac{1}{2}(\angle MF_1N + \angle MF_2N).$$

The rays F_1A and F_2B are the bisectors of the angles KF_1M and LF_2N, respectively; therefore,

$$\varphi_1 = \angle AF_1B = \frac{1}{2}(\angle KF_1L + \angle MF_1N).$$

Similarly,

$$\varphi_2 = \angle AF_2B = \frac{1}{2}(\angle KF_2L + \angle MF_2N).$$

4.31. Let φ be the angle between any of the tangent lines under consideration and the Ox axis. Consider the circle S centered at $((a+b)\cos\varphi, (a+b)\sin\varphi)$ and passing through the point $A = (a\cos\varphi, b\sin\varphi)$.

The tangent line to the ellipse at the point A has the equation

$$\frac{a\cos\varphi}{a^2}x + \frac{b\sin\varphi}{b^2}y = 1.$$

This line is perpendicular to the line AO_1, which is given by

$$y = \frac{a\sin\varphi}{b\cos\varphi}x + c.$$

Hence the circle S is tangent to the ellipse.

Next, let us show that the circle S is tangent to the lines l_1 and l_2. Suppose that the line l_1 is tangent to the ellipse at a point $(-a\cos\alpha, b\sin\alpha)$. Then it has the equation

$$\frac{-x\cos\alpha}{a} + \frac{y\sin\alpha}{b} = 1.$$

This means, in particular, that $\tan\varphi = b\cos\alpha / a\sin\alpha$. The squared distance from the origin to l_1 equals

$$\frac{b^2}{\sin^2\alpha}\cdot\frac{1}{1+\tan^2\varphi} = b^2\frac{\cos^2\varphi}{\sin^2\alpha} = b^2\cos^2\varphi\left(1 + \frac{a^2\sin^2\varphi}{b^2\cos^2\varphi}\right) = b^2\cos^2\varphi + a^2\sin^2\varphi.$$

The last expression coincides with the squared radius of the circle S. For the line l_2, we obtain a similar expression.

4.32. The normal to the ellipse at the point (x_0, y_0) has the equation

$$\frac{-y_0}{b^2}x + \frac{x_0}{a^2}y = x_0y_0\left(\frac{1}{a^2} - \frac{1}{b^2}\right).$$

It intersects the major semiaxis at the point with coordinate

$$x_1 = b^2 x_0 \left(\frac{1}{b^2} - \frac{1}{a^2}\right) = x_0 \frac{a^2 - b^2}{a^2}.$$

In addition,

$$r^2 = y_0^2 + x_0^2 \frac{b^4}{a^4} = b^2 - b^2 x_0^2 \left(\frac{1}{a^2} - \frac{b^2}{a^4}\right).$$

It is easy to verify that

$$\frac{(a^2 - b^2)(b^2 - r^2)}{b^2} = x_1^2.$$

4.33. According to Problem 4.32,

$$r_1 + r_2 = |OC_1| - |OC_2| = \frac{\sqrt{a^2 - b^2}}{b}\left(\sqrt{b^2 - r_1^2} - \sqrt{b^2 - r_2^2}\right).$$

This expression can be reduced to the form

$$a^4 r_1^2 - 2a^2(a^2 - 2b^2) r_1 r_2 + a^4 r_2^2 - 4b^4(a^2 - b^2) = 0.$$

Consider this relation as a quadratic equation in r_1. It has roots r_1 and r_3; therefore,

$$r_1 + r_3 = \frac{2a^2(a^2 - 2b^2)}{a^4} r_2.$$

4.34. According to Problem 4.33, the numbers r_i satisfy the recurrence relation

$$r_{i+2} - k r_{i+1} + r_i = 0.$$

Therefore, $r_p = a\lambda_1^p + b\lambda_2^p$, where λ_1 and λ_2 are the roots of the equation $x^2 - kx + 1 = 0$. Clearly, $\lambda_1 \lambda_2 = 1$, i.e., $r_p = a\lambda^p + b\lambda^{-p}$. Now the required formula is verified in an obvious way.

4.35. We can assume that the triangle ABC is obtained by applying the transformation

$$(2) \qquad z \mapsto \frac{z + \bar{z}}{2} + \frac{z - \bar{z}}{2} \cos\alpha = z\cos^2\frac{\alpha}{2} + \bar{z}\sin^2\frac{\alpha}{2}$$

to a regular triangle with vertices w, εw, and $\varepsilon^2 w$, where $|w| = 1$ and $\varepsilon = \exp(2\pi i/3)$. Then the semiaxes a and b of the ellipse under consideration are $1/2$ and $\cos\alpha/2$, and the distance between its foci F_1 and F_2 equals $\sqrt{a^2 - b^2} = \frac{1}{2}\sin\alpha$. A dilation with coefficient

$$\left(\frac{1}{2}\sin\alpha\right)^{-1} = \left(\sin\frac{\alpha}{2}\cos\frac{\alpha}{2}\right)^{-1}$$

maps the points F_1 and F_2 to the points $(\pm 1, 0)$. The composition of transformation (2) and this dilation is

$$z \mapsto z\cot\frac{\alpha}{2} + \bar{z}\tan\frac{\alpha}{2}.$$

Let $a = w\cot\alpha/2$. Then the polynomial with roots A, B, and C has the form

$$P(x) = \left(x - a - \frac{1}{a}\right)\left(x - a\varepsilon - \frac{1}{a\varepsilon}\right)\left(x - a\varepsilon^2 - \frac{1}{a\varepsilon^2}\right).$$

It is easy to verify that $P'(x) = 3x^2 + 3\varepsilon + 3\bar{\varepsilon} = 3x^2 - 3$; hence the roots of the polynomial P' are ± 1.

4.36. Any hyperbola can be obtained by applying an affine transformation to the hyperbola $y = x^{-1}$, for which the required assertion is obvious.

4.37. Let $ax^2 + bxy + cy^2 = (px + qy)(rx + sy)$. Then the lines $px + qy = 0$ and $rx + sy = 0$ are parallel to the asymptotes of the hyperbola under consideration. These lines are orthogonal if and only if $pr + qs = 0$, i.e., $a + c = 0$.

4.38. Let $a = \alpha + i\alpha^{-1}$, $b = \beta + i\beta^{-1}$, and $c = \gamma + i\gamma^{-1}$ be the vertices of the given triangle on the complex plane. Verify that $h = -\alpha\beta\gamma - i(\alpha\beta\gamma)^{-1}$ is the orthocenter of this triangle. Show, for example, that the number $(a - h)/(b - c)$ is purely imaginary.

4.39. Let φ be the value of the angle formed by the asymptotes that contains the hyperbola. If $\varphi \geqslant 90°$, then the required set is empty, and if $\varphi < 90°$, then it is a circle (centered at the center of the hyperbola) minus the four intersection points with the asymptotes. This can be proved by using the solution to Problem 4.24.

4.40. Suppose that $A = (a, a^{-1})$, $B = (b, b^{-1})$, and $C = (c, c^{-1})$. Then, for the values $x = -x_0, a, b, c$, we have

$$(x_0 - x)^2 + (x_0^{-1} - x^{-1})^2 = 4x_0^2 + 4x_0^{-2}.$$

Thus the numbers $-x_0$, a, b, and c are the roots of a polynomial of the form

$$x^4 - 2x_0x^3 + \cdots .$$

Hence $-x_0 + a + b + c = 2x_0$, i.e., $a + b + c = 3x_0$. Similarly, $a^{-1} + b^{-1} + c^{-1} = 3x_0^{-1}$. Therefore, the point (x_0, x_0^{-1}) serves not only as the center of the circle circumscribed about the triangle ABC, but also as its center of mass. This is only possible if the triangle ABC is equilateral.

4.41. A point X equidistant from a point A and a circle of radius R centered at O must satisfy the relation $|OX| - |AX| = R$ if the point A lies outside the circle and the relation $|OX| + |AX| = R$ if A lies inside the circle. If the point A lies on the circle, then X must lie on the ray OA.

4.42. The center of the circle passing through a point A and tangent to a circle S is equidistant from A and S; thus Problem 4.41 applies.

4.43. Each line A_tB_t is specified by the equation

$$\frac{x - 1 - t}{y - 1 - t} = \frac{1 + t + 1 - t}{1 + t - 1 + t} = \frac{1}{t}, \qquad \text{i.e.,} \quad y = 1 - t^2 + tx = -\left(t - \frac{x}{2}\right)^2 + \frac{x^2}{4} + 1.$$

For a fixed x and t ranging over all real values, y takes all values not exceeding $(x^2/4) + 1$. Thus the required set is described by the inequality $y \leqslant (x^2/4) + 1$.

4.44. Take the coordinate system in which the line l has the equation $x = 0$ and the point O has the coordinates $(1, 0)$. A point $X = (x, y)$ belongs to the required set if and only if the circle with diameter $[O, X]$ intersects the line l. This means that the distance from the center of this circle to l does not exceed the radius, i.e.,

$$\left(\frac{x+1}{2}\right)^2 \leqslant \left(\frac{x-1}{2}\right)^2 + \frac{y^2}{4}.$$

Thus the required set is given by the inequality $y^2 \geqslant 4x$.

4.45. By a similarity transformation we mean a composition of a proper motion and a homothety. Let X_1 and X_2 be two positions of the point X, and let X_1' and X_2' be the positions of the point X' at the same moments of the time. There exists a unique similarity transformation that takes X_1 to X_1' and X_2 to X_2'. At every time moment this transformation takes the point X to the corresponding point X'. Let O be the center of the similarity transformation under consideration, and OH

the altitude of the triangle XOX'. The point H is obtained from X by a similarity transformation; therefore, H moves along some straight line. Taking into account the fact that $XX' \perp OH$, we obtain the same set as in Problem 4.44.

4.46. We can assume that the center of the circle S coincides with the origin. Let the point X have coordinates $(c, 0)$. A point $A = (x, y)$ belongs to the required set if and only if S intersects the circle S_1 with diameter $[A, O]$. Let us denote the radius of S by a, the radius of S_1 by R, and the distance between the centers of S and S_1 by d. The circles S and S_1 intersect if and only if there exists a triangle with sides a, d, and R, i.e.,

$$(R-a)^2 \leqslant d^2 \leqslant (R+a)^2.$$

Taking into account the fact that $4d^2 = (x+c)^2 + y^2$ and $4R^2 = (x-c)^2 + y^2$, we obtain the inequality

$$a^2 - 2ra \leqslant cx \leqslant 2Ra + a^2,$$

which is equivalent to the inequality $(cx - a^2)^2 \leqslant 4a^2R^2$, i.e.,

$$(c^2 - a^2)x^2 - a^2y^2 \leqslant a^2(c^2 - a^2).$$

4.47. We can assume that the equation of the conic has the form

$$A(z^2 + \bar{z}^2) + Bz\bar{z} + Cz + \bar{C}\bar{z} + D = 0. \tag{3}$$

Indeed, both an ellipse and a hyperbola have equations of the form $A(z^2 + \bar{z}^2) + Bz\bar{z} = 1$ ($B < 2A$ corresponds to an ellipse and $B > 2A$ to a hyperbola), and a parabola has the equation

$$z^2 + \bar{z}^2 + 2z\bar{z} + 2iz - 2i\bar{z} = 0.$$

Let u be the center of a regular triangle with vertices $u + v\varepsilon^k$, where $k = 1, 2, 3$ and $\varepsilon = \exp(2\pi i/3)$. If this triangle is inscribed in the conic (3), then the numbers $z_k = u + v\varepsilon^k$ with $k = 1, 2, 3$ satisfy (3). Summing these three equalities, we obtain

$$A(u^2 + \bar{u}^2) + B(u\bar{u} + v\bar{v}) + Cu + \bar{C}\bar{u} + D = 0 \tag{4}$$

(we have used the relation $\varepsilon^1 + \varepsilon^2 + \varepsilon^3 = 0$). Let us substitute the value $z = z_3 = u + v$ in (3) and subtract the obtained relation from (4). The result is $\mathrm{Re}(Fv + A\bar{v}^2) = 0$, where $F = 2Au + B\bar{u} + C$. Performing similar calculations for $z = z_1 = u + v\varepsilon$, we obtain $Fv + A\bar{v}^2 = 0$. Since $v \neq 0$, we have

$$|v|^2 = |2Au + B\bar{u} + C|^2 A^{-2} \tag{5}$$

if $A \neq 0$; the case $A = 0$ corresponds to a circle. Substituting (5) in (4), we obtain the equation of the required conic.

Note that the second conic coincides with the initial one if and only if $B = 0$, which corresponds to an equilateral hyperbola.

4.48. The equation of the secant plane can obviously be written either in the form $z = ax + by + c$ or in the form $ax + by + c = 0$. In the former case, the projection of the section is specified by the equation $x^2 + y^2 = ax + by + c$, and in the latter, by the equation $ax + by + c = 0$.

4.49. Consider the intersection of the cone $ax^2 + by^2 + cz^2 = 0$, where $a \geqslant b \geqslant c$, with the sphere $\lambda(x^2 + y^2 + z^2) = \mu$. The intersection points satisfy the equation

$$(a + \lambda)x^2 + (b + \lambda)y^2 + (c + \lambda)z^2 = \mu.$$

If $\lambda = -b$, this equation specifies a pair of planes, and the section of the cone by the pair of planes then coincides with the section of the sphere by the pair of planes.

4.50. (a) Let $f = 0$ and $g = 0$ be the equations of the quadrics, and let $l = 0$ be the equation of the plane containing their common plane section. In the plane $l = 0$, the equations $f = 0$ and $g = 0$ specify the same curve; therefore, $f = \lambda g + lm$, where m is a linear function. The plane $m = 0$ also contains a common plane section of the quadrics $f = 0$ and $g = 0$.

(b) Let the quadrics under consideration have the equations $f = 0$, $g = 0$, and $h = 0$. According to (a), we can assume that $g = f + lm_1$ and $h = f + lm_2$. In this case, $g = h + l(m_1 - m_2)$. This means that the planes of the other common sections are specified by the equations $m_1 = 0$, $m_2 = 0$, and $m_1 = m_2$. The third equation is a linear combination of the first and the second; therefore, the three planes have a common line.

4.51. (a) A point A of an ellipse centered at O lies on its principal axis if and only if the circle of radius $|OA|$ centered at O is tangent to the ellipse.

The intersection of the given ellipsoid with the sphere of radius $|OA|$ centered at O is the cone

$$\left(\frac{1}{a^2} - \frac{1}{r^2}\right) x^2 + \left(\frac{1}{b^2} - \frac{1}{r^2}\right) y^2 + \left(\frac{1}{c^2} - \frac{1}{r^2}\right) z^2 = 0, \tag{6}$$

where $r = |OA|$. The required secant plane is uniquely determined as the plane tangent to the cone along the ray OA.

(b) Let OA_1 and OA_2 be the principal semiaxes of the ellipse defined as the section of the ellipsoid by the plane $x \cos\alpha + y \cos\beta + z \cos\gamma = 0$. If $A = A_1$ or A_2 (i.e., $r = r_1$ or r_2), this plane must be tangent to the cone (6). It is easy to verify that this condition is equivalent to the equality

$$\frac{a^2 \cos^2\alpha}{a^2 - r^2} + \frac{b^2 \cos^2\beta}{b^2 - r^2} + \frac{c^2 \cos^2\gamma}{c^2 - r^2} = 0.$$

4.52. On the plane Π, consider a circle centered at A. According to the assertion of Problem A14, the polar of this circle with respect to a circle centered at O is a conic with focus O, and any conic with focus O can be obtained in this way. Rotating the plane Π in the space about the axis OA, we see that a similar assertion is true of the quadrics that have axes of rotation.

(a) Consider a sphere centered at A and the family of planes parallel to a fixed line l and tangent to this sphere. The polar of this family of planes with respect to a sphere centered at O is a conic with focus O lying in the plane that passes through O and is perpendicular to l. The polar of the sphere centered at A is a quadric obtained by rotating some plane conic centered at O about the axis OA.

(b) Consider the family of planes passing through a fixed point M and tangent to the sphere centered at A. The perpendiculars from M to these planes form a right circular cone. Therefore, the perpendiculars from the point $O = F_1$ to these planes also form a right circular cone K.

The polar transformation with respect to the sphere centered at O maps the family of planes under consideration to a plane section C of an ellipsoid (the image of the sphere centered at A under the polar transformation with respect to the sphere centered at O). Clearly, C is the intersection of the cone K (having vertex O) with this ellipsoid.

Chapter 5

5.1. The map under consideration has the following properties:

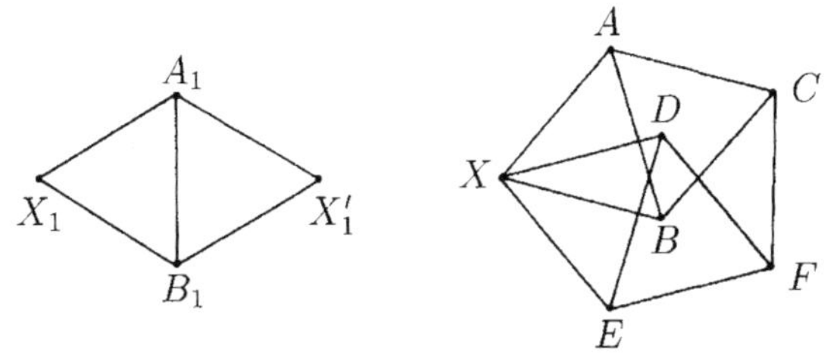

(a) (b)

FIGURE S.2

(i) it leaves all points of the hyperplane $(x, a) = 0$ fixed;
(ii) it preserves the scalar product (x, y), i.e., it is an isometry;
(iii) it takes a to $-a$, i.e., it is not the identity map.

5.2. Let $e_1, \dots, e_n$ be an orthonormal basis. Consider the vectors $\varepsilon_i = Ae_i$. By assumption, these vectors are pairwise orthogonal. Let us show that they have equal lengths. Note that $e_i+e_j \perp e_i-e_j$. Thus, by assumption, we have $(\varepsilon_i+\varepsilon_j, \varepsilon_i-\varepsilon_j) = 0$, i.e., $|\varepsilon_i| = |\varepsilon_j|$. Therefore, $|\varepsilon_1| = \dots = |\varepsilon_n| = a$. The transformation $a^{-1}A$ maps an orthonormal basis to an orthonormal basis; hence it is an isometry.

5.3. The transformation under consideration is a rotation through some angle φ about some axis l. Take the plane perpendicular to l and choose two lines l_1 and l_2 in this plane that intersect l and make the angle $\varphi/2$ with each other. The composition of the symmetries about the lines l_1 and l_2 coincides with the initial transformation.

5.4. The orthogonal complement V of the fixed point set has dimension $n-s$. The restriction of f to V has one fixed point (the origin); therefore, it can be represented as the composition of $n-s$ symmetries about hyperplanes in V. Consider the symmetries in $\mathbb{R}^n$ about the hyperplanes in $\mathbb{R}^n$ that contain those hyperplanes and the fixed point set. Their composition coincides with the transformation f.

Now suppose that the isometry f is represented as the composition of symmetries with respect to hyperplanes $H_1, \dots, H_k$. Let W be the intersection of these hyperplanes. On one hand, W lies in the fixed point set of f, and hence $\dim W \leqslant s$. On the other hand, $\dim W \geqslant n-k$. Therefore, $n-k \leqslant s$, i.e., $k \geqslant n-s$.

5.5. Suppose that the image $f(X)$ of some point X contains two points X_1 and X_1'. Consider a regular triangle XAB with side 1. Take $A_1 \in f(A)$ and $B_1 \in f(B)$; the triangles $X_1A_1B_1$ and $X_1'A_1B_1$ are then regular and have side 1 (Figure S.2 (a)), so that $|X_1X_1'| = \sqrt{3}$.

Next, take points A, B, C, D, E, and F such that the eleven segments shown in Figure S.2 (b) are all of length 1. In the images of these points, choose points $A_1, \dots, F_1$, one in each image.

For fixed points X_1 and X_1', the condition $|A_1X_1| = |A_1X_1'| = 1$ determines two possible positions of the point A_1. The possible positions of the point B_1 are the same, but $A_1 \neq B_1$, because $|A_1B_1| = 1$. Therefore, the point C_1 coincides with X_1 or with X_1'. For the points D_1 and E_1, the possible positions are the same and $D_1 \neq E_1$ for the same reason as for the points A_1 and B_1. Therefore, F_1 coincides with X_1 or with X_1'. As the result, we see that either $|C_1F_1| = 0$ or $|C_1F_1| = \sqrt{3}$. But, by assumption, the equality $|C_1F_1| = 1$ must hold. We have obtained a contradiction.

5.6. Let f be the rotation through the angle $\sqrt{2}\pi$ about some point O. Take an arbitrary point $A_0 \neq O$ and set $A_i = f(A_{i-1})$ for $i = 1, 2, \dots$. Then $A_i \neq A_j$ for $i \neq j$, because the number $i\sqrt{2} - j\sqrt{2}$ is not an integer. For $A = \{A_0, A_1, A_2, \dots\}$, we have $f(A) = \{A_1, A_2, A_3, \dots\} \subset A$ and $f(A) \neq A$.

5.7. On the complex plane $\mathbb{C}$, consider the set A consisting of all points of the form $a_0 + a_1e^i + a_2e^{2i} + \dots + a_ke^{ki}$, where $k \geqslant 0$ and $a_0, \dots, a_k$ are nonnegative integers. Let B and C be the subsets of A for which $a_0 = 0$ and $a_0 \neq 0$, respectively. Then $A = B \cup C$, $B = e^iA$ (i.e., B is obtained by rotating A through an angle of 1 radian), and $C = A + 1$ (i.e., C is obtained by translating A). It remains to verify that $B \cap C = \varnothing$.

Suppose that

$$b_1e^i + b_2e^{2i} + \dots + b_me^{mi} = c_0 + c_1e^i + \dots + c_ne^{ni},$$

where $b_i, c_j \in \mathbb{Z}$ and $c_0 \neq 0$. Then $c_0 + a_1e^i + \dots + a_re^{ri} = 0$, where $a_i \in \mathbb{Z}$ and $c_0 \neq 0$. This contradicts the transcendence of e^i (see, e.g., [**Ge**]).

5.8. The dual triangle has angles $\pi - a$, $\pi - b$, $\pi - c$. The sum of the angles of a spherical triangle is larger than π; therefore, $3\pi - (a+b+c) > \pi$, i.e., $a+b+c < 2\pi$.

5.9. The angles of a self-polar triangle equal $\pi/2$.

5.10. Consider a sphere centered at O. The required condition holds for the projection from O of the northern hemisphere to the plane tangent to the sphere at the north pole.

5.11. The proof is the same as in Euclidean geometry: if a point O is equidistant from points A and B (lines a and b) and from points B and C (lines b and c), then it is equidistant from the points A and C (the lines a and c) as well.

5.12. (a) Let O be the center of the sphere. The medians of a spherical triangle ABC intersect at a point lying on the ray OM determined by the vector $\overrightarrow{OA} + \overrightarrow{OB} + \overrightarrow{OC}$.

(b) The altitudes of a spherical triangle ABC intersect at a point lying on the ray OH determined by the vector

$$w = \overrightarrow{OA}\tan\alpha + \overrightarrow{OB}\tan\beta + \overrightarrow{OC}\tan\gamma.$$

Let us verify, for example, that the planes OAH and OBC are perpendicular, i.e., the scalar product of the vectors $\overrightarrow{OA} \times w = \overrightarrow{OA} \times \overrightarrow{OB}\tan\beta + \overrightarrow{OA} \times \overrightarrow{OC}\tan\gamma$ and $\overrightarrow{OB} \times \overrightarrow{OC}$ is zero. The lengths of the vectors $\overrightarrow{OA} \times \overrightarrow{OB}$ and $\overrightarrow{OB} \times \overrightarrow{OC}$ are equal to $\sin c$ and $\sin a$, and the angle between them is β. Thus we must verify the equality

$$\sin c \sin a \cos\beta \tan\beta - \sin b \sin a \cos\gamma \tan\gamma = 0,$$

but it follows from the law of sines.

5.13. The length of a spherical circle of radius r is $2\pi \sin r$. Therefore, the area of a spherical disk of radius r is

$$\int_0^r 2\pi \sin x\, dx = 2\pi(1 - \cos r) = 4\pi \sin^2(r/2).$$

5.14. According to the first law of cosines,

$$\cos\alpha = \frac{-\cos b \cos c + \cos a}{\sin b \sin c}.$$

Therefore,

$$2\cos^2\frac{\alpha}{2} = 1+\cos\alpha = \frac{-\cos(b+c)+\cos a}{\sin b\sin c} = \frac{2\sin\frac{a+b+c}{2}\sin\frac{b+c-a}{2}}{\sin b\sin c},$$

$$2\sin^2\frac{\alpha}{2} = 1-\cos\alpha = \frac{\cos(b-c)-\cos a}{\sin b\sin c} = \frac{2\sin\frac{b-c+a}{2}\sin\frac{a-b+c}{2}}{\sin b\sin c}.$$

5.15. According to the preceding problem,

$$\cos\frac{\alpha}{2}\cos\frac{\beta}{2} = \frac{\sin p}{\sin c}\sin\frac{\gamma}{2} \quad\text{and}\quad \sin\frac{\alpha}{2}\sin\frac{\beta}{2} = \frac{\sin(p-c)}{\sin c}\sin\frac{\gamma}{2}.$$

Therefore,

$$\cos\frac{\alpha+\beta}{2} = \frac{\sin p-\sin(p-c)}{\sin c}\sin\frac{\gamma}{2} = \frac{\cos\frac{a+b}{2}}{\cos\frac{c}{2}}\sin\frac{\gamma}{2}.$$

Similarly,

$$\sin\frac{\alpha}{2}\cos\frac{\beta}{2} = \frac{\sin(p-b)}{\cos c}\frac{\gamma}{2}, \quad \sin\frac{\beta}{2}\cos\frac{\alpha}{2} = \frac{\sin(p-a)}{\sin c}\cos\frac{\gamma}{2},$$

$$\sin\frac{\alpha+\beta}{2} = \frac{\sin(p-a)+\sin(p-b)}{\sin c}\cos\frac{\gamma}{2} = \frac{\cos\frac{a-b}{2}}{\cos\frac{c}{2}}\cos\frac{\gamma}{2}.$$

5.16. (a) Using the two preceding problems, we obtain

$$\cos\frac{\alpha+\beta+\gamma}{2} = \cos\frac{\alpha+\beta}{2}\cos\frac{\gamma}{2} - \sin\frac{\alpha+\beta}{2}\sin\frac{\gamma}{2} = \frac{\cos\frac{a+b}{2}-\cos\frac{a-b}{2}}{\cos\frac{c}{2}}\cos\frac{\gamma}{2}\sin\frac{\gamma}{2}$$

$$= -\frac{2\sin\frac{a}{2}\sin\frac{b}{2}}{\cos\frac{c}{2}}\frac{\sqrt{\sin p\sin(p-a)\sin(p-b)\sin(p-c)}}{\sin a\sin b}.$$

(b) The relations

$$\frac{\sin\frac{\alpha+\beta}{2}}{\cos\frac{\gamma}{2}} = \frac{\cos\frac{a-b}{2}}{\cos\frac{c}{2}} \quad\text{and}\quad \frac{\cos\frac{\alpha+\beta}{2}}{\sin\frac{\gamma}{2}} = \frac{\cos\frac{a+b}{2}}{\cos\frac{c}{2}}$$

imply

$$\frac{\sin\frac{\alpha+\beta}{2}-\cos\frac{\gamma}{2}}{\sin\frac{\alpha+\beta}{2}+\cos\frac{\gamma}{2}} = \frac{\cos\frac{a-b}{2}-\cos\frac{c}{2}}{\cos\frac{a-b}{2}+\cos\frac{c}{2}} \quad\text{and}\quad \frac{\cos\frac{\alpha+\beta}{2}-\sin\frac{\gamma}{2}}{\cos\frac{\alpha+\beta}{2}+\cos\frac{\gamma}{2}} = \frac{\cos\frac{a+b}{2}-\cos\frac{c}{2}}{\cos\frac{a+b}{2}+\cos\frac{c}{2}}.$$

To obtain the required equality, it suffices to multiply these two equalities and transform the sums and differences of sines and cosines in the numerators and denominators into products:

$$\sin\frac{\alpha+\beta}{2} - \cos\frac{\gamma}{2} = 2\cos\frac{\pi-\gamma+\alpha+\beta}{4}\sin\frac{\alpha+\beta+\gamma-\pi}{4},$$

$$\cos\frac{a-b}{2} - \cos\frac{c}{2} = 2\sin\frac{a-b+c}{4}\sin\frac{b+c-a}{4}, \qquad \text{etc.}$$

5.17. The first law of cosines implies $\cos c = \cos a\cos b$, and the second law of cosines implies $\cos\alpha = \sin\beta\cos a$ and $\cos\beta = \sin\alpha\cos b$.

According to the law of sines, $\sin\alpha\sin b = \sin\beta\sin a$; therefore,

$$\sin\alpha\sin b\cdot\cos a = (\sin\beta\cos a)\sin a = \cos\alpha\sin a, \qquad \text{i.e.,} \quad \tan\alpha\sin b = \tan a.$$

According to the law of sines, $\sin a = \sin\alpha\sin c$; therefore,

$$\sin a\cos c = \sin\alpha\sin c\cos c = \sin\alpha\sin c\cos a\cos b = \sin c\cos a\cos\beta,$$

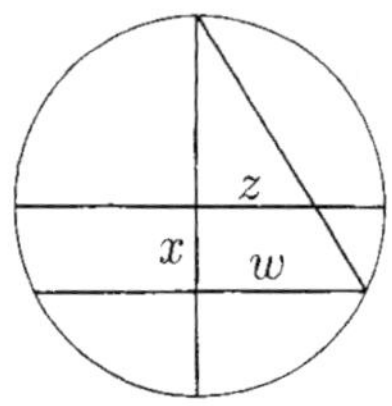

FIGURE S.3

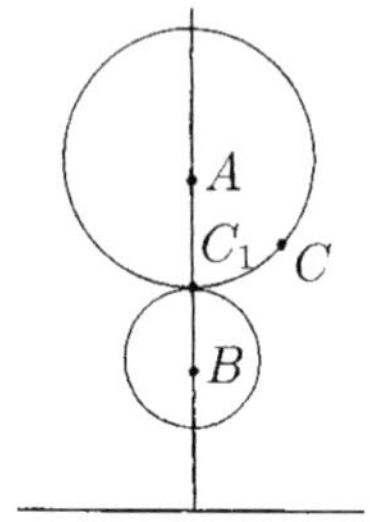

FIGURE S.4

i.e.,

$$\tan a = \tan c \cos \beta.$$

5.18. Let I be the center of the circle inscribed in a spherical triangle ABC; suppose that the inscribed circle is tangent to the side AB at a point K. Then $|AK| = p - a$, $|IK| = r$, $\angle IAK = \alpha/2$, and $\angle AKI = \pi/2$. Therefore, according to the assertion of Problem 5.17,

$$\tan r = \tan(\alpha/2) \sin(p - a).$$

It remains to note that by the assertion of Problem 5.14, we have

$$\tan \frac{\alpha}{2} = \sqrt{\frac{\sin(p-b)\sin(p-c)}{\sin p \sin(p-a)}}.$$

5.19. Let O be the center of the circle circumscribed about a spherical triangle ABC, and M the midpoint of the side AB. Then $\angle OBM = (\beta + \gamma - \alpha)/2 = \pi/2 - (\alpha - s)$. Therefore, according to the assertion of Problem 5.17, $\tan(a/2) = \tan R \sin(\alpha - s)$. It remains to note that

$$\cot \frac{a}{2} = \sqrt{\frac{\sin(\beta - s)\sin(\gamma - s)}{\sin s \sin(\alpha - s)}};$$

this formula in obtained from the formula for $\tan(\alpha/2)$ by passing to the polar triangle.

5.20. (a) The polars of the spherical lines through a given point A sweep out a great disk, namely, the polar of the point A. Hence, the polars under consideration sweep out the pair of spherical digons formed by the polars of the endpoints of the given segment.

(b) The polars of the spherical lines intersecting the given segments sweep out a set of area smaller than 4π; therefore, the sphere contains a point not lying in this set. The polar of this point intersects none of the given segments.

5.21. It is sufficient to prove that $w = 2z/(1+z^2)$ (Figure S.3). Clearly, $x^2+w^2 = 1$ and $x = (w - z)/z$. Therefore,

$$(w-z)^2 + z^2w^2 = z^2, \qquad \text{i.e.,} \quad w = \frac{2z}{1+z^2},$$

as required.

5.22. Let AB be the maximal side of a hyperbolic triangle ABC. It is sufficient to prove that $|AB| < |AC| + |BC|$. Suppose that the points A and B lie in the upper half-plane on a vertical ray. Take a point C_1 on the Euclidean segment $[A, B]$

such that $|AC_1| = |AC|$ (the distances are hyperbolic). Consider the hyperbolic circles of radii $|AC_1|$ and $|BC_1|$ centered at A and B, respectively (Figure S.4). The hyperbolic circles in the Poincaré model are Euclidean circles; therefore, the circles under consideration are tangent at the point C_1. In particular, the point C lies outside the second circle, i.e., $|BC| > |BC_1|$.

5.23. Let d be the hyperbolic distance from the point (x, y) to a point $(0, t)$. Then

$$\cosh \alpha = 1 + \frac{x^2 + (t-y)^2}{2ty} = \frac{1}{2y}\left(\frac{x^2+y^2}{t} + t\right).$$

The function $(x^2 + y^2)t^{-1} + t$ attains minimum at $t = \sqrt{x^2+y^2}$; this value of t corresponds to the point through which the perpendicular from (x, y) to the hyperbolic line under consideration passes. The minimum distance equals $\cosh^{-1}(\sqrt{x^2+y^2}/y)$, where $\cosh^{-1}$ is the inverse function to cosh.

5.25. To each point z of the unit disk, we assign the point $w = i(1+z)/(1-z)$ of the upper half-plane. We have $z = (w-i)/(w+i)$. A proper motion in the coordinates w is given by $w \mapsto (pw+q)/(rw+s)$, where $ps - qr = 1$. In the coordinates z, this transformation has the form

$$z \mapsto \frac{(ip + is - q + r)z + ip - is + q + r}{(ip - is - q - r)z + ip + is + q - r}.$$

Multiplying the numerator and denominator of this fraction by $i/2$, we obtain an expression of the required form.

5.26. In the Klein model, the assertion reduces to the corresponding assertion of Euclidean geometry.

5.27. Clearly,

$$\cosh\frac{a}{c} = \frac{1}{2}(e^{a/c} + e^{-a/c}) = \frac{1}{2}\left(\tan\frac{\alpha}{2} + \cot\frac{\alpha}{2}\right) = \frac{1}{\sin\alpha},$$
$$\sinh\frac{a}{c} = \frac{1}{2}(e^{a/c} - e^{-a/c}) = \frac{1}{2}\left(\cot\frac{\alpha}{2} - \tan\frac{\alpha}{2}\right) = \cot\alpha.$$

The third relation follows from these two.

If $0 < \alpha < \pi/2$, then each relation under consideration gives a unique expression for a; therefore, they are all equivalent.

5.28. Consider the Poincaré model in the upper half-plane. We can assume that the given line coincides with the imaginary axis and the given point lies on the unit circle (this can be achieved by applying a homothety with center at the origin). For the point $\cos\alpha + i\sin\alpha$, the angle of parallelism equals α, and the distance a from this point to the given line is evaluated by the formula

$$\cosh\frac{a}{c} = 1 + \frac{\cos^2\alpha + (1-\sin\alpha)^2}{2\sin\alpha} = \frac{1}{\sin\alpha}.$$

This relation between a and α is equivalent to the required one.

5.29. The required relations follow from the relations

$$\cosh c = \cosh a \cosh b, \qquad \sinh a = \sinh b \tan\alpha, \qquad \tanh b = \sinh a \tan\beta$$

proved above and from the identity $\cosh^2 x - \sinh^2 x = 1$ (and from the trigonometric identities). Say, $\sinh a = \sinh c \sin\alpha$ is proved by taking the square of both

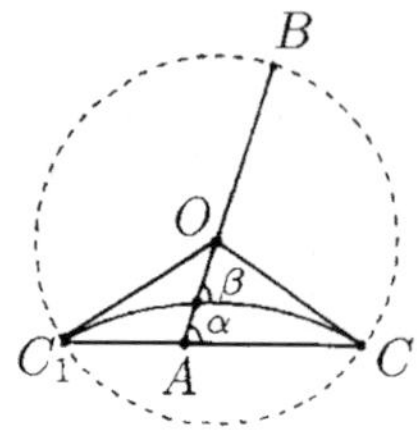

FIGURE S.5

sides of the equality $\sinh a = \sinh b \tan \alpha$ and applying the relations

$$\sinh^2 b = \cosh^2 b - 1 = \frac{\cosh^2 c - \cosh^2 a}{\cosh^2 a} = \frac{\sinh^2 c - \sinh^2 a}{\cosh^2 a}.$$

5.30. (a) Clearly,

$$\begin{aligned}\cosh a &= \cosh b \cosh c - \sinh b \sinh c \cos \alpha \\ &= \cosh b(\cosh a \cosh b - \sinh a \sinh b \cos \gamma) - \sinh b \sinh a \cos \alpha \sin \gamma / \sin \alpha,\end{aligned}$$

i.e.,

$$\cosh a(1 - \cosh^2 b) = -\cosh b \sinh a \sinh b \cos \gamma - \sinh b \sinh a \sin \gamma \cot \alpha.$$

Taking into account that $1 - \cosh^2 b = -\sinh^2 a$, we obtain

$$\cosh a = \frac{\sinh a}{\sinh b}(\cosh b \cos \gamma + \sin \gamma \cot \alpha).$$

It remains to note that $\sinh a / \sinh b = \sin \alpha / \sin \beta$.

(b) Applying (a) twice, we obtain

$$\cosh a \sin \beta = (\cosh a \sin \beta \cos \gamma + \cos \beta \sin \gamma) \cos \gamma + \cos \alpha \sin \beta, \quad \text{i.e.,}$$

$$\cosh a \sin \beta \sin^2 \gamma = \cos \beta \sin \gamma \cos \gamma + \cos \alpha \sin \gamma.$$

It remains to reduce both sides by $\sin \gamma$.

5.31. Suppose that the side AB of an angle BAC passes through the center O of the disk δ. Consider the line CC_1 in the Klein model and the line CC_1 in the Poincaré disk model (Figure S.5). The required assertion reduces to the obvious relation $\beta < \alpha$.

5.32. Use the Klein model.

5.33. The assertion is obviously implied by the formula of Problem 5.30 (b). But it can also be proved directly.

Consider two hyperbolic triangles ABC and $A_1B_1C_1$ with equal respective angles. Let us show that under a superposition of BAC and $B_1A_1C_1$, B coincides with B_1 and C with C_1. Indeed, if $B \neq B_1$ or $C \neq C_1$, then the segments $[B, C]$ and $[B_1, C_1]$ either intersect at some point M or they are disjoint. In the former case, we can assume that the point M is different from B. Then MBB_1 is a hyperbolic triangle whose sum of angles is π. In the latter case, BCC_1B_1 is a hyperbolic quadrilateral whose sum of angles is 2π. Neither can be true.

5.34. We assume that one side of the angle is the imaginary axis Oy in the Poincaré upper half-plane model. To draw a perpendicular from a point A to this line, we must draw the circle of radius $|OA|$ centered at O. The other side of the angle is an arc of a Euclidean circle, and its hyperbolic projection to the Oy axis is a bounded set.

5.35. (a), (b) Suppose that O is the center of a regular n-gon, AB is one of its sides, and M is the midpoint of this side. Then AOM is a right triangle and has angles $\alpha/2$ and π/n and legs $a/2$ and r. The required relation follows from the last relation in Problem 5.29.

A right triangle AOM with the angles specified above exists if and only if we have the inequalities $\cosh r > 1$ and $\cosh(a/2) > 1$. Both these inequalities are equivalent to $\alpha < (n-2)\pi/n$.

(c) For $\alpha = 2\pi/n$, we have $\cosh r = \cosh(a/2) = \cot(\pi/n)$. The inequality $\cot(\pi/n) > 1$ holds for $n > 4$.

5.36. Let us cut off a right triangle with legs a and b from the quadrilateral. It has angles α and β and hypotenuse c. The remaining triangle has angles $\alpha_1 = (\pi/2) - \alpha$, $\beta_1 = (\pi/2) - \beta$, and γ. According to the assertion of Problem 5.30 (b),

$$\cos\gamma = \sin\alpha_1 \sin\beta_1 \cosh c - \cos\alpha_1 \cos\beta_1 = \cos\alpha\cos\beta\cosh c - \sin\alpha\sin\beta.$$

Since $\cos\alpha\cos\beta = \cosh a \cosh b \sin\alpha\sin\beta = \cosh c \sin\alpha\sin\beta$, we have

$$\cos\gamma = ((\cosh c)^2 - 1)\sin\alpha\sin\beta = (\sinh c)^2 \sin\alpha\sin\beta = \sinh a \sinh b.$$

5.37. The area of the triangle under consideration equals

$$\int_{-1}^{1}\left(\int_{\sqrt{1-x^2}}^{\infty}\frac{dy}{y^2}\right)dx = \int_{-1}^{1}\frac{dx}{\sqrt{1-x^2}} = \pi.$$

5.38. If $w = (z-i)/(z+i)$, then $z = i(1+w)/(1-w)$. Therefore,

$$\operatorname{Im} z = \frac{1-|w|^2}{|1-w|^2} \quad\text{and}\quad |dz| = \frac{2|w|}{|1-w|^2}.$$

5.39. Let $x + iy$ be a point in the Klein model. According to the assertion of Problem 5.21, the corresponding point in the Poincaré disk model is $w = u + iv$, where

$$x = \frac{2u}{u^2+v^2+1} \quad\text{and}\quad y = \frac{2v}{u^2+v^2+1}.$$

Simple calculations show that

$$dx^2 + dy^2 - (x\,dy - y\,dx)^2 = \frac{4(1-u^2-v^2)^2(du^2+dv^2)}{(u^2+v^2+1)^4},$$

$$(1-x^2-y^2)^2 = \left(\frac{1-u^2-v^2}{u^2+v^2+1}\right)^4.$$

Therefore,

$$\frac{dx^2+dy^2-(x\,dy-y\,dx)^2}{(1-x^2-y^2)^2} = \frac{4(du^2+dv^2)}{(1-u^2-v^2)^2} = \left(\frac{2|dw|}{1-|w|^2}\right)^2.$$

According to the assertion of Problem 5.38, the last expression coincides with ds^2.

5.40. (a) In the Poincaré upper half-plane model, the side lengths of the infinitesimal rectangle under consideration are dx/y and dy/y.

In the Poincaré unit disk model, the side lengths of the infinitesimal rectangle under consideration are

$$\frac{2dx}{1-(x^2+y^2)} \quad\text{and}\quad \frac{2dy}{1-(x^2+y^2)}.$$

The area of the infinitesimal rectangle equals the product of its side lengths.

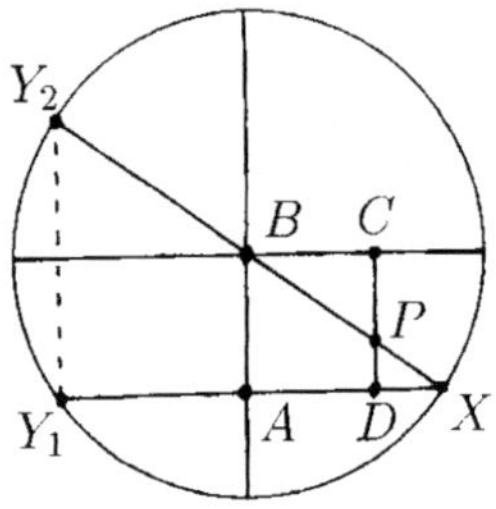

FIGURE S.6

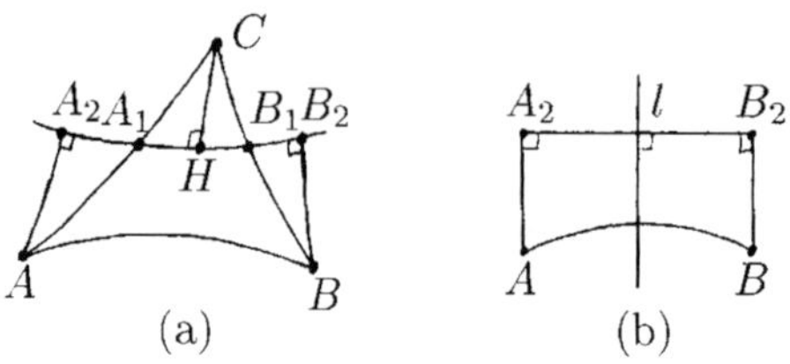

FIGURE S.7

(b) The length of the side that lies on the line $\varphi = \text{const}$ is equal to

$$\frac{1}{2}\ln\left(\frac{1-\rho}{1+\rho}\cdot\frac{1+\rho+d\rho}{1-\rho-d\rho}\right) = \frac{1}{2}\ln\left(1+\frac{d\rho}{1+\rho}\right)\left(1+\frac{d\rho}{1-\rho-d\rho}\right)$$
$$\approx \frac{1}{2}\ln\left(1+\frac{2d\rho}{1-\rho^2}\right) \approx \frac{d\rho}{1-\rho^2}.$$

The length of the side lying on the line $\rho = \text{const}$ equals

$$\frac{1}{2}\ln\left(\frac{\sqrt{1-\rho^2}+\rho d\varphi}{\sqrt{1-\rho^2}-\rho d\varphi}\right) \approx \frac{1}{2}\ln\left(1+\frac{2\rho d\varphi}{\sqrt{1-\rho^2}}\right) \approx \frac{\rho d\varphi}{\sqrt{1-\rho^2}}.$$

5.41. The circumference of a hyperbolic circle of radius r equals $2\pi\sinh r$; therefore, the area of a hyperbolic disk of radius r is

$$\int_0^r 2\pi\sinh t\,dt = 2\pi\cosh t\,\Big|_0^r = 2\pi(\cosh r - 1) = 4\pi\sinh^2\left(\frac{r}{2}\right).$$

5.42. Consider a hyperbolic triangle with side lengths $2a$, $2b$, and $2c$. Let a' be the length of the segment joining the midpoints of its two sides. We must prove that $a' < a$, i.e., $\cosh 2a' < \cosh 2a$. According to the law of cosines,

$$\cosh a' = \cosh b\cosh c - \sinh b\sinh c\cos\alpha,$$
$$\cosh 2a = \cosh 2b\cosh 2c - \sinh 2b\sinh 2c\cos\alpha.$$

We put $\cosh b = x_1$, $\sinh b = x_2$, $\cosh c = y_1$, $\sinh b = y_2$, and $\cos\alpha = p$; then the required inequality takes the form

$$2(x_1y_1 - x_2y_2p)^2 - 1 < (2x_1^2-1)(2y_1y^2-1) - 4x_1y_1x_2y_2p.$$

This inequality is equivalent to the inequality

$$x_2^2y_2^2p^2 < (x_1^2-1)(y_1^2-1).$$

The last inequality is obvious, because $x_1^2-1 = x_2^2$, $y_1^2-1 = y_2^2$, and $p^2 = \cos^2\alpha < 1$.

5.43. Consider the Klein model. If we place the vertex B in the center of the disk, then the Euclidean quadrilateral $ABCD$ will be a rectangle (Figure S.6). Let the ray AD intersect the circle at point X, and let P be the intersection point of the lines BX and CD. It suffices to prove that the segments $[A, D]$ and $[B, P]$ have equal hyperbolic lengths. But this obviously follows from the equality of the cross ratios $[A, D, X, Y_1]$ and $[B, P, X, Y_2]$, because $Y_1Y_2 \parallel AB \parallel CD$.

5.44. (a) Suppose that A_1 and B_1 are the midpoints of the sides AC and BC, respectively (Figure S.7 (a)). Let us draw a perpendicular $[C, H]$ from the vertex C to the line A_1B_1 and consider the points A_2 and B_2 symmetric to the point H with respect to A_1 and B_1, respectively. It is easy to verify that $A_2 \neq B_2$. If

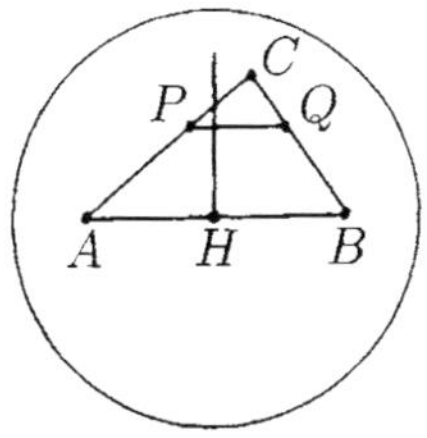

FIGURE S.8

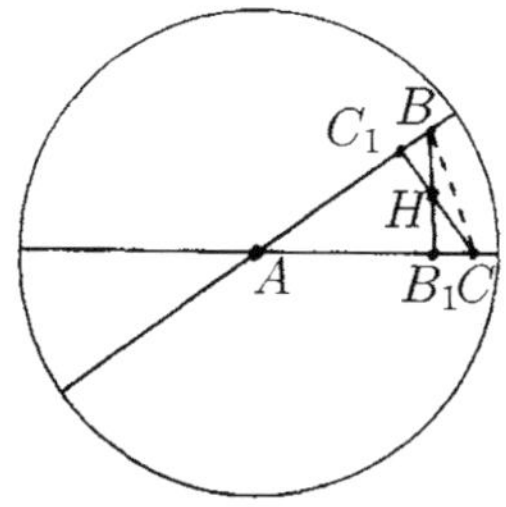

FIGURE S.9

H lies on the segment $[A_1, B_1]$, this is obvious, and if H lies on the continuation of $[A_1, B_1]$, then the equality $|A_1H| = |B_1H|$ cannot hold and so neither can the equality $|A_2H| = |B_2H|$. Thus the points A_2 and B_2 uniquely determine the line A_1B_1; therefore, it suffices to prove that the perpendicular l under consideration, which is drawn from the midpoint of the segment $[A, B]$, is perpendicular to the line A_2B_2.

The triangles A_1CH and A_1AA_2 (as well as the triangles B_1CH and B_1BB_2) are congruent right triangles; therefore, in the quadrilateral AA_2B_2B, the angles at the vertices A_2 and B_2 are $\pi/2$ and the sides AA_2 and BB_2 are equal (they are both equal to CH). Such a quadrilateral must have an axis of symmetry (Figure S.7 (b)), which must coincide with the line l. Therefore, $l \perp A_2B_2$.

(b) The case of coinciding lines is obvious; so, we assume that the lines that are isometrically mapped onto each other do not coincide. There are two possibilities.

(i) Two different segments $[X, X']$ and $[Y, Y']$ have a common point M. Then the isometry under consideration is the symmetry with respect to M.

(ii) The midpoints of all segments of the form $[X, X']$ are pairwise distinct. Let P, Q, R be the midpoints of the segments $[X, X']$, $[Y, Y']$, $[Z, Z']$. It is sufficient to show that the points P, Q, R are collinear. The symmetry about P maps the points X, Y, Z to collinear points $X'' = X'$, Y'', Z'' such that the line PX' passes through the midpoints of and is perpendicular to the segments $[Y', Y'']$ and $[Z', Z'']$. Therefore, according to (a), the line PX' is perpendicular to the lines PQ and PR. Hence P, Q, R are collinear.

5.45. (a) The intersection point of two bisectors of a triangle is equidistant from all sides of the triangle; therefore, it lies on the third bisector.

(b) Let H, P, and Q be the midpoints of the sides AB, AC, and BC of a triangle. Consider the Klein model. Let us place the point H in the center of the disk (Figure S.8). According to assertion (a) of Problem 5.44, the perpendicular to the segment $[A, B]$ through its midpoint is perpendicular to the line PQ. This means that in the Euclidean triangle ABC, the segment $[P, Q]$ is parallel to the side AB. Hence the segments $[A, Q]$ and $[B, P]$ intersect at a point K lying on the segment $[C, H]$. Indeed, there is a (Euclidean) homothety with center at K that maps the segment $[P, Q]$ onto the segment $[B, A]$; therefore, the line joining the midpoints of these segments passes through K.

(c) Consider the Klein model. If we place the point A in the center of the disk (Figure S.9), then the hyperbolic altitudes BB_1 and CC_1 coincide with the Euclidean altitudes. They intersect at some point H. The line AH is a Euclidean perpendicular to the line BC. But $[A, H]$ is a diameter of the disk; therefore, AH is also a hyperbolic perpendicular to the line BC.

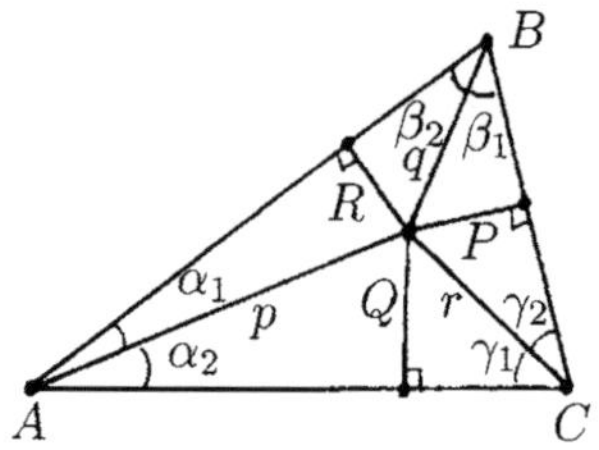

FIGURE S.10

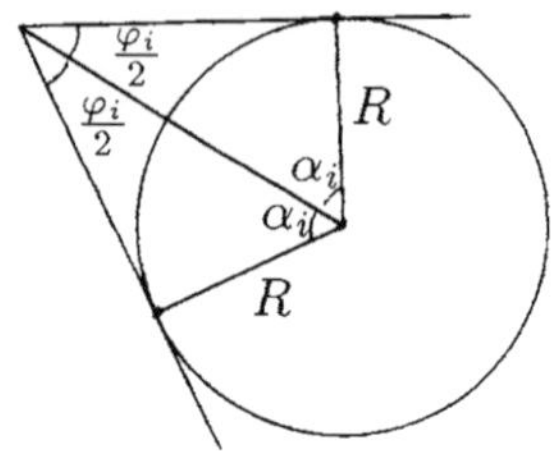

FIGURE S.11

5.46. Let us draw perpendiculars AA_2, BB_2, and CC_2 from the vertices of the triangle to the line l. Since $\angle AC_1A_2 = \angle BC_1B_2 = \varphi$, the law of sines implies

$$\sinh AA_2/\sinh AC_1 = \sin\varphi = \sinh BB_2/\sinh C_1B;$$

therefore,

$$\frac{\sinh|AC_1|}{\sinh|C_1B|} = \frac{\sinh|AA_2|}{\sinh|BB_2|}.$$

Similarly,

$$\frac{\sinh|BA_1|}{\sinh|A_1C|} = \frac{\sinh|BB_2|}{\sinh|CC_2|}, \qquad \frac{\sinh|CB_1|}{\sinh|B_1A|} = \frac{\sinh|CC_2|}{\sinh|AA_2|}.$$

It remains to multiply the obtained inequalities.

5.47. It is sufficient to prove that if the specified intervals intersect at some point M, then the required relations hold. The converse assertion follows from the uniqueness of a point dividing a segment in a given ratio.

(a) We use the notation of Figure S.10. According to the law of sines, $\sinh R = \sinh p \sin\alpha_1$ and $\sinh Q = \sinh p \sin\alpha_2$. Therefore,

$$\frac{\sin\alpha_1}{\sin\alpha_2} = \frac{\sinh R}{\sinh Q}.$$

Similarly,

$$\frac{\sin\beta_1}{\sin\beta_2} = \frac{\sinh P}{\sinh R} \quad\text{and}\quad \frac{\sin\gamma_1}{\sin\gamma_2} = \frac{\sinh Q}{\sinh P}.$$

(b) Applying the law of sines to the triangles BAA_1 and CAA_1, we obtain

$$\frac{\sinh|BA_1|}{\sinh|AB|} = \frac{\sin BAA_1}{\sin BA_1A}, \qquad \frac{\sinh|A_1C|}{\sinh|AC|} = \frac{\sin A_1AC}{\sin AA_1C}.$$

But $\sin BA_1A = \sin AA_1C$; therefore,

$$\frac{\sinh|BA_1|}{\sinh|A_1C|} = \frac{\sinh|AB|}{\sinh|AC|}\cdot\frac{\sin BAA_1}{\sin A_1AC}.$$

Now, it is easy to understand that relations (a) and (b) are equivalent.

5.48. We prove an even stronger statement, namely, the existence of an n-gon with angles $\varphi_1, \dots, \varphi_n$ circumscribed about a circle. The n-gon can be composed of right triangles with angles $\varphi_i/2$ and α_i and fixed leg R (Figure S.11). Such triangles can form an n-gon if and only if $\alpha_1 + \cdots + \alpha_n = \pi$. According to the last relation in Problem 5.29, we have $\sin\alpha_i = \cos(\varphi_i/2)/\cosh R$, i.e.,

$$\alpha_i = \arcsin(\cos(\varphi_i/2)/\cosh R).$$

Thus $\sum \alpha_i \to 0$ as $R \to \infty$, and

$$\sum \alpha_i = \sum \frac{\pi - \varphi_i}{2} = \frac{1}{2}(\pi n - \varphi_1 - \cdots - \varphi_n) > \pi$$

if $R = 0$. Therefore, there exists a number R such that $\sum \alpha_i = \pi$.

5.49. By the law of cosines,

$$\cos\alpha = \frac{\cosh b \cosh c - \cosh a}{\sinh b \sinh c};$$

therefore,

$$2\cos^2\frac{\alpha}{2} = 1 + \cos\alpha = \frac{\cosh(b+c) - \cosh a}{\sinh b \sinh c} = \frac{2\sinh p \sinh(p-a)}{\sinh b \sinh c},$$

$$2\sin^2\frac{\alpha}{2} = 1 - \cos\alpha = \frac{\cosh a - \cosh(b-c)}{\sinh b \sinh c} = \frac{2\sinh(p-b)\sinh(p-c)}{\sinh b \sinh c}.$$

5.50. (a) Using the formulas proved in the solution to Problem 5.49, we can easily verify that

$$\sin\frac{\alpha+\beta}{2} = \frac{\sinh(p-a) + \sinh(p-b)}{\sinh c}\cos\frac{\gamma}{2} = \frac{\cosh\frac{a-b}{2}}{\cosh\frac{c}{2}}\cos\frac{\gamma}{2}, \tag{7}$$

$$\cos\frac{\alpha+\beta}{2} = \frac{\sinh p - \sinh(p-c)}{\sinh c}\sin\frac{\gamma}{2} = \frac{\cosh\frac{a+b}{2}}{\cosh\frac{c}{2}}\sin\frac{\gamma}{2}. \tag{8}$$

Therefore,

$$\sin\frac{S}{2} = \cos\frac{\alpha+\beta+\gamma}{2} = \frac{\cosh\frac{a+b}{2} - \cosh\frac{a-b}{2}}{\cosh\frac{c}{2}}\sin\frac{\gamma}{2}\cos\frac{\gamma}{2}.$$

It remains to note that

$$\cosh\frac{a+b}{2} - \cosh\frac{a-b}{2} = 2\sinh\frac{a}{2}\sinh\frac{b}{2},$$

$$\sin\frac{\gamma}{2}\cos\frac{\gamma}{2} = \frac{\sqrt{\sinh p \sinh(p-a)\sinh(p-b)\sinh(p-c)}}{\sinh a \sinh b}.$$

(b) Formulas (7) and (8) can be written in the form

$$\frac{\cos\frac{S+\gamma}{2}}{\cos\frac{\gamma}{2}} = \frac{\cosh\frac{a-b}{2}}{\cosh\frac{c}{2}} \quad\text{and}\quad \frac{\sin\frac{S+\gamma}{2}}{\sin\frac{\gamma}{2}} = \frac{\cosh\frac{a+b}{2}}{\cosh\frac{c}{2}}.$$

The former implies that

$$\frac{\cos\frac{S+\gamma}{2} - \cos\frac{\gamma}{2}}{\cos\frac{S+\gamma}{2} + \cos\frac{\gamma}{2}} = \frac{\cosh\frac{a-b}{2} - \cosh\frac{c}{2}}{\cosh\frac{a-b}{2} + \cosh\frac{c}{2}},$$

i.e.,

$$\tan\frac{S}{4}\tan\left(\frac{S}{4} + \frac{\gamma}{2}\right) = \tanh\frac{p-b}{2}\tanh\frac{p-a}{2}.$$

Similarly, the latter gives

$$\tan\frac{S}{4}\cot\left(\frac{S}{4} + \frac{\gamma}{2}\right) = \tanh\frac{p-c}{2}\tanh\frac{p}{2}.$$

It remains to multiply the formulas obtained.

5.51. Let I be the center of the inscribed circle, and let K be the point of tangency of the inscribed circle with the side AB. Then

$$\tanh|IK| = \sinh|AK|\tan\angle IAK, \qquad \text{i.e.,} \qquad \tanh r = \sinh(p-a)\tan(\alpha/2).$$

Expressing $\tan(\alpha/2)$ from the formulas given in Problem 5.49, we obtain the required relation.

5.52. Let O be the center of the circumscribed circle, and let M be the midpoint of the side BC. Then $\tanh|CM| = \tanh|CO|\cos\angle COM$, i.e.,

$$\tanh\frac{a}{2} = \tanh R\cos\left(\frac{\beta+\gamma-\alpha}{2}\right).$$

In addition,

$$\frac{\beta+\gamma-\alpha}{2} = \frac{\pi}{2} - \left(\alpha+\frac{S}{2}\right).$$

Thus

$$\tanh R = \frac{\tanh\frac{a}{2}}{\sin\left(\alpha+\frac{S}{2}\right)}. \tag{9}$$

According to assertion (a) of Problem 5.49, we have

$$\cosh a = \frac{\cos\alpha+\cos\beta\cos\gamma}{\sin\beta\sin\gamma};$$

therefore,

$$\cosh^2\frac{a}{2} = \frac{\cos\alpha+\cos\beta\cos\gamma+\sin\beta\sin\gamma}{2\sin\beta\sin\gamma} = \frac{\sin\left(\beta+\frac{S}{2}\right)\sin\left(\gamma+\frac{S}{2}\right)}{\sin\beta\sin\gamma}$$

and

$$\sinh^2\frac{a}{2} = \frac{\sin\frac{S}{2}\sin\left(\alpha+\frac{S}{2}\right)}{\sin\beta\sin\gamma}.$$

Expressing $\tanh(a/2)$ from these formulas and substituting the result in (9), we obtain the required relation.

5.53. We can assume that in the Poincaré upper half-plane model, the points A, B, C, D have the coordinates $(0,a)$, $(0,b)$, $(k,\sqrt{b^2-k^2})$, $(k,\sqrt{a^2-k^2})$, where $a>b>0$. Then

$$e^{|CD|} = \frac{\sqrt{a^2-k^2}}{\sqrt{b^2-k^2}},$$

$$\cosh|AD| = 1+\frac{k^2+(a-\sqrt{a^2-k^2})}{2a\sqrt{a^2-k^2}} = \frac{a}{\sqrt{a^2-k^2}}, \qquad \cosh|BC| = \frac{b}{\sqrt{b^2-k^2}}.$$

Therefore,

$$\sinh|AD| = \frac{1}{\sqrt{a^2-k^2}}, \qquad \cosh|BC| = \frac{1}{\sqrt{b^2-k^2}}.$$

5.54. There exists only one line l parallel both to the rays AB and DC and to the rays AD and BC. Let A', B', C', D' be the endpoints of the perpendiculars from the points A, B, C, D to the line l. The lines AA' and CC' are the bisectors of the angles A and C, and the lines BB' and DD' are the bisectors of the exterior angles B and D. Therefore, the point C_b symmetric to C with respect to the line BB' lies on the line AB, and the point C_d symmetric to C with respect to DD' lies on the line AD. Let C_b' and C_d' be the projections of the points C_b and C_d to the line l. Then $|C_bC_b'| = |CC'| = |C_dC_d'|$; hence the points C_b and C_d are symmetric with respect to the line AA'. Therefore, $|AC_b| = |AC_d|$. It remains to note that $|AC_b| = |AB| + |BC_b| = |AB| + |BC|$ and $|AC_d| = |AD| + |DC|$.

5.55. Let B_a and B_c be the points symmetric to B with respect to the bisectors of the exterior angles A and C, respectively. The points B_a and B_c lie on the rays DA and DC, and $|DB_a| = |DB_c|$. Therefore, the bisector of the interior angle D is perpendicular to the segment $[B_a, B_c]$ and passes through its midpoint, and the bisectors of the exterior angles A and C are perpendicular to the segments $[B_a, B]$ and $[B_c, B]$ and pass through their midpoints. The perpendiculars to the sides of a triangle through their midpoints belong to one pencil of lines. Using similar arguments, we prove that the bisector of the interior angle B belongs to the same pencil of lines.

5.56. First, we perform a motion that superposes the points on the absolute at which the given pairs of parallel lines intersect. Then we can assume that in the upper half-plane model, the lines under consideration are rays perpendicular to the real axis. Any pair of such rays can be transformed into any other pair by a translation and a homothety.

5.57. For the elliptic and parabolic pencils, the assertion is obvious. Consider a hyperbolic pencil of lines, i.e., the pencil of lines perpendicular to some line AB, where A and B are points on the absolute. Let us draw tangent lines to the corresponding circle through the points A and B. Suppose that they intersect at a point P. Then the (Euclidean) continuations of all lines of the pencil intersect at the point P. To prove this, it is sufficient to consider a projective transformation that maps the disk into itself and takes the segment $[A, B]$ to a diameter.

5.58. In the Poincaré upper half-plane model, any horocycle can be transformed into a horocycle having an equation of the form $y = \text{const}$ by means of a motion, and any two horocycles of this form can be superposed by means of a homothety.

5.59. We can assume that in the Poincaré upper half-plane model, the points A and B have coordinates (x_1, y) and (x_2, y). Then

$$\cosh d = 1 + \frac{(x_1 - x_2)^2}{2y^2},$$

and the length l of the arc of the horocycle equals $|x_1 - x_2|/y$. Thus

$$\cosh d = 1 + \frac{l^2}{2}, \quad \text{i.e.,} \quad l = 2\sinh(d/2).$$

5.60. If two perpendiculars through the midpoints intersect at one point, then the assertion is obvious.

If two perpendiculars through the midpoints are parallel, they can be regarded as rays parallel to the Oy axis in the upper half-plane. All vertices of the triangle are then at the same distance from the real axis; therefore, the third perpendicular is a ray parallel to the Oy axis.

If two perpendiculars through midpoints are perpendicular to some line l, then all vertices of the triangle are equidistant from l; therefore, the third perpendicular is also perpendicular to the line l.

As a byproduct, we have proved that any three points lie either on one circle, or on one horocycle, or on one hypercycle.

5.61. See the solution to Problem 5.45 (c).

5.62. An arc PQ of a circle, a horocycle, or a hypercycle makes equal angles with the (hyperbolic) line PQ at the points P and Q. Let α_1, β_1, and γ_1 be the angles corresponding to the arcs BC, CA, and AB. Then $\alpha = \pi - \beta_1 - \gamma_1$, $\beta = \pi - \alpha_1 - \gamma_1$, and $\gamma = \pi - \alpha_1 - \beta_1$ (for a circle) or $\alpha = \gamma_1 - \beta_1$, $\beta = \gamma_1 - \alpha_1$, and $\gamma = \pi - \alpha_1 - \beta_1$ (for a horocycle or a hypercycle). Thus $\alpha + \beta - \gamma = \pm(\pi - 2\gamma_1)$. The angle γ_1 is constant, because it corresponds to the constant arc AB.

5.63. The diagonal AC cuts the quadrilateral $ABCD$ into two equal triangles. Let M be the midpoint of the segment $[A, C]$. Then the triangles DCM and BAM are equal; therefore, M is the midpoint of $[B, D]$. Thus the symmetry about M transforms the line AB into the line CD, and hence the perpendicular from the point M to AB is a common perpendicular to the lines AB and CD.

5.64. The isometry f maps each line l to some line $f(l)$. Consider a point X moving along a line l to the absolute. The distance from X to the line $f(l)$ does not exceed the distance from X to the point $f(X)$; therefore, the distance from X to l is bounded. This means that the lines l and $f(l)$ approach the absolute at the same points, and hence $l = f(l)$.

Any point A can be represented as the intersection of two lines. The isometry f maps each of these lines onto itself; therefore, each point A is fixed.

5.65. The triangle inequality implies that the transformation f takes a line to a line. The pairs of parallel lines are mapped to pairs of parallel lines because only parallel lines contain arbitrarily close points. Clearly, a circle of radius r centered at O is mapped to the circle of radius kr centered at $f(O)$.

Consider a triangle with vertices on the absolute. Its image under the transformation f is also a triangle with vertices on the absolute, and the circle inscribed in one triangle is mapped to the circle inscribed in the other. All triangles with zero angles are isometric; therefore, all circles inscribed in such triangles have the same radius. Clearly, this radius r is finite. As a result, we obtain $r = kr$, i.e., $k = 1$.

Addendum

A.1. Consider a nonreal line in $\mathbb{C}P^2$. Its equation has the form $l(P) + im(P) = 0$, where l and m are real-valued linear functions and the lines $l(P) = 0$ and $m(P) = 0$ do not coincide. These lines intersect at one real point, and this point belongs to the line under consideration in $\mathbb{C}P^2$.

A.2. A sphere in $\mathbb{R}^n$ has an equation of the form $\sum_{n=1}^{n}(x - a_i)^2 = R^2$. Passing to $\mathbb{C}P^n$, we obtain the equation $\sum_{i=1}^{n}(z_i - a_i z_{n+1})^2 = R^2 z_{n+1}^2$. Therefore, the intersection of the sphere with the plane at infinity is given by $z_1^2 + \cdots + z_n^2 = 0$.

A.3. Take the points $A_1(1,0,0)$, $B_1 = (0,1,0)$, $C_1 = (0,0,1)$, $M = (1/3, 1/3, 1/3)$ in $\mathbb{R}^3$. Clearly, the triangle $A_1B_1C_1$ is regular and M is its center. Choose points A, B, C on the rays OA, OB, OC (O is the origin) so that they lie in the plane $A_1B_1C_1$. The points A_1, B_1, C_1 then lie on the sides of the triangle ABC, and M is the intersection point of the segments $[A, A_1]$, $[B, B_1]$, $[C, C_1]$. This can only occur if the triangle ABC is regular and $A = (-1, 1, 1)$, $B = (1, -1, 1)$, and $C = (1, 1, -1)$.

A.4. The tangent line to the given ellipse at the point (x_0, y_0, z_0) has the homogeneous equation

$$\frac{xx_0}{a^2} + \frac{yy_0}{b^2} = zz_0.$$

If this tangent line passes through the point $(1, \pm i, 0)$, then $y_0 = \pm i b^2 x_0 / a^2$. Substituting this expression in the equation of the ellipse, we obtain

$$\frac{x_0^2}{a^2} - \frac{b^2 x_0^2}{a^4} = z_0^2, \quad \text{i.e.,} \quad \frac{z_0}{x_0} = \pm\frac{\sqrt{a^2 - b^2}}{a^2}.$$

Thus the tangent line is

$$\frac{x x_0}{a^2} + \frac{y y_0}{b^2} = \pm\frac{\sqrt{a^2 - b^2}}{a^2} z x_0.$$

Clearly, it passes through one of the foci $(\pm\sqrt{a^2 - b^2}, 0, 1)$.

Remark. Consider $I, J = (1, \pm i, 0)$ and some real points F_1 and F_2. Problem A.4 implies that all conics inscribed in the quadrilateral IF_1JF_2 have common foci, namely, F_1 and F_2.

A.5. The conic $ax^2 + bxy + cy^2 + dx + ey + f = 0$ intersects the line at infinity at points $(x_i, y_i, 0)$ such that $ax_i^2 + bx_iy_i + cy_i^2 = 0$. By Viète's theorem, $(x_1/y_1) \cdot (x_2/y_2) = c/a$. In addition, for a hyperbola with perpendicular asymptotes, the equality $a + c = 0$ holds. Indeed, this equality holds for the hyperbola $xy = k$; applying the rotation $x' = x\cos\varphi + y\sin\varphi$, $y' = -x\sin\varphi + y\cos\varphi$, we obtain $a' = a\cos^2\varphi + c\sin^2\varphi$, $c' = a\sin^2\varphi + c\cos^2\varphi$; therefore, $a' + c' = a + c$.

A.6. (a) Let O be the intersection point of the radical axes of two pairs of circles. Then the degrees of the point O with respect to all three pairs of circles are equal; therefore, the point O lies on the radical axes of the third pair of circles.

(b) We can assume that $A, B = (1, \pm i, 0)$. Then the conics are circles, and their common chords are the radical axes.

A.7. (a), (b) Let us draw lines l_1 and l_2 passing through A and intersecting the given conic. Suppose that l_1 intersects the conic at points P_1 and Q_1 and l_2, at points P_2 and Q_2. The intersection point of the lines P_1Q_1 and P_2Q_2 lies on the polar line of the point A with respect to the given conic. The intersection point of the lines P_1P_2 and Q_1Q_2 also lies on this polar line. Joining the points constructed, we obtain the required polar line.

(c) Let us construct the polar line a of the point A with respect to the given conic. The required tangent line joins the point A to the intersection point of a with the conic.

(d) Let us draw a straight line trough A, take points P and Q on this line, and construct the polar lines p and q of these points. The required tangent line joins A with the intersection point of p and q.

A.8. The ellipse $x^2/a^2 + y^2/b^2 = 1$ $(a > b)$ has foci $(\pm\sqrt{a^2 - b^2}, 0)$, and its directrices are specified by the equations $x = \pm a^2/\sqrt{a^2 - b^2}$. Let us draw the line $x = \sqrt{a^2 - b^2}$ through one focus. It intersects the ellipse at points $(x_0, \pm y_0)$, where $x_0 = \sqrt{a^2 - b^2}$. The tangent lines to the ellipse at these points are

$$\frac{x x_0}{a^2} \pm \frac{y y_0}{b^2} = 1.$$

Their intersection point has the coordinates $(a^2/x_0, 0)$. Therefore, the polar line of the focus $(\sqrt{a^2 - b^2}, 0)$ has the equation $x = a^2/\sqrt{a^2 - b^2}$.

A.9. We can assume that the given conic is a circle centered at O. We then have $a \perp OA, \ldots, d \perp OD$, and therefore,

$$[a, b, c, d] = [OA, OB, OC, OD] = [A, B, C, D].$$

A.10. Let the line BC intersect the conic at points I and J. We can assume that $I, J = (1, \pm i, 0)$. Then the conic is a circle centered at A_1, A_1B_1 and A_1C_1 are the diameters of this circle, and AB and AC are orthogonal to A_1B_1 and A_1C_1, respectively. The lines AA_1, BB_1, and CC_1 under consideration are the altitudes of the triangle $A_1B_1C_1$; therefore, they meet at one point.

A.11. (a) The sides of the triangle ACE are the polar lines of the vertices of the triangle BDF. According to Problem A10, the lines AD, CF, BE intersect at one point.

(b) The dual statement is equivalent to the assertion that the intersection points of AC with DF, of CE with FB, of EA with BD are collinear.

A.12. (a) The dual assertion (a triangle PQR is self-polar with respect to a conic C if and only if its vertices coincide with the intersection points of the diagonals and of the two pairs of opposite sides of the quadrilateral inscribed in C) directly follows from the construction of the polar line of a point with respect to a conic (see Problem A7 (a)).

(b) In homogeneous coordinates, the polar line of a point (x_0, y_0, z_0) with respect to the conic $z^2 = x^2/a^2 + y^2/b^2$ has the equation $zz_0 = xx_0/a^2 + yy_0/b^2$. In particular, the polar line of $X = (1, 0, 0)$ is YZ.

A.13. (a) Let x_1 and x_2 be the points of tangency of the conic to the tangent lines drawn from the point a. The tangent lines at x_1 and x_2 have the equations $xAx_1^T = 0$ and $xAx_2^T = 0$. Both these lines pass through a; therefore, $aAx_1^T = aAx_2^T = 0$. Hence the line x_1x_2 has the equation $aAx^T = 0$.

(b) The point x_0 satisfying the equation $x_0Bx_0^T = 0$ is dual to the line $x_0Ax^T = 0$. This line is tangent to the conic $xCx^T = 0$ at some point x_1; therefore, its equation has the form $x_1Cx^T = 0$. Thus $x_1C = x_0A$. Clearly, $x_1Cx_1^T = x_0AC^{-1}CC^{-1}Ax_0^T = x_0AC^{-1}Ax_0^T$. Hence the equalities $x_0Bx_0^T = 0$ and $x_1Cx_1^T = 0$ are equivalent if $B = AC^{-1}A$, i.e., if $C = AB^{-1}A$.

A.14. (a) In the homogeneous coordinates (x, y, z), the circle C will have the equation $(x - az)^2 + y^2 - R^2z^2 = 0$ corresponding to the matrix

$$\begin{pmatrix} 1 & 0 & -a \\ 0 & 1 & 0 \\ -a & 0 & a^2 - R^2 \end{pmatrix}.$$

Therefore, the dual conic C^* corresponds to the matrix

$$\begin{pmatrix} 1 & 0 & 0 \\ 0 & 1 & 0 \\ 0 & 0 & -1 \end{pmatrix} \begin{pmatrix} 1 & 0 & -a \\ 0 & 1 & 0 \\ -a & 0 & a^2 - R^2 \end{pmatrix}^{-1} \begin{pmatrix} 1 & 0 & 0 \\ 0 & 1 & 0 \\ 0 & 0 & -1 \end{pmatrix}$$
$$= -\frac{1}{R^2} \begin{pmatrix} a^2 - R^2 & 0 & -a \\ 0 & -R^2 & 0 \\ -a & 0 & 1 \end{pmatrix}.$$

Thus the conic C^* has the equation

$$(a^2 - R^2)x^2 - R^2y^2 - 2axz + z^2 = 0, \qquad \text{i.e.,} \quad x^2 + y^2 = \frac{a^2}{R^2}\left(x - \frac{z}{a}\right)^2.$$

(b) The equation of the conic C^* can be written in the form

$$\frac{x_1^2}{a_1^2} + \frac{y^2}{b_1^2} = 1, \qquad \text{where} \quad x_1 = x + \frac{a}{R^2 - a^2}, \quad a_1^2 = \frac{R^2}{(R^2 - a^2)^2}, \quad b_1^2 = \frac{1}{R^2 - a^2}.$$

Clearly, the polar line of the point $(a, 0)$ with respect to the unit circle has the equation $x = a^{-1}$. Now all the required statements are easy to verify.

A.15. For two conics, there exists exactly one triangle that is self-polar with respect to both of them. It can be defined either as the triangle with vertices at the intersection points of the common chords of the given conics or as the triangle whose sides pass through the intersection points of the common tangent lines to the given conics. From the first definition, we see that the triangle PQR under consideration is precisely the self-polar triangle, and the second definition implies that the intersection point of the common tangent lines to the conics lie on the sides of this triangle.

A.16. A projective involutive transformation is completely determined by the images X' and Y' of some points X and Y. Take an arbitrary point A and consider the point B at which the circumscribed circles of the triangles AXX' and AYY' intersect. Let Z and Z' be the intersection points of the circle passing through A and B with the line under consideration. It is sufficient to prove that the transformation $Z \mapsto Z'$ is projective. Let O be the intersection point of the line AB with the given line (if $O = \infty$, then the transformation $Z \mapsto Z'$ is a symmetry with respect to some point). In the coordinate system with origin at the point O, the transformation $Z \mapsto Z'$ has the form $z \mapsto c/z$.

A.17. Let O be the point through which the lines of the pencil pass. Consider two pairs a, a' and b, b' of lines corresponding to each other under the involution. Some line l intersects these lines at points A, A' and B, B', respectively. Let P be the intersection point of the circles circumscribed about the triangles OAA' and OBB'. Consider the circle centered on the line l and passing through the points O and P. This circle intersects l at points C and C'. Clearly, the lines OC and OC' are orthogonal, and the involution under consideration maps them to each other.

A.18. Let $a, a', \dots$ be the coordinates of the projections of the points under consideration. According to Menelaus' theorem,

$$(a-c)(b-a')(b'-c') = -(a'-c')(b'-a)(b-c).$$

It is easy to verify that this relation is equivalent to

$$[a, b, c', a'] = [a', b', c, a].$$

A.19. If the point O is at infinity, the required statement is equivalent to the assertion of the preceding problem.

A.20. Let us apply a projective transformation that maps the points A, B, C, D to the vertices of a square. Both assertions then become obvious.

A.21. Let $A_1, \dots, A_n$ be the given points. Consider the involutions $\sigma_1, \dots, \sigma_n$ of the circle such that σ_i takes each point X on the circle to the intersection point of the line XA_i with the circle for every i. It is required to construct the fixed points of the transformation $\sigma_n\sigma_{n-1}\cdots\sigma_1$. This transformation is projective, and its fixed points can be constructed by the Steiner method.

A.22. The points $f(g(X))$ and $g(f(X))$ coincide if and only if F and G are the intersection points of two pairs of opposite sides of a quadrilateral inscribed in the conic. This means that the point F lies on the polar line of G with respect to the given conic.

A.23. Two lines AX_1 and AX_2 are orthogonal if and only if

$$[AX_1, AX_2, AI, AJ] = -1,$$

where $I, J = (1, \pm i, 0)$. Thus, for the pencil of lines passing through the point A, the transformation $AX_1 \mapsto AX_2$ is projective. Clearly, this transformation is involutive. But for any projective involution $X_1 \mapsto X_2$ of a conic, all chords X_1X_2 pass through one point.

A.24. It is sufficient to prove that $[A_1, B_1, S_1, S_2] = [A_2, B_2, S_2, S_1]$. To this end, we must show that the signs of these cross ratios coincide and

$$\frac{A_1S_1}{B_1S_1} : \frac{A_1S_2}{B_1S_2} = \frac{A_2S_2}{B_2S_2} : \frac{A_2S_1}{B_2S_1}, \qquad \text{i.e.,} \qquad \frac{A_1S_1 \cdot A_2S_1}{B_1S_1 \cdot B_2S_1} = \frac{A_2S_2 \cdot A_1S_2}{B_2S_2 \cdot B_1S_1}.$$

The latter equality is obvious because the expressions on the right- and left-hand sides are equal to $(r_a/r_b)^2$, where r_a and r_b are the radii of the circles.

A.25. The matrix corresponding to the conic $x^2/(a-\lambda) + y^2/(b-\lambda) - z^2 = 0$ is

$$A = \begin{pmatrix} (a-\lambda)^{-1} & 0 & 0 \\ 0 & (b-\lambda)^{-1} & 0 \\ 0 & 0 & -1 \end{pmatrix},$$

and the matrix corresponding to the dual conic is

$$B = A^{-1} = \begin{pmatrix} a-\lambda & 0 & 0 \\ 0 & b-\lambda & 0 \\ 0 & 0 & -1 \end{pmatrix};$$

thus the dual conic has the equation $(a-\lambda)x^2 + (b-\lambda)y^2 = z^2$. It is easy to verify that the points $(\pm 1, \pm i, \sqrt{a-b})$ satisfy this equation.

A.26. According to the assertion of Problem A25, the conics dual to the confocal conics

$$\frac{x^2}{a-\lambda} + \frac{y^2}{b-\lambda} = z^2$$

pass through the points $(\pm 1, \pm i, \sqrt{a-b}z)$. This means that the lines

$$\pm x \pm iy + \sqrt{a-b}z = 0$$

dual to these points are tangent to the confocal conics. It is easy to verify that every such line passes through one of the points $(1, \pm i, 0)$.

A.27. Consider the family of planes containing the line l. The points dual to them form a line l^*. Let Q^* be the quadric dual to the quadric Q. It is easy to verify that $l^* \subset Q^*$ if $l \subset Q$. Indeed, any plane containing l intersects Q in two lines; therefore, it is tangent to Q, and hence the point dual to this plane belongs to Q^*. Conversely, if $l^* \subset Q^*$, then $l \subset Q$. Therefore, if the line l is not contained in Q, then the line l^* intersects the quadric Q^* in precisely two points.

A.28. The restriction of a motion of Lobachevsky 3-space to the absolute is a linear-fractional transformation over $\mathbb{C}$, and any linear-fractional transformation over $\mathbb{C}$ has a fixed point; if the transformation is not the identity map, then it cannot have more than two fixed points.

A.29. It is sufficient to prove that if $a, b, c, d \in \mathbb{C}$ are pairwise different numbers, then there exists a linear-fractional transformation f such that $f(a) = b$, $f(b) = a$, $f(c) = d$, and $f(d) = c$. The first three equalities uniquely determine f. The linear-fractional transformation $f \circ f$ has fixed points a and b and yet another fixed

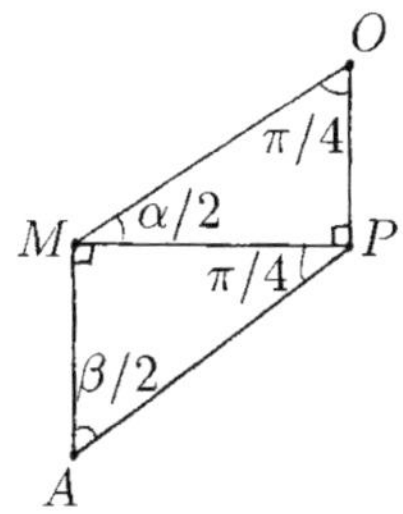

FIGURE S.12

point that cannot coincide with a or b. Thus $f \circ f$ is the identity transformation, and hence $f(d) = c$.

A.30. (a) Let $ABCD$ be the tetrahedron. According to the assertion of Problem A29, there exists a motion that maps the dihedral angle at the edge AC to the dihedral angle at the edge BD.

(b) Consider the Poincaré upper half-space model. We can assume that the vertex under consideration is at a point at infinity. Then the faces of the trihedral angle at this vertex lie on the faces of a Euclidean triangular prism with base on the absolute. The sum of the dihedral angles of a triangular prism is π.

A.31. The proof is similar to the solution of Problem A30 (b).

A.32. In the Poincaré upper half-space model, consider a trihedral angle with vertex at a point at infinity. Its faces form a triangular (Euclidean) prism with base on the absolute. The volume element in the upper half-space has the form $z^{-3}\,dx\,dy\,dz$. Therefore, the volume of the part of the trihedral angle contained in the half-space $z \geqslant c$ equals

$$\int_c^\infty Sz^{-3}\,dz = \frac{S}{2c^2},$$

where S is the Euclidean area of the section of the prism by the plane $z = c$. It is only essential that the volume is finite. This means that a horosphere cuts a trihedral angle with vertex on the absolute into two parts so that the volume of one of them (the one containing the vertex of the angle) is finite.

Let us cut off the vertices of the tetrahedron (with vertices on the absolute) from the tetrahedron itself by horospheres. As a result, the tetrahedron will be divided into five parts, each of finite volume.

A.33. Suppose that O is the center of the cube, P is the center of one of its faces, M is the midpoint of an edge of this face, and A is a vertex of this edge. Then OMP and APM are right triangles; their angles are specified in Figure S.12.

Recall that in a right triangle with leg a, we have the relation $\cosh a = \cos\alpha/\sin\beta$. Applying this relation to the leg MP in both triangles, we obtain

$$\cos\frac{\pi}{4} : \sin\frac{\alpha}{2} = \cosh|MP| = \cos\frac{\beta}{2} : \sin\frac{\pi}{4}.$$

A.34. See the solution to Problem 5.64.

A.35. The triangle inequality implies that f maps all lines to lines. Consider a triangle with vertices on the absolute. The image of this triangle (including its interior) is also a triangle with vertices on the absolute. Clearly, any two triangles with vertices on the absolute are isometric. After these remarks are made, we can apply the solution of Problem 5.65.

A.36. As a model of the Lobachevsky space, we take the model on a two-sheeted hyperboloid, i.e., we assume that the points $z_1, \dots, z_{n+2}$ lie in $\mathbb{R}^{n+1}$ and their coordinates satisfy the relation

$$x_1^2 + \cdots + x_n^2 - x_{n+1}^2 = -1$$

with $x_{n+1} > 0$. Let $[u, v] = u_1v_1 + \cdots + u_nv_n - u_{n+1}v_{n+1}$ be the pseudoscalar product in $\mathbb{R}^{n+1}$. Then $\cosh(d_{ij}) = [z_i, z_j]$.

Let us take the matrix $(z_1, \dots, z_{n+2})$ formed by the columns of coordinates of the points $z_1, \dots, z_{n+2}$ and augment it to an $(n+2) \times (n+2)$ matrix S by zeros in the first row. Let J be the diagonal matrix of order $n+2$ with diagonal $(1, 1, \dots, 1, -1)$. It is easy to verify that

$$A = ([z_i, z_j]) = S^T JS.$$

But the first row of the matrix S is zero, and hence $\det S = 0$; therefore, $\det A = 0$.

Bibliography

[A] E. Artin, *Geometric algebra*, Interscience, New York–London, 1957.

[Ba] L. Bankoff, *The metamorphosis of butterfly problem*, Math. Mag. **60** (4) (1987), 195–200.

[Be] M. Berger, *Geometry*, Vols. 1, 2, Springer-Verlag, Berlin, 1987.

[BHH] W. L. Black, H. C. Howland, and B. Howland, *A theorem about zig-zags between two circles*, Amer. Math. Monthly **81** (1974), 754–757.

[Bo1] N. Bourbaki, *Algèbre*, Paris, Hermann.

[Bo2] N. Bourbaki, *L'Architecture des mathématiques*, Les grands courants de la pensée mathématique (Cahiers du Sud), 1948, pp. 35–47).

[BQ] F. S. Beckman and D. A. Quarles Jr., *On isometries of Euclidean spaces*, Proc. Amer. Math. Soc., **4** (1953), 810–815.

[D] M. Darboux, *Sur le théorème fondamental de la géométrie projective*, Math. Ann. **17** (1880), 55–61.

[DSS] V. N. Dubrovskii, Ya. A. Smorodinskii, and E. L. Surkov, *The world of relativity*, "Nauka", Moscow, 1984. (Russian)

[Ge] A. O. Gel'fond, *Transcendental and algebraic numbers*, Gostekhizdat, Moscow, 1952; English transl., Dover, New York, 1960.

[GS] B. Grünbaum and G. S. Shepard, *Descartes' theorem in n dimensions*, Enseign. Math. **37** (1991), 11–15.

[HM] U. Haagerup and H. J. Munkholm, *Simplices of maximal volume in hyperbolic n-space*, Acta Math. **147** (1981), 1–11.

[L1] J. A. Lester, *Euclidean plane point-transformations preserving unit area or unit perimeter*, Arch. Math. **45** (1985), 561–564.

[L2] ———, *Martin's theorem for Euclidean n-space and a generalization to the perimeter case*, J. Geom. **27** (1986), 29–35.

[M] J. Milnor, *Hyperbolic geometry: the first 150 years*, Bull. Amer. Math. Soc. **6** (1982), 9–24.

[P] R. Perline, *Non-Euclidean flashlights*, Amer. Math. Monthly **103** (1996), 377–385.

[S1] G. Salmon, *A treatise on conic sections*, Chelsea, New York, 1954.

[S2] ———, *A treatise on analytic geometry of three dimensions*, Chelsea, New York, 1965.

Author Index

Subject Index

Selected Titles in This Series

(Continued from the front of this publication)

For a complete list of titles in this series, visit the AMS Bookstore at **www.ams.org/bookstore/**.